Exploring the Ocean

Volume 4

Reference Volume and Series Index

Trevor Day

OXFORD
UNIVERSITY PRESS

Consulting Editor Professor Dorrik Stow
University of Southampton Oceanography Centre, UK

Published in the United States of America by
Oxford University Press, Inc.
198 Madison Avenue
New York, NY 10016
www.oup.com

Planned and produced by
Andromeda Oxford Limited
11-13 The Vineyard
Abingdon
Oxon OX14 3PX
UK
www.andromeda.co.uk

Library of Congress Cataloging-in-Publication Data
Day, Trevor.
Exploring the ocean / Trevor Day.
p . cm.
Summary: Explores various aspects of the world's oceans including currents, tides, weather patterns, the sea floor, marine life, uses by humans, and ecological issues.
Contents: v. 1. The physical ocean -- v. 2. Life in the ocean -- v. 3. Uses of the ocean -- v. 4. Reference and set index.
ISBN 0-19-515738-9 (set : alk. paper)
1. Oceanography--Juvenile literature. 2. Marine organisms--Juvenile literature. 3. Marine biology--Juvenile literature. 4. Ocean [1. Oceanography. 2. Marine biology. 3. Ocean.] I. Title.

GC21.5.D385 2003
551.46--dc21 2002042583

Project Director Susan Kennedy
Project Manager Derek Hall
Design Tony Truscott
Editors Marion Dent, Deborah Fox, Suzanne Juby
Cartographers Tim Williams, Richard Watts
Picture Manager Claire Turner
Picture Research Alison Floyd
Production Clive Sparling

Printed in the U.A.E. by Oriental Press Limited, Dubai

USING THIS SET

This four-volume set describes the fascinating world of the ocean through informative text and colorful illustrations and maps. Each volume explores a different theme. **Volume 1** is concerned with the physical ocean: its origin and structure, and associated features such as tides, weather systems, islands, and coral reefs; **volume 2** describes the varied natural life in the ocean: evolution, marine animals and plants, food webs, ocean zones, communities of creatures, and conservation; **volume 3** looks at the way humans use the ocean: exploration from earliest times to the present day, exploitation of ocean resources, pollution, and ocean management.

In each volume, feature boxes highlight items of special interest. Cross-references at the bottom of each left-hand page point the reader to further information on related topics in other sections and volumes: the relevant volume number is given in bold type, followed by the page numbers. Each volume also has its own subject index.

Volume 4 contains a comprehensive series index, together with further reference resources, including maps, datafiles, full descriptions of the five oceans, biographies of men and women who have contributed to our knowledge of the oceans, timelines, places to visit, websites, further reading, and a glossary.

Throughout the set the following abbreviations are used for measurements:–
Length in (inches), cm (centimeters), ft (feet), m (meters), mi (miles), km (kilometers)
Area sq in (square inches), sq cm (square centimeters), sq ft (square feet), sq m (square meters), sq mi (square miles), sq km (square kilometers)
Volume cu mi (cubic miles), cu km (cubic kilometers)
Temperature F (Fahrenheit), C (Celsius)
Weight Given in U.S. tons, followed by metric tonnes

Previous page: A walrus basking on a rock in Alaska.

Right: A closeup of the prow of the *Cutty Sark*, one of the fast sailing ships known as tea clippers, now preserved at Greenwich, England.

Contents

The Arctic Ocean	4
The Atlantic Ocean	8
The Indian Ocean	14
The Pacific Ocean	18
The Southern Ocean	22
Earth's Timeline	26
How Deep?	28
Surface Exploration Timeline	30
Undersea Exploration Timeline	32
Biographies	34
Glossary	46
Further Reading	50
Websites	52
Places to Visit	53
Index	56

THE ARCTIC OCEAN

The Arctic Ocean was once an ancient sea, the Sinus Borealis, situated in the Northern Hemisphere. This sea shifted northward 200 million years ago, finally reaching the North Pole. Most of the Arctic Ocean lies within the Arctic Circle (the region north of latitude 66.5° N), but in places the ocean stretches beyond it.

The Arctic Ocean is almost entirely surrounded by land—being bordered by North America, Eurasia, and Greenland. It is always cold in the region; the ocean is often covered by ice, called pack ice, at least 10 ft (3 m) thick. The pack ice forms when sea water in the cooler parts of the ocean freezes. During the winter, pack ice covers most of the Arctic Ocean. However, in summer, surface air temperatures rise, and the ice shrinks back to cover about half of the Arctic Ocean. Currents push the ice in a clockwise direction from North American waters toward Russia, and then around again.

Between Greenland and Scandinavia, water flows into and out of the Arctic Ocean and mixes with water from the Atlantic Ocean. Warm water from the North Atlantic Drift current flows into the Atlantic Ocean and sweeps up toward the north, forming the Norwegian Current. The warm waters of the Norwegian Current keep most of the Norwegian coast free of ice throughout the year, even within the Arctic Circle. A cold current from the Arctic—the Greenland Current—sweeps toward the south. This current freezes the sea along the eastern coast of Greenland all year long, except for a brief period in summer.

The Scandinavian explorer Fridtjof Nansen identified the deep hollow that contains the Arctic Ocean (known as its basin) in the 1890s, when he took depth readings through the arctic ice. One part of the giant hollow that contains the Arctic Ocean is called the Nansen Basin. About half the Arctic Ocean sea floor consists of continental shelf. This is by far the largest proportion for any ocean.

Sea ice

Every summer, chunks of ice break off from the land in western Greenland and extreme northeastern Canada and fall into the sea. The floating chunks of ice are known as icebergs. More than 10,000 icebergs form in the region each year. Some drift far enough south to endanger ships in the Labrador Sea and just off Newfoundland in the Atlantic Ocean. In 1912 RMS *Titanic* crashed into an iceberg and sank. Two years later, the International Ice Patrol was set up. This organization monitors iceberg movement and issues warnings to ships.

Ships with specially strengthened hulls, called ice-breakers, help keep the shipping channels free from pack ice in winter and prevent ships from becoming trapped in the ice.

Mineral resources

Below shallower parts of the Arctic Ocean lie huge deposits of coal, oil, and natural gas. Oil companies extract oil from the floor of the Beaufort Sea off Alaska and northern Canada, and there are rich gas reserves near some northern Canadian islands. On the Russian side of the Arctic Ocean there are large oil and gas fields. The arctic region also contains some of the world's largest reserves of iron, copper, lead, and the radioactive metal uranium. But because the conditions in the Arctic are so inhospitable—with extreme cold, driving winds, and shifting ice—companies have so far mined little of the area's fossil fuel and mineral reserves.

Nevertheless, environmental groups and representatives of the indigenous peoples are concerned about pollution and damage to the area. They try to persuade governments to curb exploitation of the Arctic's resources.

People of the region

Very few people live in the land areas bordering the Arctic Ocean because of the extreme cold. Of the few native peoples who live in the arctic regions of North America, Scandinavia, and Siberia, some hunt birds such as geese and marine mammals such as walruses, ringed seals, and gray whales in order to provide food and sometimes clothing for themselves and their families. However, industries such as oil exploration and mineral mining are setting up their operations ever closer to the people's homes and hunting territories. Increasing numbers of indigenous people now work for mining companies in order to earn their living.

The Arctic Ocean lies at the North Pole. It is the smallest and shallowest of the world's five oceans.

Vital phytoplankton

In the Arctic Ocean tiny, plantlike organisms called phytoplankton ("phyto" is from the Greek word for "plant," and "plankton" from the Greek word for "drifting") grow from February to September, when there is enough sunlight for them to photosynthesize (make their own food substances by using the energy from sunlight). Tiny, drifting animals known as zooplankton feed on the phytoplankton, and other marine animals feed on phytoplankton and zooplankton. The animals store some of the food in their bodies as fat, and this keeps them alive during the bitterly cold winters when phytoplankton do not grow.

Fishing

On the arctic continental shelves, fishers catch species such as char and capelin, and in the deeper waters they fish for arctic cod. In the warmer waters near Scandinavia, they catch Atlantic cod, haddock, and plaice. Over the years, fishers have seriously overharvested some fish species, and as a result the number of fish has been declining. Some nations that fish in the area—including Canada, Norway, Iceland, and Russia—have tried to reverse the decline in fish numbers by catching only an agreed amount of fish each year so stocks can build up again. The agreed amounts are called quotas. This action has helped some fish stocks to recover—for example, those in the Barents Sea north of Russia.

Hunting

Since the 1970s commercial hunting of marine mammals in the arctic region has greatly declined. Whales, and most species of seals, are now protected animals. Nevertheless, in 2002, the governments of Canada and Greenland allowed hunters to kill more than 400,000 young harp seals in the areas of northern Canada and Greenland. However, conservationists (people who try to protect natural habitats and the species that live in them) doubt whether it is possible to catch such high numbers of individuals year after year, without seriously reducing the seal population.

Arctic Ocean Data

Area	3,351,811 sq mi (8,676,520 sq km)
Average depth	6,349 ft (1,935 m)
Maximum depth	18,400 ft (5,608 m) Molloy Deep
Volume	4,028,950 cu mi (16,790,582 cu km)
Major geological features	*Plates*: North American, Eurasian;
	Ridges: Alpha, Lomonosov, Nansen
Major islands (area)	
Greenland	840,000 sq mi (2,175,500 sq km)
Baffin Island	195,928 sq mi (507,451 sq km)
Victoria Island	83,897 sq mi (217,292 sq km)
Ellesmere Island	75,767 sq mi (196,236 sq km)

Tourism

More than one million tourists visit the Arctic each year. Many of them come to see some of the area's wildlife—such as seals, polar bears, and whales. Some tourists arrive at the Arctic in icebreaking ships, which plow their way through the Northwest Passage (the sea route along the arctic coast of Canada and Alaska that joins the Pacific and Atlantic Oceans) in the summer months. Others arrive by airplane.

Scientific importance

The area within the Arctic Circle holds special interest for scientists. Signs of global warming (an increase in the temperature of the planet's atmosphere) are likely to show up earlier and more distinctly in polar regions than in warmer parts of the world. This is because scientists can easily observe melting ice, caused by temperature increases. Scientists monitor the pack ice using information provided by satellites and from submarines beneath the surface of the sea. They discover whether the area and thickness of arctic pack ice is changing. A major melting of the ice would have serious environmental consequences because it would alter the flow of ocean currents. Also, at present, the arctic pack ice separates the chilly arctic air from the comparatively warm sea water below. If this barrier disappeared, a lot of heat energy from the sea would be released into the atmosphere, speeding up any global warming in the region.

Pollution

The Arctic Ocean's environment is easily damaged. In the cold conditions, biological breakdown and decay work slowly. Spilt oil, for example, takes many years to break down and disappear, rather than the few months it would take in warmer climates. When dark oil spills stain snow and ice, the stained snow and ice absorb more heat energy from the Sun than the surrounding white snow and ice (which reflect most of the Sun's rays). When this happens, the dark snow and ice melt much more quickly than they would under natural conditions. Because oil breaks down so slowly, it also poses a threat to the health of birds and mammals. Oil clogs feathers and fur, destroying their insulating properties and making it impossible for the animals to fly, swim, or walk properly. If animals swallow the oil, it can poison them.

This map shows the main underwater features of the Arctic Ocean, as well as major seas, bays and inlets, islands, and adjacent landmasses.

BERING SEA
St. Lawrence Island
BERING STRAIT
Arctic Circle
CHUKCHI SEA
Wrangel Island
EAST SIBERIAN SEA
NORTH AMERICA
EURASIA
BEAUFORT SEA
Canada Basin
Northwind Ridge
Chukchi Plateau
Mendeleyev Abyssal Plain
Mendeleyev Ridge
Wrangel Abyssal Plain
New Siberian Islands
LAPTEV SEA
Banks Island
Victoria Island
Melville Island
Prince of Wales Island
Queen Elizabeth Islands
Somerset Island
Devon Island
Ellesmere Island
ARCTIC OCEAN
Mendeleyev Basin
Makarov Basin
ALPHA RIDGE
LOMONOSOV RIDGE
North Pole
Fram Basin
NANSEN RIDGE (Arctic Mid-Ocean Ridge)
Nansen Basin
North Land
Voronin Trough
KARA SEA
Svyataya Anna Trough
Franz Josef Land
East Novaya Zemlya Trough
Novaya Zemlya
UDSON BAY
Baffin Island
Baffin Basin
BAFFIN BAY
Baffin Bay
DAVIS STRAIT
Yermak Plateau
Molloy Deep–Arctic Ocean's deepest point -18,400 ft (5,608 m)
Svalbard
GREENLAND
GREENLAND SEA
BARENTS SEA
Spitsbergen Bank
LABRADOR SEA
Greenland Basin
Mohns Ridge
Kolbeinsey Ridge
Jan Mayen
Iceland Plateau
Jan Mayen Ridge
Basin
Dumshaf Abyssal Plain
DENMARK STRAIT
SCANDINAVIA
NORTH ATLANTIC OCEAN
Iceland
Reykjanes Ridge
Faroe–Iceland Ridge
Norwegian
NORWEGIAN SEA
Faroe Islands
Shetland Islands

The Atlantic Ocean

The Atlantic Ocean is named after the Greek god Atlas. The ocean covers an area about six times that of the United States. The equator divides the Atlantic Ocean into the North Atlantic and South Atlantic.

The Atlantic Ocean's most prominent sea-floor feature is the Mid-Atlantic Ridge, a vast, undersea mountain chain. This ridge winds from north to south along the center of the Atlantic—roughly halfway between the Americas to the west and Europe and Africa to the east. Molten rock rises up from the Earth's mantle at the Mid-Atlantic Ridge. As the rock cools and solidifies, it lays down new ocean floor—adding nearly 1 inch (2.5 cm) of new sea floor each year. At the same time, old sea floor is destroyed by being pushed back down into the Earth's mantle at deep troughs, or trenches. The new sea floor spreads more quickly than it is destroyed. As a result, the Atlantic Ocean is getting larger, and Europe and the Americas are gradually moving apart. The ocean is about 40 ft (12 m) wider than when Christopher Columbus crossed it in 1492. Volcanoes erupting along or close to the Mid-Atlantic Ridge formed volcanic islands like Iceland, Surtsey, and the Azores.

In the North Atlantic, ocean currents move in a clockwise direction picking up warm water from the Gulf of Mexico and carrying it east across the Atlantic as part of the Gulf Stream. The current continues north toward the Arctic Ocean as the North Atlantic Drift. Because they carry warm water, the Gulf Stream and the North Atlantic Drift keep coastal northwest Europe unusually mild.

In the summer, the tropical North Atlantic sometimes produces warm updrafts of moist air that create fierce storms. The most violent storms are called hurricanes. In the South Atlantic, a counterclockwise current pushes cool water toward the equator. The cool water prevents hurricanes forming in the tropical South Atlantic.

Several seas (geographical regions within oceans) connect to the Atlantic Ocean basin. (A basin is a hollow in the Earth's crust filled with sea water.) The Mediterranean Sea is made up of several basins that flooded within the last five million years. The Caribbean Sea and Gulf of Mexico occupy basins that became almost enclosed by land some three million years ago. The North Sea formed within the last 18,000 years when the European continental shelf flooded.

Mineral resources

Companies mine oil and gas from the continental shelves of the Gulf of Mexico, North Sea, and off the east coast of Canada and the west coast of Africa. In the North Atlantic, mining companies excavate sand and gravel from shallow

Atlantic Ocean Data

Area	31,563,463 sq mi (81,527,400 sq km)
Average depth	12,391 ft (3,777 m)
Maximum depth	28,232 ft (8,605 m) Puerto Rico Trench
Volume	73,902,416 cu mi (307,902,776 cu km)
Major geological features	*Plates*: African, Antarctic, Caribbean, Eurasian, North American, South American; *Ridges*: Mid-Atlantic, America-Antarctic, Atlantic-Indian; *Trenches*: Cayman, Puerto Rico, South Sandwich

Major seas	**Area**	**Maximum depth**
Caribbean Sea	971,400 sq mi/2,515,900 sq km	28,232 ft/8,605 m
Mediterranean Sea	969,100 sq mi/2,509,900 sq km	16,896 ft/5,150 m
Gulf of Mexico	582,100 sq mi/1,507,600 sq km	11,496 ft/3,504 m
Hudson Bay	281,900 sq mi/730,100 sq km	364 ft/111 m
Black Sea	196,100 sq mi/507,900 sq km	7,257 ft/2,212 m
North Sea	164,900 sq mi/427,100 sq km	1,440 ft/439 m
Baltic Sea	147,500 sq mi/ 382,000sq km	295 ft/90 m

Major islands (area)	
Great Britain (mainland)	84,200 sq mi (218,077 sq km)
Cuba	42,804 sq mi (110,862 sq km)
Newfoundland	42,030 sq mi (108,855 sq km)
Iceland	39,700 sq mi (102,820 sq km)
Ireland	31,839 sq mi (82,462 sq km)

The Atlantic is the second largest of the five oceans. Scientific study of the oceans began in the North Atlantic, and we know more about the North Atlantic Ocean than any other ocean.

GREENLAND
GREENLAND SEA
BAFFIN BAY
HUDSON BAY
LABRADOR SEA
NORWEGIAN SEA
SCANDINAVIA
BALTIC SEA
Helsinki
Oslo
Stockholm
St. Petersburg
ORESUND
NORTH SEA
Rotterdam
Copenhagen
Gdansk
Felixstowe
Antwerp
KIEL CANAL
Le Havre
Hamburg
STRAIT OF DOVER
North Atlantic Drift
ST. LAWRENCE-SEAWAY
Gulf Stream
Montreal
New York
NORTH AMERICA
EUROPE
BOSPHORUS
BLACK SEA
DARDANELLES
Marseilles
Naples
Barcelona
Lisbon
Oran
Algiers
Piaeus
MEDITERRANEAN SEA
Casablanca
STRAIT OF GIBRALTAR
Alexandria
SUEZ CANAL
New Orleans
STRAITS OF FLORIDA
GULF OF MEXICO
NORTH ATLANTIC OCEAN
Las Palmas
WINDWARD PASSAGE
MONA PASSAGE
North Atlantic Gyre
CARIBBEAN SEA
Dakar
AFRICA
Colon
Freetown
PANAMA CANAL
Lagos
Equator
SOUTH AMERICA
Lobito
PACIFIC OCEAN
Rio de Janeiro
São Paulo
SOUTH ATLANTIC OCEAN
Buenos Aires
Montevideo
Cape Town
Ocean current
Important sea route
Major port or harbor
Marine protected area
Low oxygen zone
South Atlantic Gyre
SCOTIA SEA
DRAKE PASSAGE
SOUTHERN OCEAN
Palmer
BELLINGSHAUSEN SEA
WEDDELL SEA
ANTARCTICA

This map shows the main underwater features of the North Atlantic, as well as major seas, bays and inlets, islands, and adjacent landmasses.

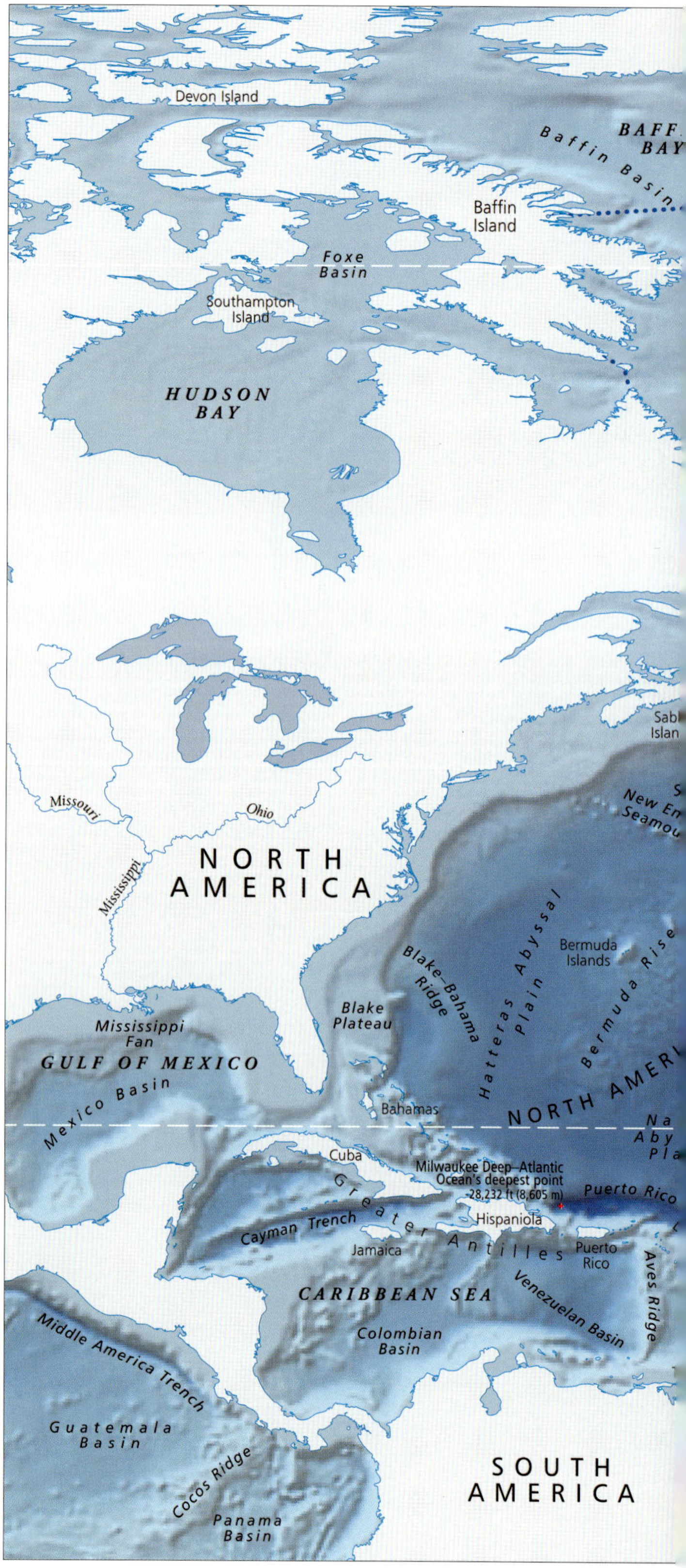

water to provide building materials. The seabed off the southwest African coast yields diamonds. There are also valuable metals such as copper and nickel at the bottom of the Atlantic Ocean, but it is too expensive to raise them.

Fishing

In the late 1990s the Atlantic Ocean provided about 25 percent of the world's marine fish catch. About two-thirds of this catch came from the northern Atlantic and off the coasts of north and west Africa. There, cold water rises, bringing valuable nutrients to the surface. This effect is known as upwelling. Fish come to feed on the tiny, drifting plants and animals called plankton that flourish where upwelling occurs, and fishers are able to reap rich rewards.

The North Atlantic is probably the world's most heavily fished region. Surface and midwater fish, such as sardines, anchovies, and herring, make up the bulk of the catches. But cod, and bottom-dwelling fish such as plaice, flounder, and haddock, are the most important commercially. Commercial fishing has drastically reduced the stocks of many species.

There are very rich fisheries off the west African coast. However, governments of the countries in whose territorial waters they occur—such as Namibia and Mauritania—earn foreign currency by selling fishing rights to other states. As a result, large fishing fleets from places such as Spain and Portugal take most of the fish, and there are few left for local fishers to catch.

North Atlantic Ocean Data

Area	15,938,626 sq mi (41,080,840 sq km)
Average depth	11,666 ft (3,556 m)
Maximum depth	28,232 ft (8,605 m) Puerto Rico Trench
Volume	35,064,624 cu mi (146,080,848 cu km)
Major geological features	*Plates*: African, Caribbean, Eurasian, North American; *Ridge*: Mid-Atlantic; *Trenches*: Cayman, Puerto Rico

GREENLAND
GREENLAND SEA
Mohns Ridge
Jan Mayen
Kolbeinsey Ridge
Iceland Plateau
Jan Mayen Ridge
Basin
NORWEGIAN SEA
Norwegian
Arctic Circle
Iceland
Faroe–Iceland Ridge
Faroe Islands
Shetland Islands
LABRADOR SEA
Imarssuak Seachannel
Reykjanes Ridge
Maury Seachannel
NORTHWEST ATLANTIC MID-OCEAN CANYON
Rockall Bank
Rockall Trough
NORTH SEA
BRITISH ISLES
Porcupine Bank
Porcupine Abyssal Plain
Newfoundland
West European Basin
Flemish Cap
Grand Banks of Newfoundland
Biscay Abyssal Plain
EUROPE
Newfoundland Basin
Newfoundland Rise
Iberian Basin
IBERIAN PENINSULA
Corsica
Sardinia
Balearic Islands
MEDITERRANEAN SEA
Azores
Azores–Cape St Vincent Ridge
MID-ATLANTIC RIDGE
Corner Seamounts
Madeira Islands
NORTH ATLANTIC OCEAN
Great Meteor Tablemount
Canary Basin
Canary Islands
Cape Verde Abyssal Plain
Tropic of Cancer
CAPE VERDE BASIN
AFRICA
Cape Verde Islands
Gambia Abyssal Plain
MID-ATLANTIC RIDGE
Demerara Abyssal Plain
Guyana Basin
Sierra Leone Rise
Sierra Leone Basin
Guinea Basin

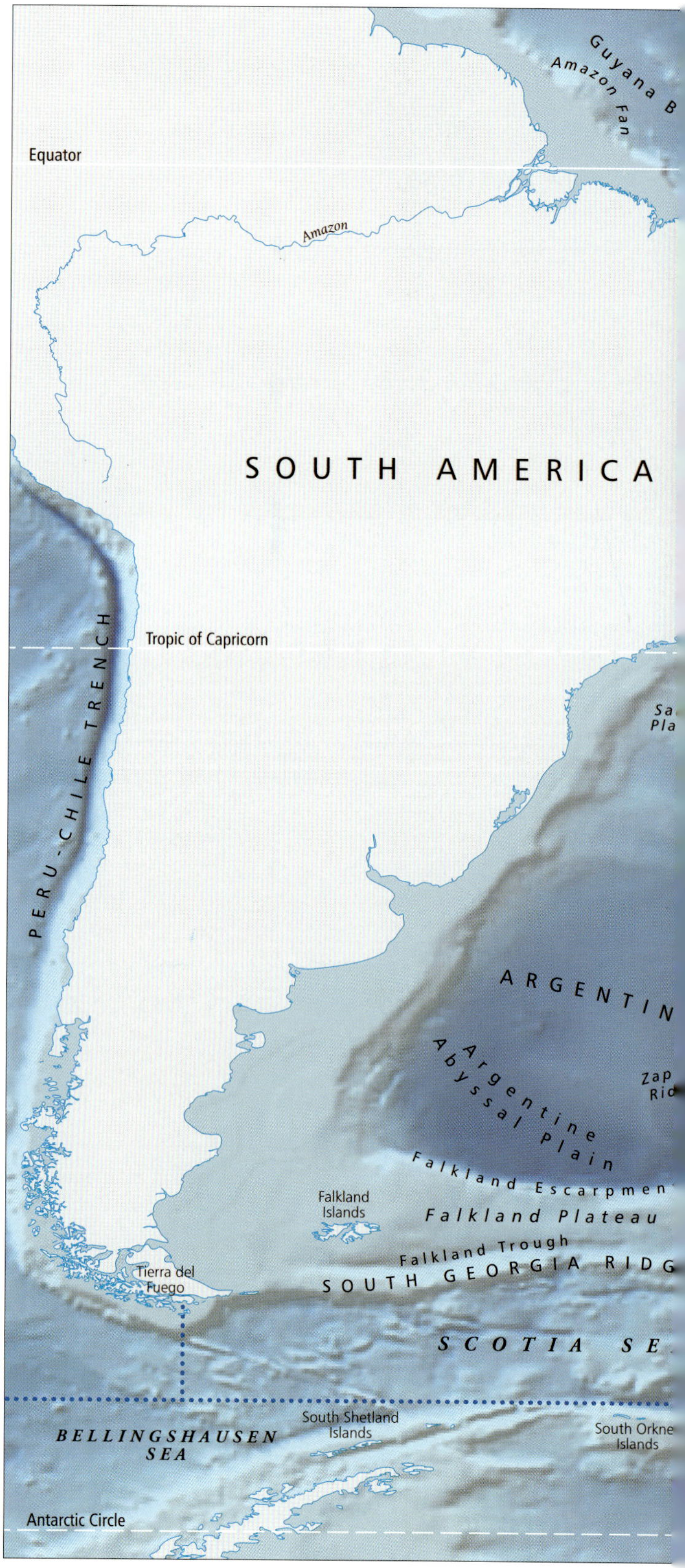

This map shows the main underwater features of the South Atlantic, as well as major seas, islands, and adjacent landmasses.

Tourism

The many and varied coastlines of the Atlantic offer a huge range of scenery, climate, and activities from watersports to exploring local culture and history. The seas around the Gulf of Mexico, Caribbean Sea, and Mediterranean Sea attract more than 200 million vacationers a year alone.

Pollution and conservation

The shipping routes between North America and northwest Europe are among the most heavily polluted in the world. Sewage, pesticides, and industrial waste such as poisonous metals are often dumped into the sea, or find their way there through rivers, causing damage to marine life. Pollutants in fish and shellfish may be eaten by humans. Many governments are now taking measures to reduce pollution—for example, Germany has 27 rapid-response, anti-pollution vessels that can reach an oil or chemical spill in German coastal waters within 48 hours.

Ships empty ballast tanks (water-filled tanks that regulate the ship's buoyancy) at sea and may discharge chemical pollutants into the sea water. Major oil spillages sometimes occur at sea, and oil tankers frequently leak quantities of oil. Oil kills creatures at sea and washes ashore to pollute beaches.

Along the west African coast, developers clear tropical rainforest and mangrove. This causes more fresh water to run into the sea. At the same time, the absence of tree roots to bind the soil makes the coastline erode more quickly.

South Atlantic Ocean Data

Area	15,624,837 sq mi (40,446,560 sq km)
Average depth	13,126 ft (4,000 m)
Maximum depth	27,313 ft (8,325 m) South Sandwich Trench
Volume	38,837,792 cu mi (161,824,928 cu km)
Major geological features	*Plates*: African, Antarctic, South American; *Ridges*: Mid-Atlantic, America-Antarctic, Atlantic-Indian; *Trench*: South Sandwich

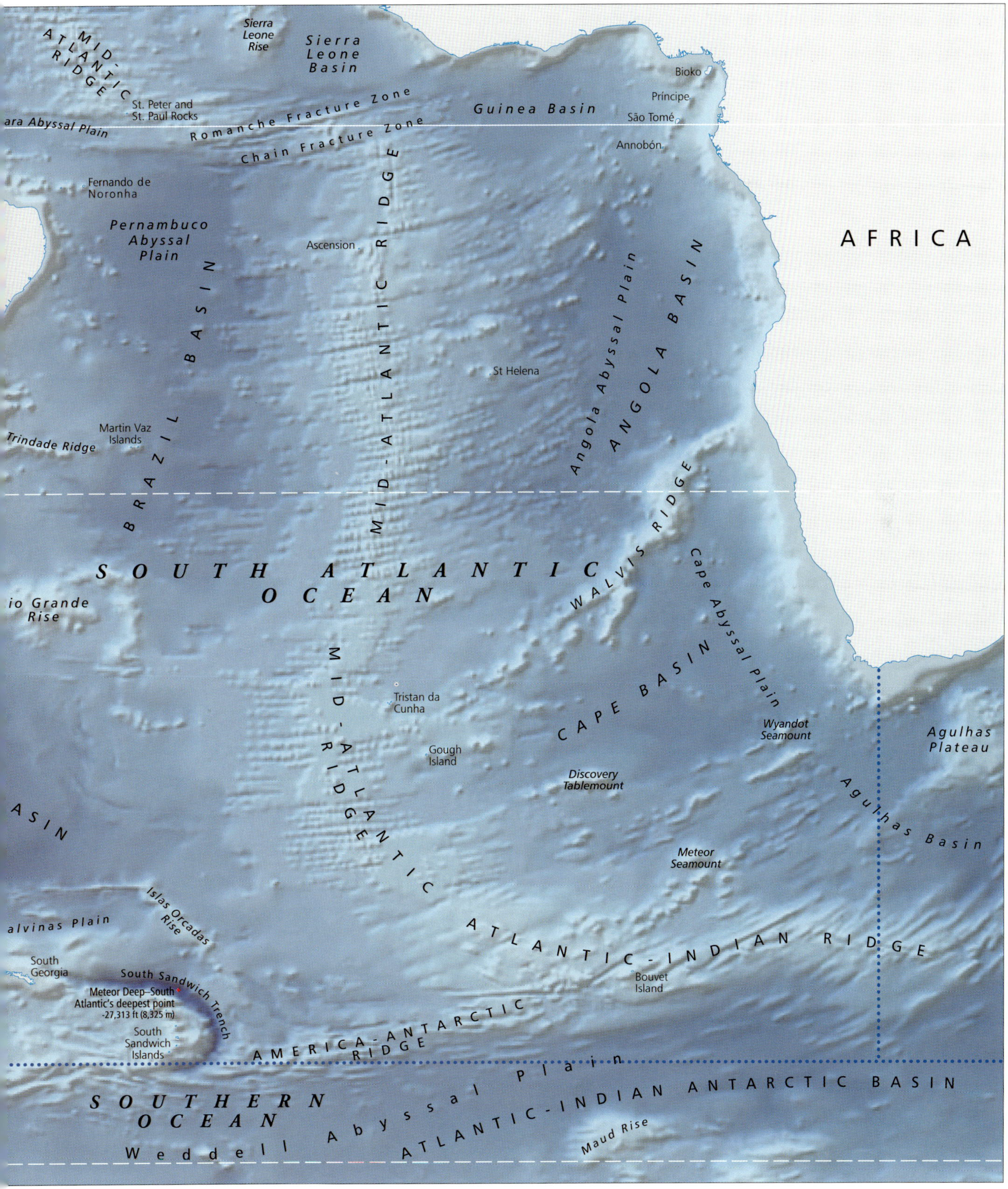
MID-ATLANTIC RIDGE
Sierra Leone Rise
Sierra Leone Basin
Bioko
Príncipe
São Tomé
Annobón
St. Peter and St. Paul Rocks
Romanche Fracture Zone
Chain Fracture Zone
Guinea Basin
ara Abyssal Plain
Fernando de Noronha
Pernambuco Abyssal Plain
Ascension
AFRICA
BRAZIL BASIN
MID-ATLANTIC RIDGE
Angola Abyssal Plain
ANGOLA BASIN
St Helena
Martin Vaz Islands
Trindade Ridge
WALVIS RIDGE
SOUTH ATLANTIC OCEAN
io Grande Rise
Cape Abyssal Plain
CAPE BASIN
Tristan da Cunha
Gough Island
MID-ATLANTIC RIDGE
Wyandot Seamount
Agulhas Plateau
Discovery Tablemount
Agulhas Basin
ASIN
Meteor Seamount
Islas Orcadas Rise
alvinas Plain
ATLANTIC-INDIAN RIDGE
South Georgia
South Sandwich Trench
Meteor Deep–South Atlantic's deepest point -27,313 ft (8,325 m)
Bouvet Island
AMERICA-ANTARCTIC RIDGE
South Sandwich Islands
SOUTHERN OCEAN
Weddell Abyssal Plain
ATLANTIC-INDIAN ANTARCTIC BASIN
Maud Rise

The Indian Ocean

Millions of years ago the southern supercontinent, known as Gondwana, began to break up. One large fragment broke in two. A piece drifted north to eventually form Africa, and the other remained in the south to form Antarctica and Australia. Australia eventually separated from Antarctica about 50 million years ago. India then broke away from the African piece. It drifted northeast, tearing fragments off Africa. Some of those fragments now lie scattered in the Indian Ocean as the islands of the Seychelles.

The Indian Ocean sea floor is dominated by the mid-ocean ridge system that runs north to south in the shape of an upside down Y. Hotspots (places where molten rock have burnt through the Earth's crust) have left volcanoes or ridges on the sea floor. The Kerguelen hotspot created the longest straight ridge on the planet—the Ninety East Ridge.

The Indus and Ganges-Brahmaputra River systems in India and Bangladesh carry sediment from the Himalayas to the Bay of Bengal. Together, they have laid down a fan of

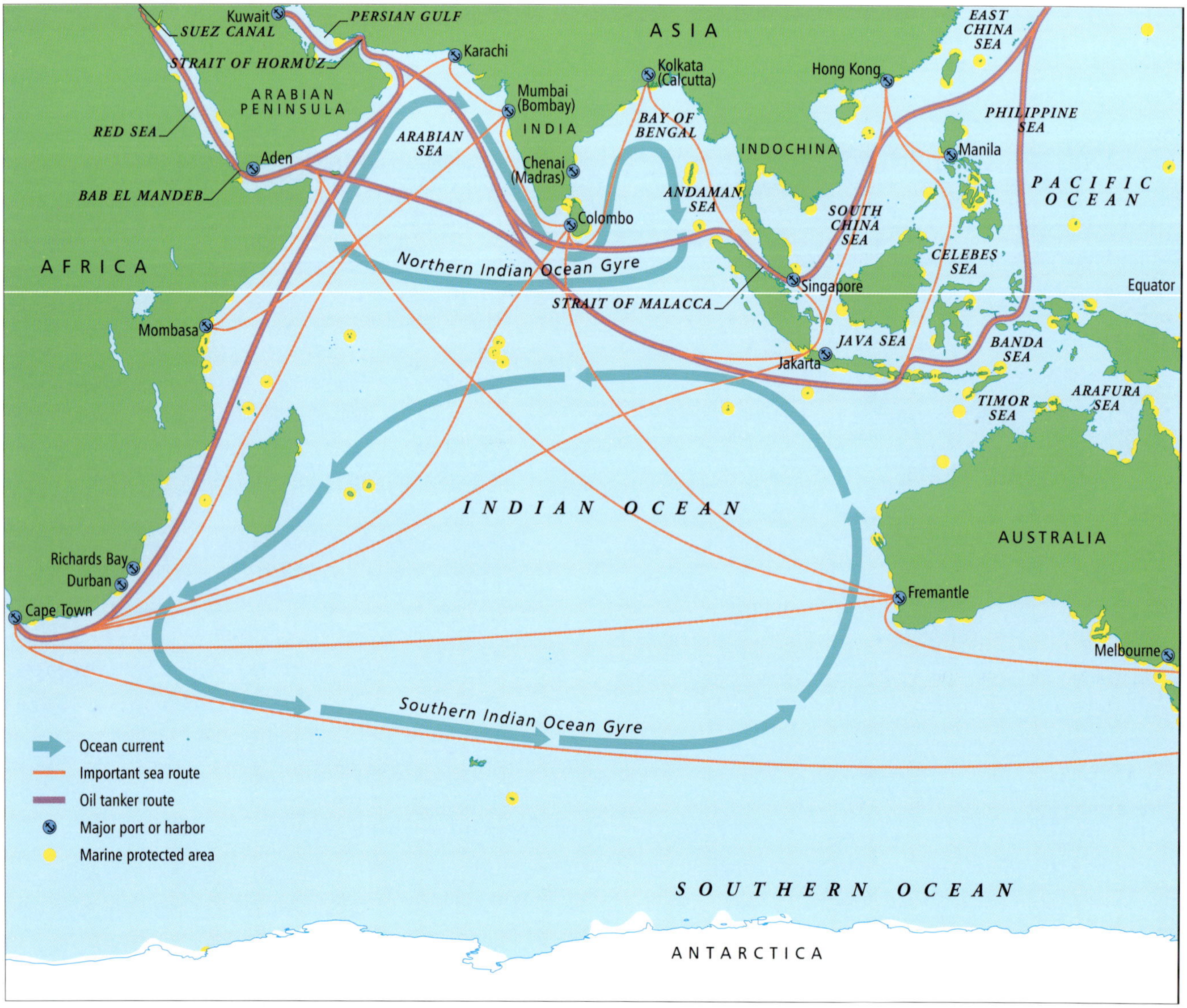

sediment that spreads over 800 mi (1,290 km) from the coast along the seabed in this part of the Indian Ocean.

Monsoons and cyclones

The Indian Ocean is totally enclosed on its northern side. Landmasses warm and cool more rapidly than the sea, so when the Asian continent warms rapidly in summer and cools quickly in winter this causes dramatic changes in the Indian Ocean's winds and currents. During the Northern Hemisphere's summer, the south Asian mainland warms up quickly. As warm air rises, it draws in seasonal winds called monsoons that blow from the southwest toward India. They bring moisture-laden air from the southern Indian Ocean and unload torrential rain onto the Indian subcontinent.

The summer winds generate a north-flowing current off the Somali coast of Africa, which drives a spiral, clockwise current, or gyre. As the winds blow surface water away from the coasts of Somalia and the Arabian Peninsula, an upwelling of cool, nutrient-rich water rises from below. The nutrients help phytoplankton to grow; many other marine organisms depend on the phytoplankton for food.

In winter, between November and April, the lands cool and warm air rises over the ocean. The monsoon winds reverse and blow from the northeast. This causes the flow of the northern gyre to reverse. Coastal upwelling then stops, and the growth of phytoplankton slows.

In low-lying areas bordering the Indian Ocean, such as Bangladesh, tropical cyclones (the name for hurricanes in this region) sometimes sweep huge waves ashore, flooding the land and causing great loss of life.

Mineral resources

About 40 percent of the world's offshore oil comes from the Indian Ocean, particularly off the coasts of Saudi Arabia, Iran, India, Indonesia, and Western Australia. Tankers laden with oil and petroleum products travel along the major sea routes from the Persian Gulf and Indonesia eastward to Japan and other East Asian countries, and westward to Africa, South America, and Europe. About half the world's oil is exported from the Persian Gulf.

The people of Indonesia, Myanmar, and Thailand mine the mineral cassiterite (a source of tin). The mineral has gradually worn away from the land and settled in shallow water. Zircon (valued as a gemstone and used in certain industrial processes) is found on beaches in India and Sri Lanka. South Africans dig up glauconite (a rock rich in potash, which is used as a fertilizer) from their beaches.

The Indian Ocean is the third largest ocean on Earth. Its origins go back 130 million years.

Contrasting nations

Some of the nations bordering the Indian Ocean's northern coast—notably Pakistan, India, and Bangladesh—are among the poorest and most densely populated in the world. These countries contrast strongly with the richer nations to the east, such as Australia, and the oil-rich states to the northwest, such as Saudi Arabia and Oman. The countries bordering the eastern coast of Africa are relatively undeveloped. Many people in this region produce just enough to feed themselves and their families. South Africa is the exception; it is a fairly rich, industrialized country.

Fishing

The Indian Ocean is the third largest ocean, but only about 10 percent of the world's marine-fish catch by weight comes from it. There are two main reasons for the realtively low catch. First, most continental shelves in the Indian Ocean are quite narrow. This limits the number of fish that can feed in the nutrient-rich shallow water where most phytoplankton grow. Second, when monsoon winds reverse in winter, they stop the upwelling of cold water. This reduces the growth of phytoplankton—the source of food for fish.

Overfishing in the coastal waters of India, Pakistan, Bangladesh, Sri Lanka, and the Maldives has reduced fish populations. On the northeastern side of the Indian Ocean—particularly in Indonesia—farmers rear fish, shellfish, and grow seaweed. Island fishers use lines baited with hundreds of hooks or purse seines (a net placed around a school of fish and drawn in) to catch tuna. Many smaller island states—such as the Maldives and Seychelles—rely on tuna-fishing, both for home consumption and for export. Fishing fleets from Russia, Japan, South Korea, and Taiwan also fish for tuna, and trawl for shrimp.

Pollution and conservation

The coral reefs of the Indo-West Pacific—the patchwork of islands that lie between the Indian and Pacific Ocean basins—have the biggest variety of marine life of any shallow marine ecosystem in the world. But the beautiful sandy

Tropic of Cancer
Equator
Tropic of Capricorn
AFRICA
ARABIAN PENINSULA
INDIA
RED SEA
PERSIAN GULF
GULF OF OMAN
GULF OF ADEN
ARABIAN SEA
BAY OF BENGAL
ANDAMAN SEA
Indus
Ganges
Brahmaputra
Indus Fan
Ganges Fan
Socotra
Arabian Basin
Lakshadweep (Laccadive Islands)
Maldive Islands
Sri Lanka
Andaman Islands
Andaman Basin
Nicobar Islands
Chagos-Laccadive Plateau
Carlsberg Ridge
Somali Abyssal Plain
Somali Basin
Coco-de-Mer Seamounts
Madingley Rise
Seychelles
Mascarene Plateau
Amirante Trench
Farquhar Group
Agalega Islands
Comoro Islands
Mascarene Basin
Davie Ridge
Madagascar
Mascarene Abyssal Plain
Mauritius
Réunion
Mauritius Trench
Rodriguez
Rodriguez Fracture Zone
Bassas da India
Europa
Chagos Archipelago
Chagos Trench
MID-INDIAN RIDGE
MID-INDIAN BASIN
Nikitin Seamounts
NINETYEAST RIDGE
Cocos Basin
Cocos Islands
Osborn Plateau
Madagascar Basin
Madagascar Plateau
Mozambique Plateau
Mozambique Abyssal Plain
Natal Basin
SOUTHWEST INDIAN RIDGE
INDIAN OCEAN
Broken Ridge
Amsterdam
St. Paul
SOUTHEAST
Agulhas Plateau
Agulhas Basin
CROZET BASIN
Crozet Plateau
Prince Edward Islands
Crozet Islands
Kerguelen Islands
Kerguelen Plateau
Heard Island
William's Seamounts
Atlantic-Indian Ridge
Conrad Rise
Enderby Abyssal Plain
Valdivia Abyssal Plain
ATLANTIC-INDIAN BASIN
INDIAN
SOUTHERN OCEAN

This map shows the main underwater features of the Indian Ocean, as well as major seas, bays and inlets, islands, and adjacent landmasses.

beaches and clear, coral-rich waters attract tourists who sometimes damage the reefs by treading on the coral or brushing past it when swimming. The pollution from badly managed tourist resorts can also damage, or even kill, coral. In parts of the Red Sea, too many visitors are overwhelming local beaches and reefs. Now, many countries encourage tourists to take special care when they dive on reefs, or are even prohibiting tourists from visiting certain reefs.

Many Madagascan, Indonesian, Malaysian, and Sri Lankan coral reefs have already been badly damaged. In Indonesia, vast regions of mangrove have been cleared for firewood and to provide land for rice cultivation or fish and shrimp farming. Tankers crossing the northern Indian Ocean often discharge oil when washing out their tanks. Together, these discharges are greater than the oil released in most major oil spills. This low-level oil pollution probably also contributes to the damage caused to coral and mangroves.

Indian Ocean Data

Area	26,064,036 sq mi (67,469,536 sq km)
Average depth	12,720 ft (3,877 m)
Maximum depth	23,376 ft (7,125 m) Java (Sunda) Trench
Volume	62,780,380 cu mi (261,590,400 cu km)
Major geological features	*Plates*: African, Arabian, Eurasian, Indo-Australian, Antarctic; *Ridges*: Mid-Indian, Ninety East, Southeast Indian, Southwest Indian; *Trench*: Java (Sunda)

Major seas	**Area**	**Maximum depth**
Andaman Sea	218,100 sq mi/564,880 sq km	14,000 ft/4,267 m
Red Sea	174,900 sq mi/453,000 sq km	7,434 ft/2,266 m
Persian Gulf	88,800 sq mi/230,000 sq km	240ft/73 m

Major islands (area)	
Madagascar	226,658 sq mi (587,040 sq km)
Sumatra	165,000 sq mi (427,350 sq km)
Sulawesi (Celebes)	72,987 sq mi (180,036 sq km)
Java	48,900 sq mi (126,650 sq km)

THE PACIFIC OCEAN

The Pacific Ocean covers an area 15 times the size of the United States. It is the largest geographical feature on Earth, stretching from the edge of the Arctic Ocean to the top of the Southern Ocean. At its widest point it extends about 11,000 mi (17,700 km), from the Americas in the east to Asia and Australasia in the west. The equator divides the Pacific Ocean into the North Pacific and the South Pacific.

The Pacific contains nearly half the world's sea water. The Mariana Trench in the West Pacific is the deepest part of all the oceans in the world. If all the land in the world were put into the Pacific Ocean, it would barely fill it.

In 1520 the Portuguese explorer Ferdinand Magellan gave this ocean the name "Pacific" from the Portuguese words "mar pacifico" meaning "the calm sea." Magellan thought the ocean was free of violent storms, but he was quite wrong. Typhoons and tropical cyclones (the terms used for hurricanes in the Pacific) besiege this ocean. Earthquakes, volcanoes, or landslides trigger giant waves called tsunamis.

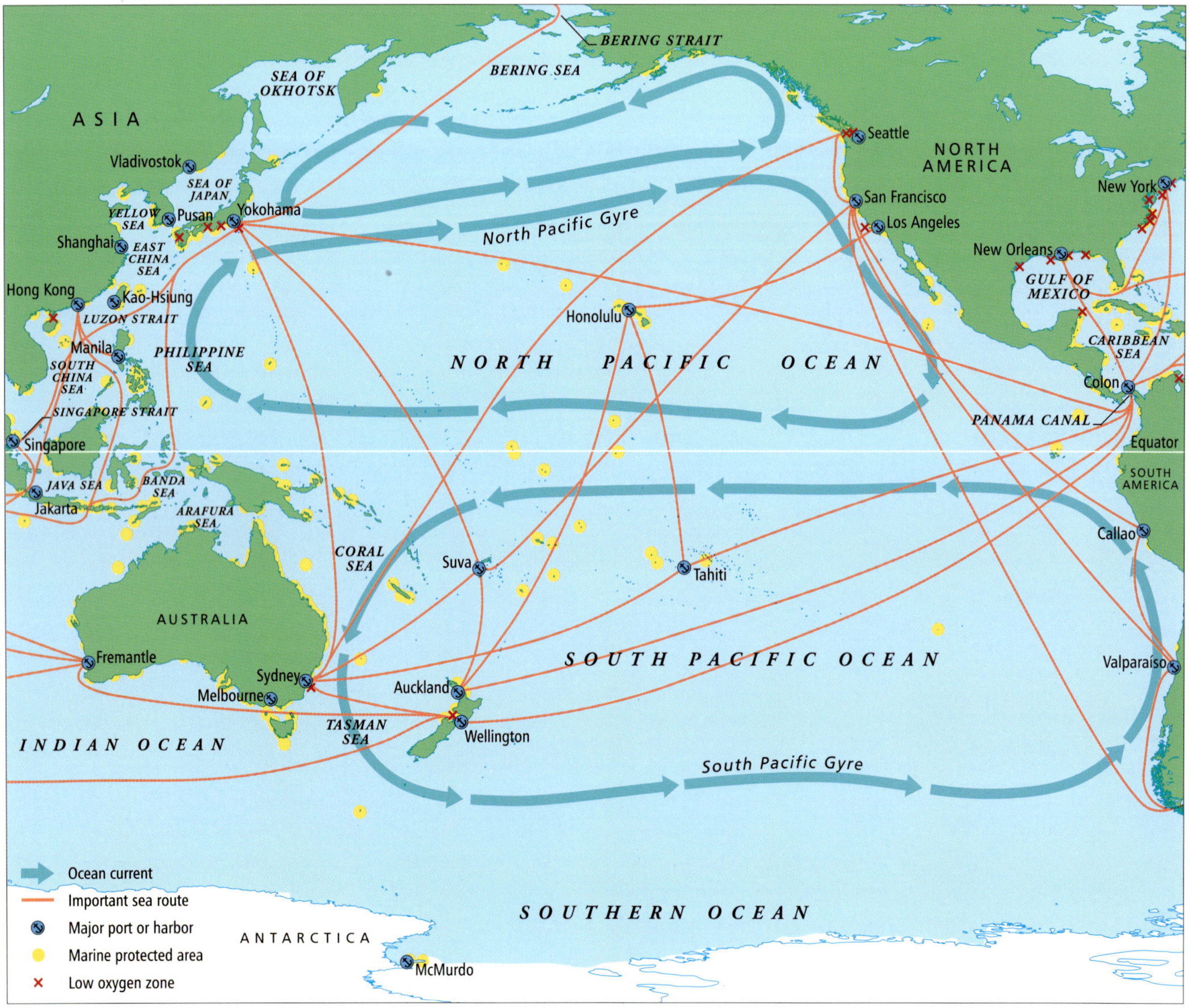

In the North Pacific, a massive spiral of water called a gyre circulates in a clockwise direction and regulates surface currents. In the South Pacific, another gyre flows counter-clockwise. Between the two, on and near the equator, is a complicated system of other currents and countercurrents (ones flowing in the opposite direction).

Ring of Fire

At the edge of the Pacific Ocean lies the "Ring of Fire." This is a giant circle of volcanically active islands and mountain ranges on continental landmasses. Nearby, the Pacific's massive tectonic plates are sliding beneath landmasses. As the plates move, they snag and then release, causing earthquakes. The earthquakes set off tsunamis. When the plates slide under landmasses they melt, and molten rock (magma) bubbles up through weaknesses in the overlying rock and erupts as lava from volcanoes.

Today, new sea floor is still being created in the Pacific. Lava rises up from a rift in the ocean floor that runs along a ridge known as the East Pacific Rise. The plates on either side of the rift move apart by about 6 in (15 cm) each year. Lava fills the gap left by the moving plates, creating new sea floor. However, more sea floor is being destroyed than created. Old sea floor—carried along by plate movement—slides into the trenches of the Pacific. The result is that the Pacific is gradually shrinking.

Hundreds of islands

The Pacific Ocean is dotted with islands. Chains of volcanic islands run roughly southeast to northwest. The oldest volcanoes are extinct and lie underneath the sea at the northwest end of the chains. They are known as seamounts. The newest volcanoes lie to the southeast end. Islands in some volcanic chains gradually sink because of Earth movements. Many islands develop coral reefs, beginning life as fringing reefs (where the coral reef develops along the shoreline) and then becoming barrier reefs as the island sinks. When the island sinks beneath the ocean, it may become an atoll.

Mineral resources

The Pacific Ocean's continental shelves are rich in minerals. Construction companies excavate sand and gravel from shallow water, particularly off Southeast Asia. There is gold on Alaskan beaches and tin in the shallow seabeds of Southeast Asia. Lumps of mixed metals called polymetallic nodules lie scattered across the sea floor of the deep Pacific, particularly between Hawaii and Mexico, but it is too expensive for companies to mine the metals. The United States, Peru, Australia, New Zealand, and China mine offshore oil and gas reserves on their continental shelves.

The Pacific Ocean is the widest, deepest, and oldest of the five oceans.

Countries of the region

Bordering the Pacific Ocean are Russia, China, Japan, Indonesia, and Australasia to the west, and North, Central, and South America to the east. In 2001 the Pacific Ocean contained six of the world's top-ten busiest ports: Singapore, Shanghai (China), and four ports in Japan. Most of the countries bordering the Pacific are quite highly industrialized. Exceptions include Guatemala to the east, and Cambodia, Vietnam, New Guinea to the west, and many of the South Pacific islands.

Fishing

In the late 1990s the Pacific Ocean provided more than 60 percent of the world's marine fish catch by weight. Most of this came from the northwest and southeast regions, with the Japanese sardine, the South American sardine, and the Peruvian anchovy making up a large part of the catch.

Fishers catch highly valuable tuna with seine nets or long lines carrying hundreds of baited hooks. Drift nets—often called "walls of death" because they catch everything in the water including endangered fish and mammals—are widely banned, but some illegal drift-net fishing continues.

Tourism

The spectacular coral reefs of the Central, South, and West Pacific regions are a major attraction for tourists and divers.

Pollution and conservation

More than half of the world's coral reefs lie in the Pacific Ocean and in the seas of the Indo-West Pacific. In the Central and Southwest Pacific many reefs are more or less undamaged, despite people fishing and diving around them. Some of the earliest marine sanctuaries—such as the Great Barrier Reef Marine Park—are established there. Over-harvesting of coral and coral-reef fish, and marine pollution such as oil spillages, have damaged many reef ecosystems in the Northwest Pacific, from the Philippines to Japan.

This map shows the main underwater features of the Pacific Ocean, as well as major seas, islands, and adjacent landmasses.

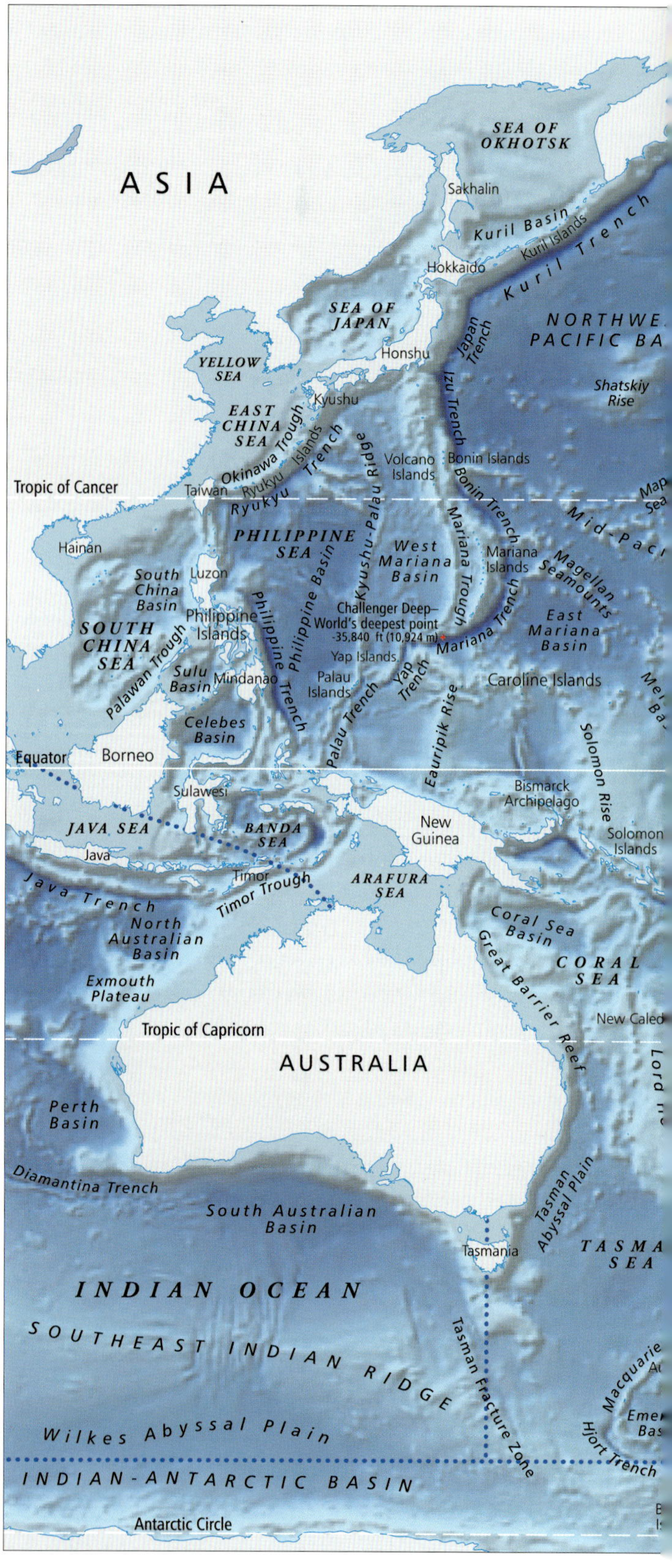

Some of the South Pacific islands within Melanesia, Micronesia, and Polynesia face similar problems. In the Northwest and South Pacific, developers have cleared many mangrove swamps to provide space for coastal towns and farms and to make way for fish- and shellfish-farming. However, the young of commercially important fish and shellfish develop in mangrove communities. When people remove mangroves, they damage the success of fisheries over a wide area because they are destroying the nursery grounds of many marine species.

Pacific Ocean Data

Area	58,957,258 sq mi (152,617,160 sq km)
Average depth	13,874 ft (4,229 m)
Maximum depth	35,840 ft (10,924 m) Mariana Trench
Volume	154,960,672 cu mi (643,375,552 cu km)
Major geological features	*Plates*: Antarctic, Cocos, Indo-Australian, Nazca, North American, Pacific, Philippine; *Ridges*: East Pacific Rise, Pacific-Antarctic; *Trenches*: Aleutian, Bonin, Izu, Japan, Kermadec, Kuril, Mariana, Middle America, Peru-Chile, Philippine, Ryukyu, Tonga

Major seas	**Area**	**Maximum depth**
South China Sea	1,148,500 sq mi/2,974,600 sq km	23,812 ft/7,258 m
Bering Sea	873,000 sq mi/2,261,000 sq km	12,913 ft/3,936 m
Sea of Okhotsk	537,500 sq mi/1,392,000 sq km	11,040 ft/3,365 m
Sea of Japan	391,100 sq mi/1,012,900 sq km	10,016 ft/3,053 m
East China Sea	256,600 sq mi/664,600 sq km	24,629 ft/7,507 m
Yellow Sea	113,500 sq mi/294,000 sq km	190 ft/58 m

Major islands (area)	
New Guinea	306,000 sq mi (792,500 sq km)
Borneo	280,100 sq mi (725,450 sq km)
Honshu	87,805 sq mi (227,413 sq km)
South Island, New Zealand	58,384 sq mi (151,213 sq km)
North Island, New Zealand	44,204 sq mi (114,487 sq km)
Luzon	40,680 sq mi (105,360 sq km)
Mindanao	36,775 sq mi (95,247 sq km)

BERING SEA
Aleutian Basin
Aleutian Islands
Aleutian Trench
Emperor Trough
Chinook Trough
Tufts Abyssal Plain
Queen Charlotte Islands
Juan de Fuca Ridge
Cascadia Basin
NORTH AMERICA
Hess Rise
Northwest Hawaiian Ridge
Musicians Seamounts
NORTHEAST PACIFIC BASIN
Midway Islands
HAWAIIAN RIDGE
Patton Escarpment
Cedros Trench
Blake Plateau
GULF OF MEXICO
Mexico Basin
Bahamas
Hawaiian Islands
Necker Ridge
Revillagigedo Islands
Middle America Trench
Cuba
Cayman Trench
Jamaica
CARIBBEAN SEA
NORTH PACIFIC OCEAN
Northwest Christmas Island Ridge
Mathematicians Seamounts
Clipperton
Albatross Plateau
Guatemala Basin
Magellan Rise
Cocos Ridge
CENTRAL PACIFIC BASIN
Line Islands
Galápagos Rift
Panama Basin
Gilbert Islands
Galápagos Islands
Carnegie Ridge
Phoenix Islands
SOUTH AMERICA
Peru-Chile Trench
Tuvalu (Ellice Islands)
Tokelau
Manihiki Atoll
Manihiki Plateau
Marquesas Islands
EAST PACIFIC RIDGE
Galápagos Rise
Samoa Islands
Fiji Islands
Fiji Plateau
Tonga Islands
Tonga Trench
Society Islands
Tuamotu Archipelago
Tuamotu Ridge
Peru Basin
Cook Islands
Nazca Ridge
Lau Ridge
Austral Islands
Austral Ridge
Pitcairn Island
Easter Island
Sala y Gómez Ridge
Chile Basin
Kermadec Islands
Kermadec Trench
Louisville Ridge
SOUTH PACIFIC OCEAN
Juan Fernández Is
Chile Rise
SOUTHWEST PACIFIC BASIN
Chatham Rise
Chatham Islands
Bounty Trough
SOUTHEAST PACIFIC BASIN
Humboldt Abyssal Plain
PACIFIC-ANTARCTIC RIDGE
SOUTHERN OCEAN
Bellingshausen Abyssal Plain
PACIFIC-ANTARCTIC BASIN

The Southern Ocean

The Southern Ocean, like the Arctic Ocean, is partially frozen all year round. Unlike the Arctic Ocean, which is an ocean surrounded by continents, the Southern Ocean is an ocean surrounding a continent—Antarctica. The South Pole lies near the center of this continent.

Antarctica is covered with a vast sheet of ice that is, on average, 5,750 ft (1,600 m) thick. The continent and its major ice shelves—the large areas of ice extending from the land into the sea—contain about 90 percent of all the ice that remains permanently on Earth from year to year without melting. The huge weight of the ice on Antarctica pushes down its continental shelf. The antarctic continental shelf is more than 1,500 ft (460 m) deep on average, compared to the global average of about 435 ft (133 m). Beyond the continental shelf, parts of the mid-ocean ridge system surround the basin of the Southern Ocean.

Two sets of currents, flowing in opposite directions, circle the Antarctic continent. The East Wind Drift hugs Antarctica and flows from east to west. Farther north, the Antarctic Circumpolar Current streams from west to east. This current is the only one that flows right around the world unimpeded by landmasses. It is also the world's largest current by volume, carrying one hundred times the volume of all the world's rivers combined.

Antarctica is the coldest place on Earth. In 1983 Russian scientists on Antarctica proved this when they recorded a temperature of -128.6° F (-89.2° C). Coastal temperatures average about -22° F (-30° C) and only rise above freezing in the antarctic summer, between December and March. The Southern Ocean surrounding Antarctica has some of the most violent weather on Earth. Very strong winds and large waves can occur at most times of the year. Ships can become trapped in the ice if the sea freezes.

A natural boundary

The geographical boundary of the Southern Ocean is 60° S—this is the northerly limit of the ocean. However, the boundary is a political one; the natural limit of the Southern Ocean is mostly north of 60° S, and reaches as far as 48° S. This natural boundary is where warm and cool currents mix in a region of water called the Antarctic Convergence. The convergence separates warmer waters from the north from cold antarctic waters to the south. The convergence is important for antarctic life because it pushes nutrient-rich water to the surface of the Southern Ocean—this process is called upwelling. Upwelling encourages the growth of phytoplankton on which many other sea creatures feed, and thus fuels marine food chains.

Sea ice

In winter, the antarctic sea ice stretches as far as 55° S in the Atlantic and 65° S in the Pacific. This is five times the area covered by ice in summer. Several thousand large icebergs (floating chunks of ice) break away from the glaciers (slow-moving rivers of ice) and ice shelves (ledges of ice) each year. They may drift for thousands of miles over many months before melting in warmer waters.

Mineral resources

Geologists suspect there are major deposits of minerals under the antarctic ice. They have discovered coal and iron ore deposits, for example, in antarctic mountain ranges. Oil and gas reserves lie beneath the continental shelf. At present, it is too expensive and difficult to exploit Antarctica's mineral and fossil fuel reserves. In 1991 many nations signed a treaty to ban antarctic fossil fuel and mineral exploitation for at least 50 years.

Everyone is a visitor

There are no native peoples in Antarctica, but scientists and tourists visit the region. In 1998–99 there were 27 permanent scientific stations on Antarctica and its nearby islands. Antarctica is becoming a popular tourist destination. In 2001 more than 15,000 tourists visited the antarctic region to see the spectacular icescapes and look at the marine wildlife. The animal life of the region includes many different species of penguins, other seabirds such as skuas, rare gulls, and albatrosses, and seals and humpback whales.

Fishing

The extreme cold, deep waters, and harsh weather conditions make commercial fishing in the Southern Ocean a

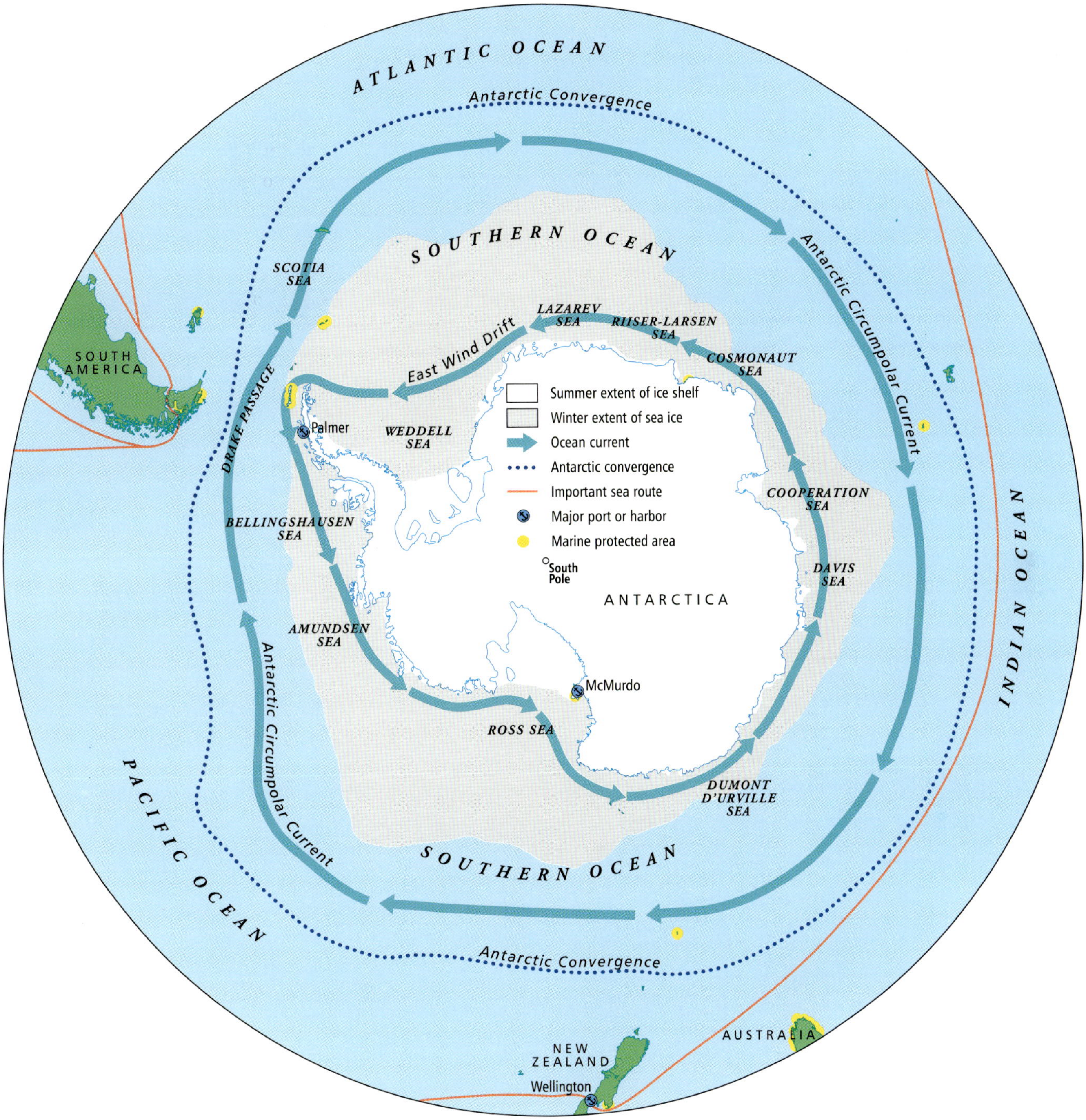

The Southern Ocean surrounds the coldest continent in the world—Antarctica. It is the fourth largest of the five oceans.

challenging occupation. In the late 1990s less than 1 percent of the global catch of seafood was from the waters in and around the Southern Ocean. Over 75 percent of the recorded catch was krill. Patagonian toothfish (also called black hake or antarctic sea bass) accounted for the other 25 percent. Many fish are caught illegally—that is, without permits.

International agreements made in 1999 to end such unregulated fishing have not been effectively enforced, and illegal fishing continues.

Hunting

People have ruthlessly pursued and slaughtered the seals and whales of the Southern Ocean for their fur, blubber, or meat. These mammals are now protected in Southern Ocean waters. Nevertheless, Japanese whalers still hunt a type of baleen whale called minke whale for what are described as "scientific purposes." The meat from the captured whales is then sold commercially.

Scientific interest

Antarctica's relatively untouched environment is particularly interesting to scientists who study the history of the Earth's climate. For example, they can analyze air bubbles trapped in the ice that froze over thousands of years ago and so record Earth's changing carbon dioxide levels. The levels of carbon dioxide have risen and fallen in the last few thousand years, but they have steadily increased in the last 200 years. Atmospheric levels of lead pollution and radioactive fallout (from nuclear explosions) have risen since 1945.

Scientists fear that polar regions will be affected by global warming (an overall rise in the Earth's surface temperature) sooner than tropical regions because a slight change in temperature can quickly prevent ice from forming or speed up the melting of ice. Global warming could trigger unpredictable changes in weather patterns and the flow of ocean currents. Scientists closely study Antarctica's ice sheets for early warnings of climate change. However, findings from the studies do not always give scientists a clear picture of what is happening. For example, by early 2002, although one major ice shelf had shrunk to half its previous size, other parts of Antarctica were colder and had increased in size because more ice had formed.

Southern Ocean Data

Area	8,102,165 sq mi (20,973,318 sq km)
Average depth	11,188 ft (3,410 m)
Maximum depth	23,737 ft (7,235 m) (South Sandwich Trench)
Volume	17,165,720 cu mi (71,518,408 cu km)
Major geological features	*Plate*: Antarctic; *Ridges*: America–Antarctic, Atlantic–Indian, Macquarie, Pacific–Antarctic, Southeast Indian

Pollution and conservation

In the 1980s British scientists discovered a hole in the ozone layer above Antarctica. The ozone layer is part of the atmosphere—the envelope of gases that surrounds the Earth. The ozone layer helps to filter out ultraviolet radiation from the Sun, which is harmful to life on Earth. But some countries release pollutants—such as the propellants used in aerosol spray cans—into the atmosphere. These pollutants destroy part of the ozone layer, causing a hole to form. When this happens, higher-than-normal levels of dangerous ultraviolet light from the Sun stream through the hole.

One of the effects of this is to harm phytoplankton near the sea's surface. When phytoplankton are damaged, and their growth checked, it can have serious consequences for marine life, because so many animals depend on phytoplankton for food.

In the 1990s the ozone hole got bigger. But because many countries have reduced or stopped releasing ozone-depleting pollutants, the ozone hole may have stopped growing.

The Antarctic Treaty

Antarctica and the Southern Ocean are protected by strict environmental legislation—stricter than for any other part of the planet. In 1959, 12 nations signed an Antarctic Treaty, and by 2000 a further 32 nations had joined them. The Treaty members agree to keep Antarctica free of military forces, nuclear weapons, and nuclear power. They have agreed that the area is a special place to be set aside for scientific research.

This map shows the main underwater features of the Southern Ocean, as well as major seas, islands, and adjacent landmasses.

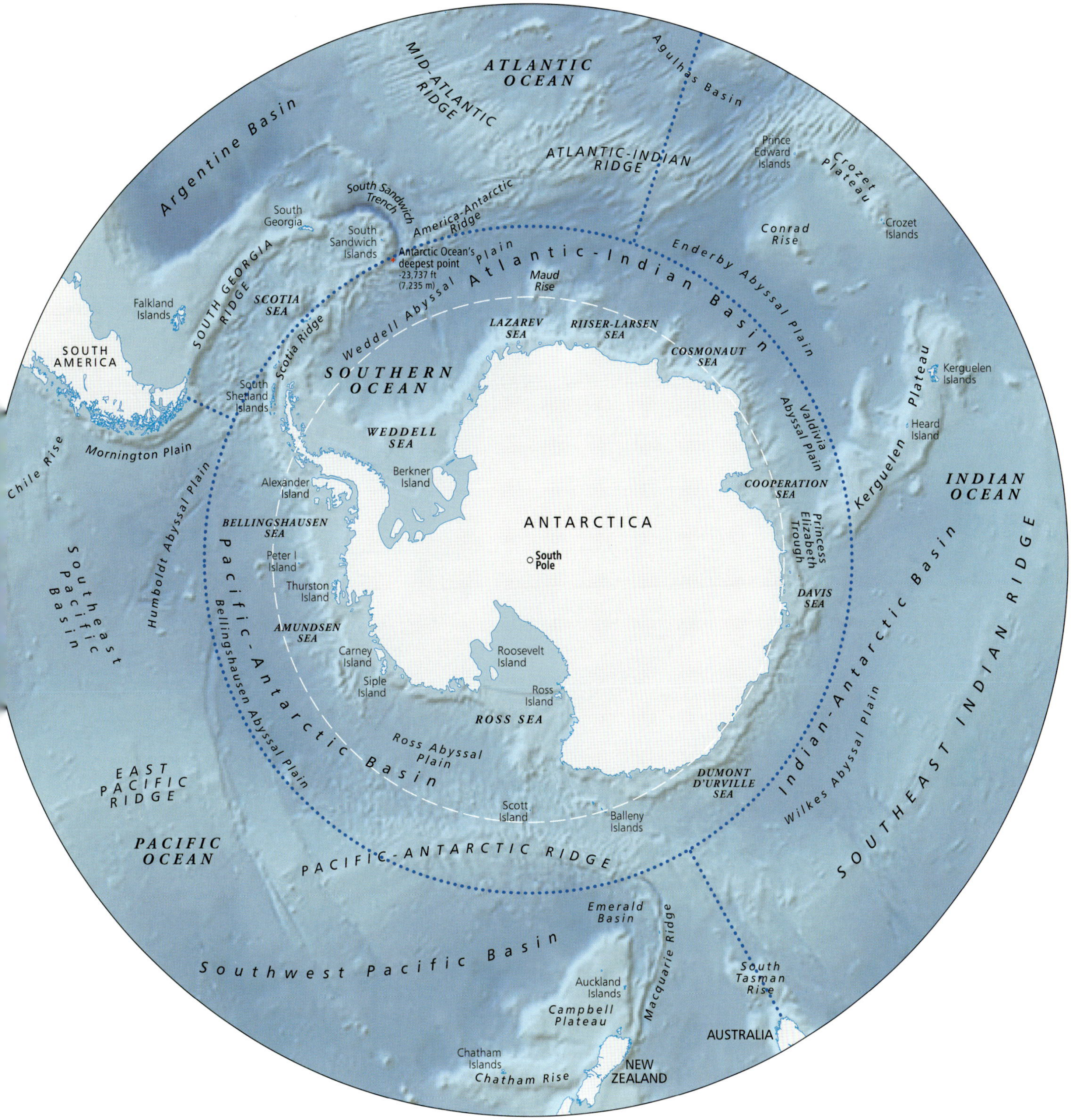
ATLANTIC OCEAN
MID-ATLANTIC RIDGE
Agulhas Basin
Argentine Basin
ATLANTIC-INDIAN RIDGE
Prince Edward Islands
Crozet Plateau
Crozet Islands
Conrad Rise
South Sandwich Trench
America-Antarctic Ridge
South Georgia
South Sandwich Islands
Antarctic Ocean's deepest point -23,737 ft (7,235 m)
SOUTH GEORGIA RIDGE
SCOTIA SEA
Falkland Islands
Weddell Abyssal Plain
Atlantic-Indian Basin
Enderby Abyssal Plain
Maud Rise
LAZAREV SEA
RIISER-LARSEN SEA
COSMONAUT SEA
SOUTH AMERICA
Scotia Ridge
South Shetland Islands
SOUTHERN OCEAN
Kerguelen Islands
Kerguelen Plateau
Heard Island
Valdivia Abyssal Plain
WEDDELL SEA
Berkner Island
Mornington Plain
Chile Rise
Alexander Island
COOPERATION SEA
INDIAN OCEAN
ANTARCTICA
Humboldt Abyssal Plain
BELLINGSHAUSEN SEA
Princess Elizabeth Trough
South Pole
Peter I Island
Southeast Pacific Basin
Thurston Island
DAVIS SEA
Pacific-Antarctic Basin
Bellingshausen Abyssal Plain
AMUNDSEN SEA
Carney Island
Roosevelt Island
Siple Island
Ross Island
ROSS SEA
Indian-Antarctic Basin
SOUTHEAST INDIAN RIDGE
Ross Abyssal Plain
EAST PACIFIC RIDGE
DUMONT D'URVILLE SEA
Wilkes Abyssal Plain
Scott Island
Balleny Islands
PACIFIC OCEAN
PACIFIC-ANTARCTIC RIDGE
Emerald Basin
Macquarie Ridge
Southwest Pacific Basin
South Tasman Rise
Auckland Islands
Campbell Plateau
AUSTRALIA
Chatham Islands
Chatham Rise
NEW ZEALAND

Earth's Timeline

This chart shows the major events that have occurred during the evolution of the Earth, although scientists do not always agree on the exact dates. Many of the changes take place over very long periods of time. Astronomers and physicists study the stars, planets, and other objects in space to understand when and how the Earth was formed.

Geologists and geophysicists investigate Earth's rocks and their formation to reveal how the Earth has changed since it came into existence. Paleontologists examine fossils and geneticists study genetic material. Together, the findings of all the different scientists help to give us a picture of the history of the Earth and its animal and plant life.

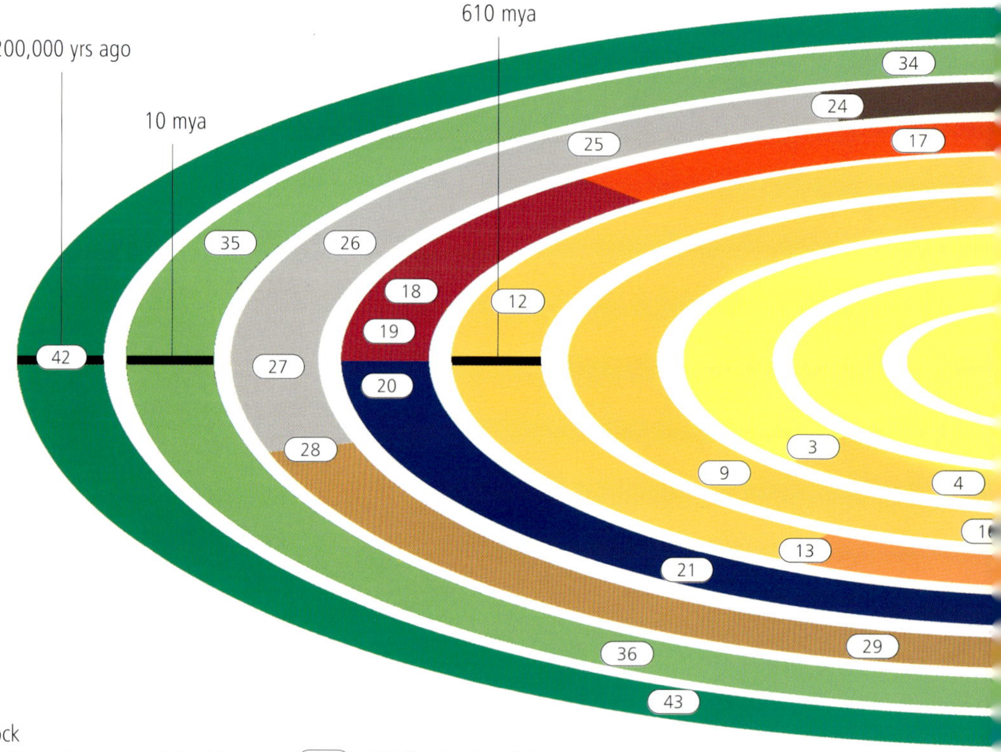

Billion years ago

1 **13** Formation of the Universe
2 **6** Solar system begins to form
3 **4.6** Earth forms as red-hot, molten rock
4 **4.4** Mars-sized body hits the Earth and may have created the Moon
5 **4.1** Earth's oldest rocks form
6 **4** Earth cools and first oceans form
7 **3.5** Simple microbes (earliest life forms) appear
8 **2.5** Photosynthetic microbes (cyanobacteria) release oxygen
9 **2** Oxygen levels in the atmosphere rise
10 **1.8** First complex-celled microbes (organisms with a nucleus) appear
11 **1.4** Early supercontinent Rodinia forms

Million years ago

12 **650** Ediacaran fauna ("the age of jellyfish")
13 **570** Supercontinent Rodinia breaks up; shallow seas become common
14 **550** Marine invertebrates with shells or skeletons become abundant and diverse during the Cambrian period
15 **480** First corals form
16 **460** First jawless fish appear
17 **455** Early land plants develop
18 **425** First jawed fish appear
19 **420** First land animals appear
20 **405** First insects appear
21 **375** First sharks appear
22 **360** Large, treelike ferns, clubmosses, and horsetails form the first land forests; first amphibians appear
23 **315–295** Swamp forests start to form major coal deposits
24 **250** Permian-Triassic mass extinction; many marine species die out
25 **245** Supercontinent Pangaea forms from the joining together of continents
26 **225** First dinosaurs appear
27 **210** Pangaea begins to break up; first mammals appear
28 **205** Triassic-Jurassic mass extinction; about one-third of marine species die out; global warming triggered by a meteorite impact may be the cause

29 **170** North Atlantic opens up

30 **160** First modern fish (teleosts) appear

31 **140** Southern supercontinent Gondwanaland begins to split up; South America separates from Africa; South Atlantic opens up

32 **90** Sinus Borealis moves north to become the Arctic Ocean

33 **65** Cretaceous-Tertiary mass extinction; dinosaurs die out; meteorite impact is the likely cause

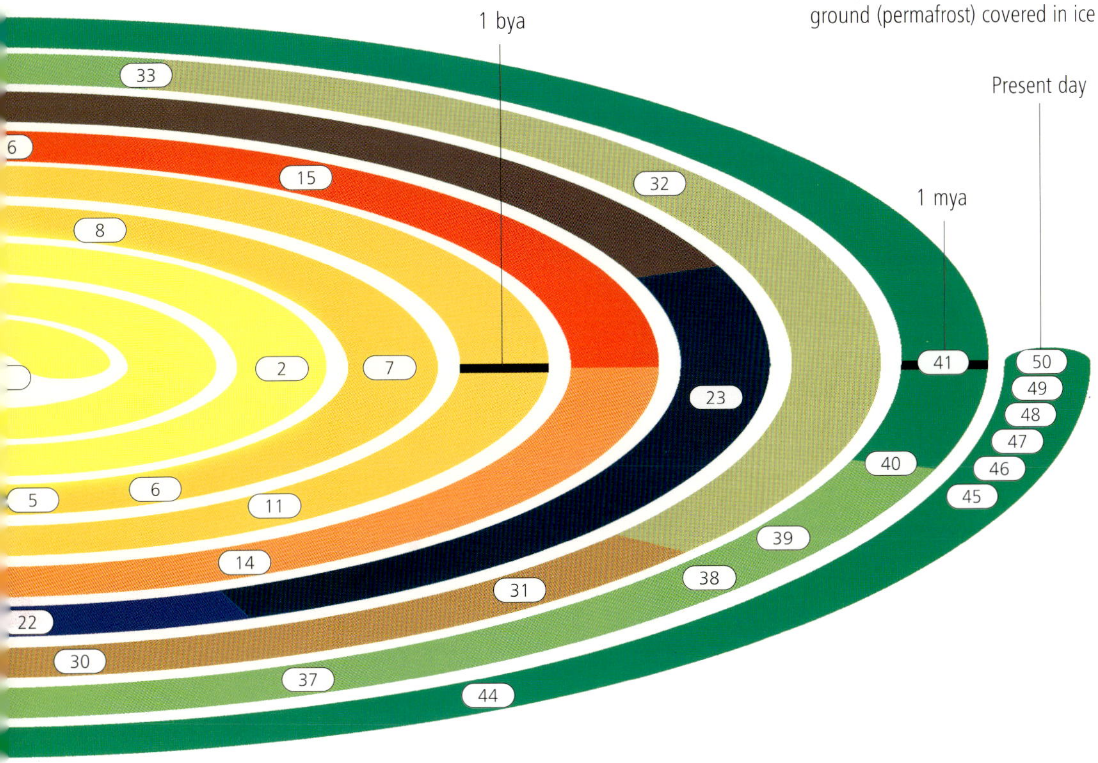

34 **55** First whales appear

35 **20** Australia moves north; Indian subcontinent collides with Eurasia, forming the Himalayas

36 **7** Mediterranean Sea becomes separated from the Atlantic Ocean and begins to dry out

37 **4.5** Mediterranean Sea reconnects with the Atlantic Ocean and sea water rushes in

38 **3.2** Isthmus of Panama connects North and South America

39 **3** Humanlike apes using stone tools

40 **1.8** Extensive northern glaciation begins; by 1.6 mya ice sheets will cover about one-third of Northern Hemisphere's land surface

41 **1** Fire used by early humans

Thousand years ago

42 **200** First modern humans (*Homo sapiens*)

43 **120** Bering land bridge between Siberia and Alaska becomes submerged

44 **70** Aboriginal people from Timor (part of Indonesia) colonize Australia by crossing the sea in boats

45 **20** Most recent ice age (glacial), with ice coverage at its maximum; Bering land bridge is well established; North Sea is mostly frozen ground (permafrost) covered in ice

46 **15** Current interglacial (warm period) begins

47 **12** Agricultural crops are grown in the Fertile Crescent (lower Nile Valley, eastern Mediterranean coast, Syria, and Mesopotamia)

48 **6** Egyptian papyrus-reed boats being used in the Mediterranean Sea

Recent

49 **200 years ago** Industrial Revolution in Europe and North America; atmospheric carbon dioxide levels begin to rise

50 **1990s** The warmest decade on record

How Deep?

How deep can marine creatures and underwater craft go? Environmental factors such as lack of light, chillingly cold temperatures, and crushing, high pressures limit the depths to which manned and unmanned craft, humans, and other creatures can descend—as the chart on the right shows.

1 Free diver: 531 ft (162 m). In January 2000 the Cuban diver Pipin Ferraras reached this depth without any pressure-protection or air supply. The pressure there is 16 times that at the surface. Ferraras trained for months before he undertook this feat.

2 Scuba diver: 925 ft (282 m). In 1994 Jim Bowden reached this depth in the Zacaton Sinkhole, Mexico. To avoid nitrogen poisoning, he breathed a helium-rich air mixture.

3 NEWTSUIT: 2,000 ft (609 m). To resist high pressures, the diver is enclosed in an armor-plated suit with power-assisted movable joints. He or she breathes air at normal surface pressure.

4 Bathysphere: 3,028 ft (923 m). In 1934 engineer Otis Barton and naturalist William Beebe descended in this steel spherical diving vessel. They were the first people to view deep-sea creatures in their natural environment.

5 Nuclear-powered submarine NR-1: 3,000 ft (914 m). This is a U.S. Navy research submarine. Military submarines may dive deeper, but the navy will not say how deep.

6 Giant tubeworms: 7,300 ft (2,225 m). In 1977 scientists in the submersible *Alvin* discovered the worms alongside hot-water vents in the eastern Pacific.

7 Sperm whale: 7,875 ft (2,400 m). Sperm whales at this depth are probably hunting giant squid. By storing oxygen in their muscles, slowing their heart rate, and reducing their activity, whales can stay submerged for 90 minutes.

8 *Alvin* three-person submersible: 14,765 ft (4,500 m). The recent, deep-diving version of this well-traveled submersible is one of a long series of *Alvin* designs that began in 1964.

9 *Nautile* three-person submersible: 19,685 ft (6,000 m). This French submersible was used in the 1997 movie *Titanic*.

10 Autonomous underwater vehicle (AUV) *Rover:* 19,685 ft (6,000 m). This self-contained vehicle, operated by the Scripps Institution of Oceanography, probes the sea floor and takes sediment samples.

11 Remotely operated vehicle (ROV) *Jason*: 19,685 ft (6,000 m). The robot vehicle, operated by Woods Hole Oceanographic Institution, is a versatile platform for taking samples and sensing the surroundings with light and sound. Operators connect it to another ROV called *Medea*.

12 Deep-water clams: 20,670 ft (6,300 m). An expedition in the submersible *Shinkai-6500* (see below) discovered the clams close to hot-water vents off the coast of Japan. The clams gain food from symbiotic bacteria that grow inside their bodies.

13 *Shinkai-6500* three-person submersible: 21,415 ft (6,527 m). In 1989 the Japanese craft reached this depth during a test dive.

14 Deep-sea anglerfish: 27,460 ft (8,370 m). The deepest recorded fish. The anglerfish is large-mouthed and has a reduced skeleton and weak muscles. Rather than hunting for prey, it attracts them with a bioluminescent lure attached to its head.

15 ROV *Kaiko*: 35,798 ft (nearly 10,912 m). In 1995 this robot vehicle explored the Challenger Deep in the Mariana Trench and came very close to breaking the deep-diving record. The ROV filmed deep-sea worms and amphipods (a type of crustacean) on the ocean floor.

16 Bathyscaphe *Trieste*: 35,800 ft (10,912 m). In 1960 Jacques Piccard and Don Walsh made this record-breaking dive in the Challenger Deep.

17 Amphipod crustacean: 35,840 ft (10,924 m). Numerous microbes, as well as invertebrates, live in the deepest part of the world's ocean. The pressures here are more than 1,000 times greater than at the sea surface, and the sea temperature is bitterly cold at 32–34° F (0–1° C).

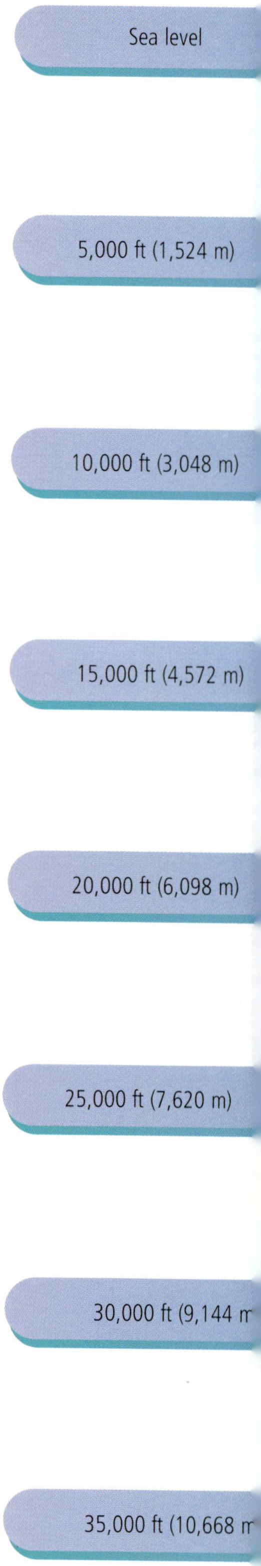

1
2
3
4
5
6
7
8
NAUTILE
9
12
10
11
13
14
16
17
15

Surface Exploration Timeline

Some of the most important events that have occurred worldwide during the exploration of the surface waters of the ocean are listed on these two pages. The events described here include some of the major advancements in the design of navigational equipment and charting, as well as the pioneering voyages of seafarers, scientists, and others whose discoveries helped to chart sea routes between nations and provide us with much of our knowledge about the sea.

ABOUT 4000 BCE Egyptians use papyrus-reed boats in Mediterranean Sea

ABOUT 2500 BCE Egyptian traders sail down the Red Sea to Punt (Somalia)

ABOUT 2500 BCE Extensive sea trade between people of the Persian Gulf and the Indian subcontinent

ABOUT 1500 BCE Southwest Pacific seafarers sail beyond sight of land to colonize uninhabited South Pacific islands

ABOUT 1100 BCE Phoenicians in merchant ships and war galleys travel throughout the Mediterranean Sea

ABOUT 600 BCE Phoenician expedition funded by Egyptian Pharaoh Necho II sails around Africa, from the Red Sea to the Mediterranean Sea

ABOUT 550 BCE Phoenician explorer Hanno sails to the Gulf of Guinea on Africa's west coast

ABOUT 330 BCE Greek explorer Pytheas (*c.* 380–300 BCE) sails from Massilia (Marseilles, France) to Britain, Ireland, Norway, and possibly Iceland. Pytheas notes that the height of the North Star varies according to latitude

ABOUT 220 BCE Greek geographer Eratosthenes (*c.* 275–194 BCE) publishes a chart of the known world. It includes lines that are early forms of latitude and longitude

ABOUT 150 CE Greek geographer and astronomer Ptolemy, also called Claudius Ptolemaeus, (*c.* 100–170 CE) compiles a map of the Roman world. It includes degrees of latitude and longitude

ABOUT 982 Norse chieftain Eric the Red (*c.* 950–1004) sails west from Iceland and discovers Greenland

ABOUT 1000 Leif Eriksson (about 975–1020), son of Eric the Red, sails from Greenland to Newfoundland and establishes a settlement there

ABOUT 1000 Chinese seafarers use the magnetic compass to assist navigation

ABOUT 1250 Middle Eastern and European seafarers use the graduated compass (marked with 32 points for precise reading) to assist navigation

ABOUT 1280 European seafarers in the Mediterranean Sea use harbor-finding charts (portulans) that incorporate measures of direction (bearings) to known landmarks

1405–1433 Chinese Admiral Zheng He (*c.* 1371–1434) commands fleets of junks that travel widely across the northern Indian Ocean as far as the east coast of Africa

1416 Prince Henry the Navigator of Portugal (1394–1460) founds his school of navigation that begins a "golden age of exploration"

ABOUT 1480 European sailors use the navigational instrument the astrolabe to determine latitude by measuring the height of the Sun or the North Star above the horizon

1487–1488 Portuguese navigator Bartholomeu Dias (*c.* 1450–1521) sails from the Atlantic Ocean to the Indian Ocean by traveling around the tip of Africa (Cape of Good Hope)

1492 Portuguese explorer Christopher Columbus (1451–1506) crosses the Atlantic Ocean, from Spain to the Bahamas, to make the European rediscovery of the Americas

1497–98 Portuguese explorer Vasco da Gama (1460–1524) sails from the Atlantic Ocean around Africa to India, opening up a sea-trade route to Asia

1519–22 Portuguese nobleman Ferdinand Magellan (1480–1521) leads the first expedition to sail around the world. Magellan is killed in 1521 and master navigator Sebastian del Cano completes the voyage in 1522

1534–36 French navigator Jacques Cartier (1491–1557) explores the Gulf of St. Lawrence, and St. Lawrence River, Canada, in his search for the Northwest

Passage between the Atlantic and Pacific Oceans

1569 Flemish map maker (cartographer) and mathematician Gerhardus Mercator (1512–94) publishes the Mercator projection—depicting a map of the world on a flat surface—as a navigational chart for seafarers

1609–10 English navigator Henry Hudson (1565–1611) explores the Hudson River and the Hudson Bay in his search for the Northwest Passage

1675 The Royal Observatory is established at Greenwich, London, England. Britain and some other countries adopt the Greenwich Meridian as the zero line of longitude

1735 Englishman John Harrison (1693–1776) makes the first very accurate ship's timepiece, or chronometer. Chronometers make it possible for ships to correctly measure longitude at sea

1768–79 Englishman Captain James Cook (1728–79) undertakes three major exploratory voyages of the Pacific Ocean. During them, he produces accurate charts and gathers scientific data on winds, currents, tides, and sea temperatures

1770 Benjamin Franklin (1706–90), colonial postmaster (responsible for managing early postal systems) and later American statesman, publishes the first navigational chart of the Gulf Stream

1831–36 Naturalist Charles Darwin (1809–92), aboard HMS *Beagle,* charts Pacific and Atlantic coasts and islands of South America. Darwin's observations lead to his theory of coral atoll formation and theory of evolution by natural selection

1855 U.S. Navy officer Matthew Fontaine Maury (1806–73) publishes *The Physical Geography of the Sea.* The book includes a wealth of information useful to seafarers and marine scientists. It is the first classic text in modern oceanography

1872–76 The British *Challenger* expedition, led by Sir Charles Wyville Thomson (1830–82), is the first largescale survey of the Pacific, Indian, and Atlantic Oceans

1884 The Greenwich Meridian is adopted worldwide as the zero line of longitude

1888 The Marine Biological Laboratory is set up at Woods Hole, Massachusetts

1893–96 Norwegian explorer and oceanographer Fridtjof Nansen (1861–1930) plots the flow of arctic pack ice, establishes that the Arctic contains a true ocean, and confirms there is no north polar continent

1895–98 American Joshua Slocum (1844–1909) in his boat *Spray* becomes the first person to sail around the world single-handedly

1903 The laboratory that later becomes the Scripps Institution of Oceanography is founded in San Diego, California

1903–1906 Norwegian explorer Roald Amundsen (1872–1928) heads the first expedition to successfully navigate the Northwest Passage

1908 U.S. inventor and businessman Elmer Ambrose Sperry (1860–1930) patents a gyroscopic compass design. Unlike a magnetic compass, it always points to true north

1930 The Woods Hole Oceanographic Institution is established at Cape Cod, Massachusetts

1946–47 Norwegian Thor Heyerdahl (1914–2002) crosses the Pacific Ocean in the balsa-wood raft *Kon-Tiki*

1946–47 Radio navigation systems developed during World War II—the U.S. LORAN (**Lo**ng **R**ange **A**id to **N**avigation) and the British DECCA—enter widespread service

1978 The first oceanographic satellite, *SEASAT-A,* is launched. Its success demonstrates the usefulness of satellite remote sensing in studying the oceans

1993 The 24th Navstar satellite is placed in orbit, completing the global positioning system (GPS). The system allows seafarers to establish their position easily, accurately, and inexpensively

1994 The Law of the Sea Treaty enters into force. This treaty deals with national claims of ownership. The treaty also governs how nations should use and safeguard the oceans

2000 The *Argo* project is launched as part of the Integrated Global Observing Strategy (IGOS). By 2003, about 3,000 *Argo* buoys are planned to monitor the temperature, saltiness, and currents of the oceans. They will regularly transmit their data via a satellite, *Jason-1*

2002 The European remote-sensing satellite *ENVISAT* is launched. Its functions include: monitoring marine pollution, wind, and sea wave patterns, and recording changes in sea temperature and ice cover

UNDERSEA EXPLORATION TIMELINE

Some of the most important events that have occurred worldwide during underwater exploration of the ocean are listed on these two pages. The events described here include some major developments in the design of diving equipment, submarines, and submersibles (research vessels that can go under water). Also described are some of the key scientific discoveries that have come about through exploration of the deep ocean waters and the sea floor.

ABOUT 4500 BCE Mesopotamian divers are diving to depths of over 33 ft (10 m) to gather shellfish

ABOUT 350 BCE Greek philosopher Aristotle (*c.* 384–322 BCE) describes a diving bell. It consists of a container lowered under water and filled with air that divers can enter

1620 CE Dutch inventor Cornelis Drebbel (1572–1633) successfully tests an oar-driven submarine in the River Thames, London, England

1674 English scientist Robert Boyle (1627–91) publishes his findings on how temperature, saltiness, and pressure in the sea alter with depth

1690 English scientist, astronomer, and mathematician Edmond Halley (1656–1742) designs and tests a diving bell. Within the decade, divers use similar bells to recover objects from wrecks

1775 American David Bushnell invents a one-person submarine, the *Turtle.* In the following year his brother, Ezra Lee, uses the submarine to try to detonate an explosive charge against the hull of a British ship

1817–18 An expedition led by Scottish explorer Sir John Ross (1777–1856) hauls up worms and starfish from a depth of 5,900 ft (1,800 m) in the Arctic Ocean

1837 German inventor Augustus Siebe (1788–1872) tests a diving helmet and suit. The Siebe suit and helmet become the standard deep-diving equipment of the 19th century

1839–43 British explorer Sir James Clark Ross (1800–62), nephew of Sir John Ross, leads an expedition to the Southern Ocean that dredges up marine life from the sea floor at depths greater than 13,100 ft (4,000 m)

1854 Sir Edward Forbes (1815–54) publishes *Distribution of Marine Life.* Despite the findings of the Rosses, Forbes argues that life cannot exist below about 2,000 ft (600 m), the so-called "azoic zone"

1855 U.S. naval officer Matthew Fontaine Maury (1806–73) publishes the first depth chart of an ocean basin. It shows the depths of the North Atlantic Ocean floor based on soundings (depth readings)

1858 The first transatlantic telegraph cable is laid

1868–70 Scottish scientist Charles Wyville Thomson (1830–82), in expeditions aboard HMS *Lightning* and HMS *Porcupine*, collects marine organisms from great depths—conclusively disproving Forbes's azoic zone theory

1872–76 Charles Wyville Thomson leads the *Challenger* expedition that makes an extensive study of the Pacific, Atlantic, and Indian Oceans. The expedition takes 492 deep-water soundings (depth readings) and 133 dredges from the seabed. They discover metallic nodules and plumb the Mariana Trench to a depth of 26,850 ft (8,185 m)

1912 German meteorologist and geologist Alfred Wegener (1880–1930) suggests that the continents float slowly about on the surface of the globe (Wegener called this theory "continental drift")

1920 Alexander Behm (1880–1952) bounces sound waves off the bottom of the North Sea in an early test of echo sounding, later called sonar

1925–27 German expedition in the research vessel *Meteor* makes detailed oceanographic studies of the Atlantic Ocean. It is the first scientific cruise to use sonar to map the ocean floor

1934 Naturalist William Beebe (1877–1962) and engineer Otis Barton descend to a depth of 3,028 ft (923 m) in a bathysphere (a spherical diving chamber lowered by cable). They are the first people to observe deep-sea life in its natural habitat

Mid-1930s Marine geologist and geophysicist Maurice Ewing (1906–74) develops seismic surveying (controlled explosions and analysis of returning sound waves) to map seabed sediment and the oceanic crust

1943 Frenchmen Jacques Cousteau (1910–97) and Émile Gagnan invent the aqualung for divers—a special mouthpiece attached to an air cylinder. It becomes the most successful form of scuba (**s**elf-**c**ontained **u**nderwater **b**reathing **a**pparatus)

1948 Swiss engineer Auguste Piccard (1884–1962) builds the bathyscaphe (a deep-diving chamber suspended beneath a buoyancy and propulsion unit)

1949 The Lamont (later Lamont Doherty) Geological Observatory is set up at Columbia University, New York

1951 British scientists in the research ship *Challenger II* use sonar to find the deepest part of the ocean, the Challenger Deep in the Mariana Trench, at a depth of about 6.8 mi (11 km)

1952 American geologist and map maker Marie Tharp identifies a rift valley in the Mid-Atlantic Ridge

1956 American marine geologists Maurice Ewing and Bruce Heezen (1923–77) announce the existence of the global mid-ocean ridge system

1958 The nuclear-powered USS *Nautilus* is the first submarine to pass under the ice at the North Pole

1960 Jacques Piccard, Auguste Piccard's son, and U.S. Navy Lieutenant Don Walsh dive 35,800 ft (10,912 m) deep in the bathyscaphe *Trieste*. No one has ever dived deeper

1960 The nuclear-powered USS *Triton* becomes the first submarine to travel completely around the globe under water without surfacing

1960 North American geologists Harry Hess (1906–69) and Robert Dietz (1914–95) propose the theory of sea-floor spreading (a mid-ocean ridge laying down new sea floor)

1962 Two French divers spend seven days at a depth of about 33 ft (10 m) in Jacques Cousteau's *Conshelf I*, an underwater diving platform

1964 Submersible *Alvin* enters service for the Woods Hole Oceanographic Institution

1967 The wreck of the Tudor flagship *Mary Rose* is located near Portsmouth, England, using a wide sonar beam to create a 3-D image of the ocean floor

1968 U.S. drilling ship *Glomar Challenger* extracts deep-sea cores of sediment and oceanic crust. They provide evidence in support of sea-floor spreading which, in turn, helps explain plate tectonics theory (the view that the Earth's crust consists of several giant, slow-moving plates)

1969 Armored diving suit, JIM (named after the deep-diving pioneer Jim Jarrett) is developed for use at depths below 1,000 ft (300 m)

1974 **F**rench-**A**merican **M**id-**o**cean **U**ndersea **S**tudy (FAMOUS) team maps the Mid-Atlantic Ridge

1977 Marine geologist Bruce Heezen and map maker Marie Tharp publish *The Floor of the Oceans*—a map that gives an impression of what an ocean would look like if it was drained of water

1977 American scientists on board the submersible *Alvin* discover unusual life forms, such as 3-ft (1-m) long worms, around deep-sea hot-water (hydrothermal) vents in the eastern Pacific Ocean

1985 Robert Ballard and an American-French team use remotely operated vehicles (ROVs) to locate the wreck of the RMS *Titanic* in about 2.5 mi (4 km) of water. The following year the team explores the wreck using ROV *Jason Junior* attached to submersible *Alvin*

1990 The U.S. drilling ship *Joides Resolution* retrieves a deep-sea sediment sample estimated at 170 million years old

1995 The Japanese ROV *Kaiko* dives to a depth of 35,798 ft (nearly 10,912 m), almost equaling *Trieste*'s descent made 35 years earlier

1995 The U.S. Navy releases satellite altimetry data (based on bouncing radio waves from a satellite to measure the level of the surface of the sea). This enables geographers and geophysicists to begin compiling a detailed 3-D digital chart of the ocean floor

1997 U.S. geophysicists David Sandwell and Walter Smith publish a new map of the global ocean floor based on satellite altimetry data combined with sonar survey data

2000 Robert Ballard's Institute of Exploration team uses the ROV *Little Hercules* to find a very well-preserved ship in the Black Sea. The vessel is 1,500 years old

2002 The underwater laboratory *Aquarius-2000* continues its program of research studying Florida's coral reef ecology

BIOGRAPHIES

Throughout the long and exciting history of ocean exploration many different scientists, navigators, and voyagers have contributed to our vast knowledge of the sea and the ocean floor. A brief life history and the major achievements of some of these people can be found on the following pages.

Agassiz, Alexander (1835–1910)
American naturalist and industrialist Alexander Agassiz was the son of Louis Agassiz (1807–1873), professor of zoology and geology at Harvard University. Alexander Agassiz was born in Switzerland. He came to the United States in 1849 and studied engineering and natural history at Harvard. A successful and generous copper mine owner, he used his fortune to fund scientific expeditions and university research.

In 1877 Agassiz carried out the first of his detailed oceanographic observations of the Pacific and Caribbean. As a result of his investigations, he suggested that these oceans were connected until the rise of the Panama Isthmus (the land connecting North and South America) during the Cretaceous period (about 135 to 65 million years ago).

Agassiz investigated the life cycles of spineless animals such as corals and starfish. He analyzed the sea urchins brought back from Charles Wyville Thomson's 1872–76 *Challenger* expedition, and Agassiz's engineering expertise helped to improve the equipment used to collect samples on this and other oceanographic expeditions. Agassiz encouraged, and partly provided the money for, the oceanographic center that became the Scripps Institution of Oceanography at La Jolla, California.

Prince Albert I of Monaco (1848–1922)
Born into Monaco's royal family, Albert became a navigator, scientist, and sponsor of oceanographic expeditions. Between 1884 and 1915 he led or sponsored research cruises that made major advances in marine biology, oceanography, and the design of sampling equipment. In 1868 Albert joined the French Navy and fought in the Franco-Prussian War. After the war, in 1873, he purchased his first sea-going ship. By the early 1880s his vessels cruising the Mediterranean Sea and North Atlantic were equipped with the latest oceanographic equipment. One of Albert's experiments involved floating bottles and barrels in the sea to study the movements of currents. By the late 1880s he had proved the existence of the North Atlantic Gyre using these methods.

In 1889 Albert became head of the principality of Monaco but still found time for oceanography. He coordinated the publication of the first depth chart of all oceans, completed in 1905. The following year Albert established the Oceanographic Institute in Paris, and in 1910 the Oceanographic Museum of Monaco. Some of Albert's ideas were well ahead of his time. He warned about the dangers of overfishing, proposed that aircraft could be used to spot schools of fish, and that marine sanctuaries should be set up to conserve wildlife.

Amundsen, Roald (1872–1928)
Norwegian explorer who commanded the *Gjöa* on the first complete sea journey through the Northwest Passage in 1903–1906 (previous journeys were partly on foot over ice). He became the first person to reach the South Pole in 1911. He published his account of the voyage of the *Gjöa* in *Amundsen's North West Passage* (1908).

Amundsen then attempted to be the first person to reach the North Pole but abandoned the endeavor on hearing that Robert Peary had reached it first in 1909. His book *The South Pole* (1913) tells the story of his dash to the South Pole. Using dog teams and skis, he approached from the Bay of Whales on the Ross Sea, arriving on December 14, 1911, 35 days before the British explorer Robert F. Scott.

Amundsen contributed much to the scientific and geological knowledge of Antarctica before returning to the Arctic in 1918 and setting out in the *Maud* to sail along the whole northern coast of Europe and Asia. Amundsen then turned to air exploration and, after a number of unsuccessful attempts, flew over the North Pole in 1926 in an inflatable airship, or blimp, built and piloted by the Italian explorer, Umberto Nobile. In 1928 Umberto Nobile's airship crashed into the Arctic Ocean. Amundsen was killed when his own airship crashed while attempting to rescue Nobile.

Ballard, Robert (1942–)
American marine geologist and oceanographer. Ballard combines his scientific research and underwater exploration with publishing books and setting up projects intended to educate the public and involve young people in scientific investigation. A graduate of the University of California, where he gained degrees in chemistry and geology, Ballard became head of the Deep Submergence Laboratory at the Woods Hole Oceanographic Institution in the early 1970s. He participated in the development of the manned deep-sea submersible *Alvin*, which scientists used to map parts of the Mid-Atlantic Ridge. Scientists in *Alvin* observed hot-water (hydrothermal) vents while submerged near the Galápagos Islands in 1977. These are vents in the sea floor that discharge water high in mineral content and heated by magma in the Earth's crust. Unique communities of marine creatures develop around such vents.

Ballard led expeditions that located the wrecks of the *Titanic*, *Lusitania*, and the German battleship *Bismarck*. In the 21st century his exploration institute in Mystic, Connecticut, has been exploring the Mediterranean and the Black Sea and has found evidence of a settlement that may date to 3000 BCE. He has also founded the Jason Project, a U.S. venture that allows young people to participate in scientific expeditions.

Battuta, Ibn (1303–1365)
Arab explorer, born at Tangier, Morocco, who spent 30 years intermittently traveling an estimated 75,000 mi (120,000 km), including extensive journeys across the northern Indian Ocean. He began by traveling to Makkah (Mecca) in 1325–26. His first journey took him through northern Africa, Egypt, and the Persian Gulf.

He returned to Makkah for a year before setting out again down the coast of East Africa and across to the Gulf of Oman. His third journey took him to Asia Minor. He crossed the Black Sea and wandered in the Ukraine, eventually returning to Constantinople (Istanbul). In the late 1320s and early 1330s Battuta traveled to India by way of Afghanistan, where he spent eight years at the court of the Sultan of Delhi.

He later sailed to China as an envoy. On his way to China he endured storms and piracy. During the journey he visited the Maldive Islands, the west coast of India, Ceylon (present-day Sri Lanka), Sumatra, Myanmar, and Cambodia. Battuta returned home overland by way of Makkah, with a detour to Sardinia.

After settling back in Tangier, he briefly visited Spain. Finally, in about 1351, he made his last journey across the Sahara to visit Timbuktu and the lands of the Niger. He dictated a story of his travels, *The Rihlah,* to an assistant called Ibn Juzayy, and this has remained a largely reliable source for the geography and social history of the region in the medieval period.

Boyle, Robert (1628–1691)
Anglo-Irish chemist and physicist from an aristocratic background who pioneered experimentation using scientific principles. He carried out most of his research in his laboratory in the city of Oxford and gave the first precise definition of a chemical element, a chemical reaction, and of chemical analysis. Boyle designed a vacuum pump (a pump used to take all the air from a container) and showed that air is necessary to transmit sound, for a candle to burn, and for life in air-breathing animals. He formulated Boyle's law, which states that the volume of a gas at a constant temperature is inversely proportional to its pressure. For example, if the pressure of a gas doubles, its volume will be reduced by half, and vice versa.

Boyle conducted many experiments. At sea, he measured sea water saltiness, temperature, and depth and described his findings in three important essays. He also performed chemical analyses on gold and silver, and he devised a way of analyzing the minerals in water. He also realized that plant extracts, notably litmus, would indicate the presence of acid and alkaline substances by changing color. Boyle was probably the first chemist to collect a sample of gas.

Carson, Rachel (1907–64)
American writer, marine biologist, and conservationist. As a child, Carson had a fascination for all forms of natural life. After working for the U.S. Bureau of Fisheries as a radio scriptwriter, she joined their staff as an aquatic

biologist in 1936. She wrote an influential trilogy of books about the sea—*Under the Sea-Wind* (1941), *The Sea Around Us* (1951), and *The Edge of the Sea* (1955).

Carson became chief editor of the U.S. Fish and Wildlife Service but resigned in 1952 to devote herself to science writing. Her bestselling book, *Silent Spring* (1962), warned of the dangers of the indiscriminate use of pesticides and strongly influenced the environmental and conservation lobbies that emerged in the late 20th century. In her writing she showed a deep understanding of nature and put forward well-researched scientific arguments about how human activities were threatening the natural environment.

Columbus, Christopher (1451–1506)
Italian-born navigator and explorer who made four voyages to the New World funded by the Spanish monarchy. Columbus was born in Genoa and went to sea at an early age before settling in Lisbon in 1476, where he became a chart maker and master mariner. He believed that it was possible to reach the Far East from Europe by sailing westward. He set out on his first voyage in 1492 in the *Santa Maria,* accompanied by two other ships, *Niña* and *Pinta.* On October 12 they landed on a small island in the Bahamas, reaching Cuba and Haiti a few weeks later.

Columbus returned to Spain in March 1493, where he was honored. He sailed west again from 1493 to 1496, landing in Guadeloupe, Montserrat, Antigua, Puerto Rico, and Jamaica. In 1498 he explored Trinidad and the mainland of South America, but he became involved in quarrels with the colonists, and he traveled back to Spain in chains. His last voyage was from 1502 to 1504 when he reached Honduras and Nicaragua. He was then marooned on Jamaica but eventually rescued. He died in poverty in Spain in 1506 at age 55.

Cook, James (1728–1779)
English explorer, navigator, and mapmaker. He joined the Royal Navy in 1755 and, during the Seven Years' War (1756–63), surveyed the St. Lawrence River in Canada and the coasts of Newfoundland and Labrador. In 1766 Cook provided detailed and accurate observations of an eclipse of the Sun. He was commissioned to take an expedition to Tahiti to observe the passage of Venus across the Sun and to search for the fabled southern continent, *Terra Australis.* He sailed in 1768, observed the passage of Venus, and then explored and charted New Zealand and the east coast of Australia, which he claimed as a British colony. He returned to England in 1771.

On his second voyage in 1772–75 he charted Easter Island, New Caledonia, and the Marquesas Islands of the South Pacific. He circumnavigated Antarctica, but sea ice prevented him getting close enough to the landmass to see it. In fact, Cook was convinced that a great southern continent did not exist. His third and last voyage began in 1776. He visited the South Pacific and then sailed up the west coast of North America, carrying out a coastal survey and hoping to find the western entrance to the Northwest Passage (the northern sea route linking the Pacific and Atlantic Oceans). His way was blocked by ice in the Bering Strait, but he surveyed the Russian coast on the eastern side of the Strait before returning to Hawaii in early 1779. He was killed there when he tried to intervene in a dispute between islanders.

James Cook's major legacy was in charting many of the coasts and islands of the Pacific Ocean and in advancing the use of accurate scientific observation in the study of the sea.

Cousteau, Jacques (1910–1997)
French documentary film maker, diving pioneer, and co-inventor of the aqualung. In 1930 Cousteau enroled in the French naval academy at Brest with the intention of becoming a pilot, but a serious car accident thwarted this ambition. At Brest, and later as a naval artillery instructor at Toulon, Cousteau devised makeshift goggles and snorkels for underwater use. During World War II he spied for the French Resistance. In 1943 Cousteau worked with engineer Émile Gagnan to create a form of underwater breathing apparatus called the aqualung, a type of scuba (**s**elf-**c**ontained **u**nderwater **b**reathing **a**pparatus). Their device delivered air to the diver at the same pressure as the surrounding water, making underwater breathing easier and safer. With an aqualung, divers can swim freely about under water using an air supply carried on their backs.

After the war, with naval backing, Cousteau set up the Undersea Research Group. It developed the aqualung and later experimented with underwater habitats for divers. In 1950 Cousteau acquired the minesweeper *Calypso* and converted it into an oceanographic vessel. His diving exploits were vividly described in his 1953 book, co-authored with diver Frédéric Dumas, *The Silent World.* In 1956 his exploits were featured in a full-length film of the same name. Cousteau was criticized for lacking scientific credibility and his early emphasis on adventure rather than environmental concerns. Nevertheless, by the late 1960s, Cousteau was a staunch campaigner for marine conservation. He introduced millions of people to the underwater world through his documentaries (at least 115), his writing (he co-authored more than 100 books), and though his co-invention, the aqualung.

Darwin, Charles (1809–1892)
British naturalist and scientist who studied medicine and theology before undertaking a five-year cruise aboard the *Beagle* (1831–36) as the ship's naturalist. During the voyage to South America and the Pacific, Darwin methodically collected data and specimens and began to develop his ideas about evolution. The theory of evolution by natural selection was presented to the Linnaean Society in London in 1858, jointly with Alfred R. Wallace, who had arrived at similar conclusions. Darwin published his influential work, *On the Origin of Species by Means of Natural Selection or the Preservation of Favoured Races in the Struggle for Life,* in 1859.

Darwin's observations on fossils, variations within species, and relationships between closely related island species—such as the finches of the Galápagos Islands—led him to conclude that living organisms evolve (slowly change) over many generations under the influence of natural selection. He believed that the individuals within a population that are best suited to their environment are the ones most likely to survive and produce offspring, and that their successful characteristics will be inherited.

Darwin's theory caused bitter dispute because it conflicted with the biblical story of Creation. Yet it gradually came to be accepted by most scientists as a fundamental biological principle. Darwin also wrote a very important work on barnacles, and he proposed a theory on the formation of coral reefs around islands. He suggested that a simultaneous sinking of the sea floor and outward growth of coral combine to form first a fringing reef, then a barrier reef, and finally an atoll.

Dietz, Robert (1914–1995)
The American geologist Robert Dietz played a major part in the record-breaking dive of the bathyscaphe (deep-diving submersible) *Trieste.* He was also the co-developer of the theory of sea floor spreading (how new sea floor is created at mid-ocean ridges and gradually moves away from the ridge). Most of Dietz's research was done at the Scripps Institution of Oceanography (SIO). Dietz's early research involved describing and mapping submarine canyons and sea floor sediments. In the 1940s and 1950s he worked jointly for the U.S. Navy and SIO and helped map important features of the Pacific Ocean basin. Dietz undertook pioneering underwater exploration firsthand and was one of the first scientists to dive with an aqualung to survey the seabed.

He played a key role in convincing the U.S. Navy to help in the record-breaking bathyscaphe dive to the bottom of the Mariana Trench in 1960. This feat is described in Robert Dietz and Jacques Piccard's book, *Seven Miles Down* (1960). In the same year Dietz published an article in the science journal *Nature* that described sea floor spreading. This idea, independently arrived at by Harry Hess (an American geologist based at Princeton University), was a major contribution to plate tectonics theory. Dietz was also one of the first people to recognize the importance that meteorite impacts played in Earth's evolution. Dietz has a Pacific seamount, an antarctic mountain, and an asteroid named after him.

Earle, Sylvia (1935–)
American marine biologist and oceanographer Sylvia Earle has combined a distinguished academic career with a great deal of practical research under the sea. In 1970 she lived for 14 days in an underwater chamber with four other oceanographers in order to study undersea habitats in the Virgin Islands. She has pioneered the use of scuba (**s**elf-**c**ontained **u**nderwater **b**reathing **a**pparatus) for biological research. In 1979, using a special suit called a JIM suit, she made a record-breaking, untethered dive (one in which the diver is not attached to the surface) to a depth of 1,250 ft (380 m). Earle has made a special study of marine algae, especially seaweeds, and has collected over 20,000 samples. Her discovery of undersea dunes off the Bahamas led to a greater understanding of how this part of the ocean was formed.

In the early 1980s Earle co-designed a low-cost, underwater submersible. It could reach depths of 3,000 ft (910 m). In the 1990s she worked with Japanese scientists to design a craft that could dive to 36,000 ft (11,000 m). Earle was the first woman to serve as chief scientist at the National Oceanic and Atmospheric Administration, the U.S. government agency that manages fisheries and marine sanctuaries and conducts oceanographic research. A tireless worker for marine conservation, in 1995 she published the popular book *Sea Change: A Message of the Oceans.* Earle has also directed the "Sustainable Seas Expeditions," a five-year project to explore and document with photographs the geology and natural life of the seas in U.S. waters.

Eriksson, Leif (about 980–1020)

Icelandic explorer and son of Norwegian explorer, Eric the Red, Leif Eriksson grew up in Greenland, where his exiled father had established a settlement. In around 1000, Eriksson led an expedition across the Atlantic Ocean in search of a fertile land west of Greenland, first glimpsed by a Viking seafarer some years before. Leif's expedition reached a place they called Helluland, meaning "land of stone slabs," which was almost certainly Baffin Island. They then sailed south, reaching first Markland ("land of woods") and then a lush landscape they called Vinland, "land of the vine," due to the many trees and bushes bearing berries and grapes. Historians believe that Eriksson's Vinland lay on some part of North America, probably Newfoundland.

In 1960 Norwegian explorers Helge Ingstad and Anne Stine Ingstad followed the route of Viking voyages recorded in medieval Icelandic manuscripts and located the remains of eight buildings and hundreds of Viking artifacts at L'Anse aux Meadows in Newfoundland. In 1965 a 15th-century map of Vinland was published. It depicted Viking exploration of North America centuries before Columbus discovered the New World.

da Gama, Vasco (1460–1524)

Portuguese navigator who became the first European to journey by sea to India, in 1497. He established a sea route for trade with the Indies, enabling Portugal to grow rich from the spice trade. Da Gama sailed around the Cape of Good Hope, landed at and named the port of Natal in South Africa, sailed up the east coast to Mozambique, and crossed the Indian Ocean. He landed at Calicut in India (now Kozhikode in Kerala) in 1498. In 1502 da Gama led another expedition to India, during which he founded a Portuguese colony in Mozambique. He then sailed on to attack Calicut in revenge for the murder of Portuguese settlers. He was made viceroy of India in 1524 but died shortly afterward.

Hardy, Alister (1896–1985)

As a child, Hardy showed a talent as a natural history artist. After serving with the Royal Engineers during World War I, Hardy studied zoology at Oxford University before joining the Fisheries Laboratory at Lowestoft, England, as a naturalist. His early experiments with a fast-towed net helped locate the tiny, drifting plankton on which herring fed. This information helped commercial fishers. In 1924 Hardy joined the four-year *Discovery* expedition to Antarctica as chief zoologist and observed and drew antarctic marine life.

On his return he joined University College, Hull, England, as a professor of zoology and oceanography, where he developed the continuous plankton recorder that bears his name. This device allows a non-specialist to take plankton samples continuously from a moving ship. The plankton recorder trails in the water and captures plankton between two layers of mesh. When analyzed in the laboratory, the section of mesh can be linked to the location where the plankton were caught. Since the 1930s, data from annual surveys using this method has benefited the fishing industry, and changes in plankton composition reveal long-term trends in climate change. Hardy also wrote the influential two-volume work, *The Open Sea,* and in 1957 he was knighted for his achievements in marine biology.

Harrison, John (1693–1776)

British creator of accurate timepieces that revolutionized the measurement of longitude at sea. (A reading of longitude provides sailors with their east-west location anywhere on the globe.) The son of a carpenter, Harrison became a cabinet maker. He built his first longcase (grandfather) clock entirely of wood in 1713, at age 20. By the mid-1720s he was making pendulum clocks that were accurate to within one second a month.

In 1707 a fleet of British naval ships lost 2,000 lives in shipwrecks caused by an inaccurate reading of longitude. As a result, in 1714, the British Government offered £20,000 to anyone who could devise a way of measuring longitude to within half a degree (two minutes of time) on a return trip from England to the West Indies (Caribbean). Harrison took up the challenge. One means of measuring longitude involved accurately comparing local time at sea with an international standard such as Greenwich Mean Time. The time difference gives a measure of longitude. This requires precise timekeeping, and clocks at sea were notoriously inaccurate. Harrison changed that. He completed his first marine timekeeper, or ship's chronometer, in 1735. Although it was large and needed regular adjustment, it was very accurate. Over the next 35 years Harrison designed four more chronometers, each one smaller and more accurate.

Captain Cook successfully tested Harrison's fifth chronometer during his second Pacific expedition (1772–75). Harrison was cheated of his prize for a long time, however. The judges, a panel of scientists, were in favor of complicated scientific methods of longitude measurement that involved measuring the height of the Moon and bright stars in the sky and the angles between them. They wanted a scientist to win the prize. Eventually, in 1773, when Harrison was about 80, he received some of the prize money after an appeal by his son, William Harrison, and the King's astronomer, Stephen Demainbray.

He, Zheng (Ho, Cheng) (about 1371–1434)
Zheng He was born at K'unming in the Yunnan province of China. In 1381 the Chinese Ming dynasty drove the Mongol Yuan dynasty out of Yunnan and Zheng He was captured and drafted into the army. He had a distinguished army career and made influential friends at court. In 1405 he was appointed commander-in-chief of missions to the western oceans. He set out, with several hundred ships and nearly 28,000 troops, to successfully defend the trading city of Malaika in present-day Malaysia. Zheng He's ships were large compared with European ships of the time and had advanced design features, such as multiple rudders for precise steering and watertight compartments that helped stop the ship from flooding if it was holed.

Zheng He made seven voyages between 1405 and 1433, traveling more than 30,000 mi (50,000 km). He visited Southeast Asia, India, Arabia, and Africa.

Prince Henry, the Navigator (1394–1460)
A Portuguese prince, Henry fought a military campaign in Morocco and then became interested in exploring Africa. He established a base for expeditions at Sagres in southwestern Portugal in 1416; this was later extended to include an observatory and a school for the study of geography and navigation. Under Henry's guidance, the school made notable contributions to the art of navigation. It compiled tables to help determine latitude, and it produced detailed nautical charts. Expeditions from the school explored Madeira, the Cape Verde Islands, the Azores, and the west coast of Africa almost as far as present-day Sierra Leone.

The rapid growth of the slave trade in the mid-1440s accelerated exploration. During the following decade many Portuguese vessels sailed down the west African coast. Exploration from Henry's school laid the foundations for the establishment of Portugal's colonial empire and its rise to prominence as a political and economic power in the 16th century.

Hensen, Victor (1835–1924)
The biologist Victor Hensen was born in Schleswig, Germany. He studied medicine and became a professor of physiology (the study of how living creatures function) at Kiel's Institute of Anatomy in 1868. Hensen made discoveries about animal and human hearing and sight, and several parts of the ear are named after him.

His spare-time interest was the study of marine life. He began to develop ways of estimating fish stocks, and from the 1870s was concerned with using nets to catch plankton—the food of fish such as herring—and calculating plankton abundance. Hensen coined the term "plankton" (from the Greek word for "drifting") in 1887. Hensen originally included swimmers such as fish and squid in his definition of plankton, but later the term included only organisms that drift or float in the water.

In 1889 Hensen led the "Plankton Expedition" on the steamer *National.* The expedition's cruise of the Atlantic Ocean sampled plankton at various depths and showed that most tropical waters were not rich in nutrients (substances that plankton need for growth) and so contained less plankton than temperate waters. The findings of the expedition involved 36 scientists analyzing the data over 30 years. Hensen's work took plankton study from mere description to detailed analysis.

Hess, Harry (1906–1969)
American geologist and geophysicist, Hess originated the theory that the sea floor is spreading. He was a professor of geology at Princeton University in the 1930s, specializing in the study of chains of islands with active volcanoes. During World War II Hess commanded an attack transport ship that was equipped with sonar (echo sounding) gear, enabling him to collect data on ocean depth. Hess measured the contours of the ocean floor in the Pacific and discovered hundreds of flat-topped mountains that looked as though they had been formed by erosion. These features, named "guyots" by Hess, were later recognized as the remains of once-active volcanoes.

In the early 1950s marine geologists Maurice Ewing, Bruce Heezen, and Marie Tharp realized that the Mid-Atlantic Ridge forms part of an undersea mountain system extending across the globe. The ridge system contains a rift valley up to 1 mi (1.6 km) deep and 30 mi (48 km) wide. Hess proposed that hot magma rises through vents in the rift and cools to form new oceanic crust. The new crust then pushes older material sideways, causing the sea floor to move away from the rift on both sides.

This theory gained support in the early 1960s when scientists investigated "magnetic reversals" in the ocean floor surrounding the rifts. The Earth's magnetic field reverses every few million years and rocks formed during each period have a recognizable magnetic alignment. Researchers found the pattern of alignments on either side of a rift were a mirror image of each other. This proved that new material was being laid down at the rift and spreading out on either side. The theory of sea floor spreading explained the movement of continents, which Alfred Wegener proposed in 1912 in his earlier theory of continental drift.

Heyerdahl, Thor (1914–2002)
Norwegian explorer, archeologist, and anthropologist Heyerdahl studied zoology, geography, and anthropology at Oslo University and carried out anthropological research in the Marquesas Islands in 1937–38. He wanted to discover how the islands' inhabitants had arrived there. He suggested that the Pacific islands had been settled by South American voyagers traveling from the east, rather than by travelers journeying from the west, as was commonly thought. In 1947, with five companions, he sailed a raft called the *Kon-tiki* from Peru to the Tuamotu Islands to discover if the journey was possible. Heyerdahl built the raft using only materials that would have been available when the original journeys were believed to have taken place. The difficult 4,300-mi (6,900-km) journey took 101 days. Although Heyerdahl proved the journey was possible, modern genetic studies suggest the early colonizers were of Asian descent and did indeed journey from the west.

In the 1960s Heyerdahl argued that Egyptian peoples could have colonized the Americas. In 1969 he supported his theory by sailing an Ancient Egyptian-style papyrus-reed boat, *Ra*, across the Atlantic. In 1977 he sailed from Iraq to India across the Persian Gulf, in a reed boat called *Tigris*, to imitate early navigators. Heyerdahl's voyages were an attempt to show that ancient civilizations could have traveled widely across the oceans. His theories undoubtedly stimulated further exploration and enormous debate in the academic world, but some seem to have been disproved by modern scientific evidence.

Hudson, Henry (1565–1611)
English explorer and navigator Hudson was employed by the Muscovy Company during 1607–8 to find the Northeast Passage (a sea route along the northern coasts of Europe and Asia) to China, but he was unsuccessful. He tried again in late 1609, for the Dutch East India Company, but extreme weather conditions forced him to turn back and he sailed instead to the east coast of North America, hoping to find a Northwest Passage. He reached New York Bay and sailed up the river, which was subsequently named the Hudson River in his memory. This voyage established Dutch territorial claims in the region. Hudson set out on a further search for a Northwest Passage in 1610, sailing from London and financed this time by English adventurers. He traveled between Greenland and Labrador through what is now called Hudson Strait into Hudson Bay, where he was

forced to spend the winter. In the spring, his starving crew mutinied, and he was set adrift with his son and seven men in a small boat. They were never seen again.

Iselin, Columbus O'Donnell (1904–1971)
A key figure in the development of modern oceanography in the United States, Iselin was born in New Rochelle, New York, the son of a wealthy banker. Iselin was a keen sailor and built his first full-size boat when only 11 years old. He studied marine science at Harvard University. On graduating, he organized an expedition to the Labrador Current in his schooner. The expedition plotted iceberg movements and confirmed those predicted by mathematical calculations. The International Ice Patrol used the expedition's findings to help predict the movement of icebergs that might endanger shipping.

In 1930 Iselin became captain of *Atlantis*, the first research vessel of the Woods Hole Oceanographic Institution (WHOI). Later, in the 1930s, Iselin became the foremost expert on the Gulf Stream. Between 1940 and 1950, as director of the WHOI based at Cape Cod, Massachusetts, he guided the institution through its expansion in size and importance. Iselin was one of the first marine scientists to recognize the great effect of rivers on coastal waters and the importance of studying the atmosphere and ocean as a combined system.

In later life he gave thought to how people could exploit the oceans without damaging them. He also contributed ideas to the initial drafting of the Law of the Sea, the international treaty that governs ownership and use of the oceans. His work during World War II helped sailors to interpret sonar readings more effectively. This meant better detection of German U-boats and probably saved thousands of Allied lives.

Magellan, Ferdinand (1480–1521)
The Portuguese navigator who led the first round-the-world voyage, Magellan was born in northern Portugal. As a child, he learned of the exploits of famous Portuguese seafarers such as Christopher Columbus and Vasco da Gama. As a young man, Magellan served with Portuguese naval fleets that sailed across the Indian Ocean as far as the Spice Islands (Moluccas) in Indonesia. Magellan rose to the rank of captain, and by age 30 he was an experienced navigator. Emperor Charles V of Spain funded an expedition for Magellan to discover a route to the Spice Islands by traveling westward. During the journey, Magellan knew he would have to cross the New World (the recently rediscovered American continent), or sail around it. In September 1519 Magellan's fleet of five small ships and 265 men left Sanlucar de Barrameda, Spain. Their route took them through the fearsome seas near the tip of South America. By March 1521 the expedition had crossed the Pacific Ocean and reached the Philippines. Nineteen members of the crew died of sickness and starvation on the way. On April 27 Magellan was killed when he became involved in a dispute between Philippine islanders. Sebastian del Cano, Magellan's master navigator, completed the voyage, returning to Spain in September 1522 with only one vessel and 18 crew, but having completed a round-the-world voyage in the process—one of the greatest feats in maritime history.

Marsigli, Count L. F. (1658–1730)
A soldier, scientist, and writer, Marsigli was born in Bologna, Italy, and spent his early life studying science and traveling in order to collect natural history data. During his time as a soldier he became an expert mapmaker and designer of military fortifications. After leaving the army, Marsigli devoted the rest of his life to carrying out investigations into the structure of mountains and the natural history of seas, lakes, and rivers.

In 1712 Marsigli founded an influential academy of science in Bologna. In 1724 he published a work on oceanography, *Histoire Physique de la Mer* (*The Natural History of the Sea*), in which he examined the geological formation of ocean basins and described the plants and fish that lived in the Mediterranean Sea. He also wrote a work on the River Danube, *Danubius* (1726), in which he described the riverbed, the water, the animal and plant life, and the nature of the rocks of the surrounding land. Marsigli was a fellow of the London Royal Society and a member of the Paris Academy of Science.

Maury, Matthew Fontaine (1806–1873)
American oceanographer, meteorologist, and naval officer, Maury was born near Fredericksburg, Virginia. Maury's active career in the navy was cut short by a stagecoach accident in 1839, which left him lame. In 1841 he

became superintendent of the naval Depot of Charts and Instruments where he studied ship's logbooks and became convinced of the importance of gathering data on winds and currents from a wide range of sources to help seafarers navigate the oceans. He arranged for ships to be equipped with specially designed logbooks to record this data, and he began to compile wind and current charts of the oceans, which he published in 1848. His original intention had been to assist sailing ships face the growing competition from steamships.

Maury was among the first to spearhead international collaboration in oceanographic and meteorological research. He helped organize the first International Meteorological Conference held in Brussels in 1853. His book *The Physical Geography of the Sea* (1855) contained the first map of an ocean basin (the North Atlantic) and is regarded as the first classic work of modern oceanography. Maury also produced charts of the Pacific and Indian Oceans. In 1861 the Civil War broke out and Maury resigned his post in order to work for the confederate government as head of coast, harbor, and river defenses. While in this post, he designed an electric torpedo. After the war, he spent some time in Mexico before becoming a professor of meteorology at Virginia Military Institute in 1868. Many oceanographers have dubbed Maury "The Father of Oceanography."

Munk, Walter (1917–)

Austrian-born meteorologist and oceanographer Walter Munk studied physics at the California Institute of Technology and then went to work at the Scripps Institution of Oceanography with Harald Sverdrup. During World War II, Munk worked with Sverdrup on the prediction of surf and tide conditions to facilitate Allied landings. In 1947 Munk sailed to the Arctic to study submarine operations and later researched the workings of ocean currents.

While working as part of the team of oceanographers monitoring atomic tests in Bikini Lagoon in the Marshall Islands of the South Pacific, he was concerned about the possibility of atomic tests generating tsunamis (giant waves). This prompted him to develop a system to warn of impending tsunamis in the early 1950s.

In 1952–53 Munk took part in a successful series of expeditions to explore the deep Pacific, and he went on to research current and wave patterns, including the effects of wind and gravity on ocean circulation. Munk developed some of the first computer programs to analyze the size of sea waves and track individual waveforms. Having established wave-recording stations at various locations, he was able to trace waves across the Pacific to their origin in the Indian Ocean.

During the late 20th century Munk developed instruments to measure tides in the open sea and to measure water behavior within waves. He also set up experiments to test the way in which sound can be transmitted through water. This research helped the investigations into global warming, as sound travels faster through warm water. However, the research met some opposition from biologists and members of the public. They were concerned about the effect of excessive sound signals on marine life, especially on species such as whales and dolphins, which use sound as a means of communication. After a halt in transmissions to conduct checks on the likely effects on marine mammals, there are plans to re-establish transmissions to monitor global warming of the oceans.

Nansen, Fridtjof (1861–1930)

Norwegian oceanographer, polar explorer, and statesman, Nansen led the *Fram* expedition (1893–96) which established that the Arctic was a true ocean. In 1880 Nansen entered the University of Christiania (now Oslo, Norway) to study physics and mathematics but soon changed to zoology (the study of animals). In 1888, after completing a doctorate on parasites, he organized a successful expedition to cross Greenland on skis and prove that central Greenland was ice-bound all year. Soon after his return, he sought funding for a trip to the Arctic to show (as he believed) that an ice-covered ocean, not a continent, lay at the North Pole. Funded by Norway and Britain, Nansen designed a research vessel, the *Fram*, with a hull designed to rise out of the ice rather than be crushed by it. Nansen's intention was to allow the ship to become locked in arctic ice near Siberia, drift close to the North Pole, and emerge near Scandinavia. Nansen achieved this successfully, and the *Fram* emerged from the ice after three years.

Meanwhile, Nansen and a companion had tried to trek to the North Pole and had to spend the winter on the ice. Nansen took deep soundings through the ice that showed that the Arctic contained an ocean. The expedition's observations led to further investigations into how winds steer ocean currents and how seas separate into different layers. Between 1900 and 1914 Nansen took part in Atlantic and Arctic Ocean expeditions and co-authored several classic oceanographic papers. From 1905 Nansen took on various political roles, including finding ways to resettle World War I refugees. In 1922 he was awarded the Nobel Prize for Peace.

Piccard, Auguste (1884–1962)
Swiss-born, Belgian engineer Auguste Piccard, with his twin brother Jean, pioneered the study of the middle atmosphere (a layer above a height of about 8 mi or 13 km) using hot-air balloons. In 1932 they achieved a height of 55,500 ft (16,916 m). Piccard subsequently worked closely with Albert Einstein on measuring levels of cosmic radiation. In the 1940s Piccard turned to oceanic exploration, designing diving vessels that could withstand the high pressures in deep water.

Piccard constructed a bathyscaphe (a deep-diving submersible) and later worked on the designs of other diving craft with his son, Jacques. In 1953 Auguste made a record-breaking descent in a bathyscaphe to 10,330 ft (3149 m) off the west coast of Africa. In 1960 Jacques and naval lieutenant Don Walsh descended in the bathyscaphe *Trieste* to 35,800 ft (10,912 m) in the Marianas Trench in the western Pacific Ocean. This is almost the deepest place in the world. It is the deepest descent to date by any diving craft.

Ptolemy (Ptolemaeus), Claudius (about CE 100–170)
Greek-born astronomer, geographer, and mathematician, Ptolemy lived in Alexandria, Egypt, and wrote about many topics, including geography, astronomy, mathematics, music, and physics. His 13-volume work, *Almagest*, developed the astronomical and mathematical theories of the Greeks. The work included a catalog of stars with their precise latitude, longitude, and magnitude, as well as Ptolemy's theory concerning the movements of the planets and stars. He imagined there was a planetary system in which the Moon, Mercury, Venus, the Sun, Mars, Jupiter, and Saturn all moved around Earth. This so-called Ptolemaic system was believed to be true until the mid-16th century, when Polish astronomer Copernicus proved that all the planets orbit the Sun.

Ptolemy's *Geographia*, consisting of a world map and 26 regional maps, strongly influenced the development of European map making, known as "cartography." The text was rediscovered in the 15th century after Greek manuscript copies were brought to Italy from Constantinople (present-day Istanbul).

Renaissance mapmakers reproduced *Geographia* and distributed it widely, using the newly developed art of printing. *Geographia* included three different ways of drawing maps, provided coordinates for some eight thousand different places, and explained ways of calculating latitude and longitude. It was regarded as an authoritative source for navigational purposes. His maps showed, incorrectly, no southern sea route around Africa. They dissuaded many 15th century navigators from seeking a sea passage from the Atlantic to the Indian Ocean around the tip of Africa, until the explorer Bartholomeu Dias did so successfully in 1487–88.

Pytheas (about 380–300 BCE)
A Greek explorer and trader from Massilia (modern Marseilles), Pytheas sailed to northwestern Europe around 325 BCE. He recorded his journey in a narrative, *On the Ocean*, which has not survived, but details of his voyage are recorded in other works. Pytheas' journey offers the earliest descriptions of Brittany, the British Isles, and the eastern coasts of the North Sea. He is thought to have circumnavigated Britain and sailed north for six days to an island he called Thule, which may have been modern Iceland. Pytheas reported that the people of Thule produced honey, drank milk, and kept their grain in barns.

Pytheas also described tin production in Cornwall, England, and visited an island that was a source of amber—it was probably Heligoland, a small island in the North Sea. Pytheas made detailed astronomical observations to chart his journey. He noted that the days became longer the farther north he traveled, and that the midsummer Sun never sank below the horizon north of Thule. He was fascinated by the big tidal range he saw in the northern latitudes, and he noted that they coincided with different phases of the Moon. He also described a congealed sea (a sea in the early stages of freezing), which may have been sea ice forming near Iceland.

Ross, Sir James Clark (1800–1862)
British polar explorer and nephew of Scottish polar explorer Sir John Ross (1777–1856). James Ross commanded the *Erebus* in 1818 and sailed in search of the Northwest Passage (a northern sea route connecting the Atlantic and Pacific Oceans) with his uncle, Sir John Ross. James then made a number of voyages to the Arctic between 1819 and 1827 with the British navigator Sir William Parry, during which he made a study of Inuit life. He sailed in search of the Northwest Passage again in 1831 and, while on this voyage, he discovered the north magnetic pole, located on Prince of Wales Island.

Between 1839 and 1843 Ross commanded an expedition to study the Earth's magnetic field in Antarctica, and it was during this expedition that he discovered the Ross Sea, Ross Island, and the Ross Shelf—all of which were subsequently named after him. His crew made soundings to depths greater than 13,100 ft (4,000 m)—a record at the time—and dredged up worms and starfish from record-breaking depths. He also explored Victoria Land and much of North Graham Land, recording his experiences in *A Voyage of Discovery and Research in the Southern and Antarctic Regions* (1847). His last voyage was in 1848–49, when he sailed to the Arctic in search of lost British explorer, Sir John Franklin, but Ross failed to find Franklin's ill-fated expedition.

Shokalsky, Yuly Mikhaylovich (1856–1940)
Russian geographer and oceanographer Shokalsky pioneered explorations of the arctic seas of Russia and of the Black Sea, among many achievements. Shokalsky was born in St. Petersburg, Russia, and attended the naval college there. In 1878 he enrolled in the naval academy and, on graduation, the Central Physics Laboratory employed him as a marine meteorologist. Between 1883 and 1908 he worked for the Russian Navy studying lakes, rivers, and coastal waters, and as a teacher of geography and "marine description" at the naval college. In 1900 he began devising an oceanographic study of the Arctic Ocean that lasted 15 years. During the study, icebreaking expeditions discovered Severnaya Zemlya, a group of islands near the North Pole. In 1908 Shokalsky joined the Naval Academy as Russia's first professor of oceanography. His book *Oceanography*, first published in 1917, won him international recognition. Shokalsky's study of the Black Sea—long delayed because of World War I and the Russian Revolution—finally began in 1923. It was completed in 1935, after 53 expeditions. During his career, Shokalsky published more than 1,300 scientific articles, and served in various official and academic posts, for example, president of the Geographic Society of the U.S.S.R. (1917–31). In 1929 the American Geographical Society awarded Shokalsky the Cullum Geographical Medal, an award given only to the most eminent geographical researchers.

Sverdrup, Harald (1888–1957)
Norwegian geographer and meteorologist Harald Sverdrup studied geography and astronomy at Oslo University. He then investigated and wrote about atmospheric and oceanic physics, including the North Atlantic trade winds. Sverdrup was scientific director of Roald Amundsen's polar expedition on the *Maud* from 1918 to 1925. As a result he wrote many articles on meteorology, magnetism, atmospheric electricity, and the physical oceanography of currents and tidal surges. In 1936 Sverdrup became director of the Scripps Institution of Oceanography (SIO) in La Jolla, California. The SIO carried out the first highly detailed study of currents in the Gulf of California and pioneered university courses in oceanography. Sverdrup was senior author of the classic textbook entitled *The Oceans: Their Physics, Chemistry and General Biology*, published in 1942.

During World War II the Scripps Institution of Oceanography and the University of California, Los Angeles, established a training school for military meteorologists. The school trained military personnel to predict weather and tide conditions so the best time could be chosen for Allied landings in Europe, North Africa, and across the Pacific Ocean.

After World War II Sverdrup studied the ways in which wind and sea currents interact together in the ocean and laid the groundwork for later studies on ocean circulation. In 1948 he returned to Norway to direct the Norwegian Polar Institute and to take on numerous academic roles, including the position of professor of geophysics at Oslo University. In his research he brought together two sciences—oceanic physics and biology—to help explain the growth of phytoplankton in spring, a major oceanographic event in most temperate waters.

Tharp, Marie (1920–)
The American geologist and mapmaker Marie Tharp began her academic career with a degree from Ohio University in English and music—an unlikely start for a scientist. Tharp joined the geology department of the University of Michigan in 1943 as part of a wartime initiative to attract women to positions left vacant by men who had been drafted. Tharp earned a geology degree and, later, a math degree while working in the petroleum industry. In 1948 she joined the geology department at Columbia University, working with Maurice Ewing and Bruce Heezen on a project that aimed to map the Atlantic ocean floor using data from echo soundings.

In the course of producing maps showing the shape of the sea floor, she identified the path of the rift valley in the Mid-Atlantic Ridge. Howard Foster, working beside Tharp, plotted oceanic earthquake locations. The earthquakes were seen to form an almost continuous line down the Mid-Atlantic Ridge. Ewing and Heezen went on to suggest, in 1956, the discovery of a global rift system on the ocean bed.

Tharp and Heezen also produced physiographic maps (partial, three-dimensional representations) of the ocean floor—showing the floor as it would look from above if the water was removed. A map of the North Atlantic was published in 1956 and one of the Indian Ocean in 1964. The maps increased scientific investigation into the theory of continental drift and also aroused public interest. This led *National Geographic* magazine to commission Tharp, Heezen, and Austrian artist Heinrich Berann to map the rest of the world's oceans. Their *World Ocean Floor*, published in 1977, became the standard representation of the world's oceans, and versions of this were used in oceanographic textbooks until the 1990s.

In 1997 Tharp was nominated by the Library of Congress as having made major contributions to the field of mapmaking.

Thomson, Sir Charles Wyville (1830–1882)
The Scottish naturalist, marine biologist, and deep-sea explorer Charles Wyville Thomson studied medicine at the University of Edinburgh, Scotland. He then taught botany and zoology, before being appointed professor of the natural history departments at the Universities of Cork and Belfast in Ireland. His book *The Depths of the Sea* (1873) recorded the results of deep-sea dredging expeditions he undertook during 1868–70 and provided evidence that invertebrate life (animals without backbones) lived at greater depths than most people had previously thought possible. He also noted differences in deep-sea temperatures, which helped in the understanding of oceanic circulation.

Thomson was director of scientific work on the HMS *Challenger* expedition from 1872 to 1876, which sailed around (circumnavigated) the globe. It covered about 79,000 mi (127,140 km) in total. The expedition—the most important in oceanography's history—carried out more than 130 bottom dredges, 150 net trawls, 490 deep-ocean soundings, and captured and identified 4,000 new marine species. Thomson wrote an account of the cruise, which was published in 1877. He was knighted and became a professor of natural history at the University of Edinburgh.

Wegener, Alfred (1880–1930)
German geologist, meteorologist, and explorer. In 1912 Wegener put forward the theory of continental drift, stating that the continents are not fixed in position but drift slowly about on the Earth's surface. Wegener had discovered that the coastlines of present-day continents could be fitted together like pieces in a jigsaw puzzle. He then gathered evidence to show similarities in rock types, geological structures, fossils, and types of vegetation on different continents to back up his theory.

Wegener's ideas about continental drift were not well supported by other geologists at first. Then, in the 1960s, geologists collected evidence to support the theory that the sea floor was indeed spreading, as first proposed by Harry Hess. Their evidence came from analyzing the magnetic pattern in rocks on both sides of spreading ridges, and by drilling through the sea bottom sediment to estimate the age of the sea floor. The mechanism of sea floor spreading provided an explanation for continental drift.

Wegener made a number of exploratory journeys to Greenland, where he established meteorological stations and attempted to discover the thickness of the ice sheet and the rate of drift of Greenland. He died of ill-health and over-exertion at age 50, during his last expedition to Greenland.

GLOSSARY

Words in SMALL CAPITALS refer to other entries in the glossary

Abyssal plain The broad, flat expanse of the ocean floor that lies at a depth of between 13,000 ft and 20,000 ft (4,000–6,000 m).
Acid rain Acidic, polluting rain. It occurs when oxides of nitrogen and sulfur dioxide—produced naturally from VOLCANOES and also by coal-fired power stations and car exhausts—are released as gases into the air and mix with water in the atmosphere, which then falls as rain.
Algae Simple forms of single- or multi-celled plant. Algae include SEAWEEDS and most types of plant PLANKTON (PHYTOPLANKTON).
Anthropology The study of people and their cultures.
Asteroid A small piece of rock orbiting the Sun left over from the birth of the SOLAR SYSTEM. *See also* METEORITE.
Asthenosphere A slowly moving part of the MANTLE on which PLATES travel.
Atmosphere The envelope of gases that surround the Earth.
Autonomous underwater vehicle (AUV) A robotic, underwater craft that operates without being connected to another vessel. *See also* REMOTELY OPERATED VEHICLE.
Bacteria Simple, microscopic single-celled ORGANISMS that lack a nucleus. *See also* PROKARYOTE.
Barrier island A long, narrow ISLAND built up from sand that has been moved by WAVES. The barrier island is separated from the mainland by a LAGOON.
Barrier reef A CORAL REEF that forms parallel to the SHORE but is separated from it by a wide strip of water (called a coral LAGOON).
Basin *See* OCEAN BASIN.
Bathyscaphe A vessel designed to operate at great depths under water. It consists of a gasoline-filled float, below which hangs a round, steel cabin. *See also* BATHYSPHERE.
Bathysphere A diving vessel consisting of a steel sphere with viewing ports, lowered from a ship by a cable. *See also* BATHYSCAPHE.
Beach The gently sloping region of the SHORE where sand, pebbles, or other loose particles settle.
Benthic Of the sea floor, or relating to ORGANISMS that live in or on the sea floor. *See also* PELAGIC.
Bioluminescence Light produced by living organisms.
Buoy A float. Most buoys are navigation or warning markers anchored to the seabed, but some drift freely and carry environmental monitoring apparatus.
Comet A giant snowball of ice and rock hurtling through space. Comets that travel close to the Earth are debris from the birth of the SOLAR SYSTEM. *See also* ASTEROID and METEORITE.
Commensalism A close association (SYMBIOSIS) between two SPECIES in which one benefits but the other neither gains nor is harmed.
Community The collection of ORGANISMS that live in a particular place, or HABITAT. *See also* ECOSYSTEM.
Continent One of the seven large, unbroken masses of land into which the Earth's surface is divided: Asia, Africa, Europe, North and South America, Australia, and Antarctica.
Continental drift The movement of CONTINENTS over the Earth's surface.
Continental margin The CONTINENTAL SHELF, SLOPE, and RISE at the edge of a continental landmass.
Continental rise The gentle rise at the base of a CONTINENTAL SLOPE, formed by accumulation of SEDIMENT.
Continental shelf The submerged, plateaulike edge of a continental landmass. It gently slopes, usually reaching a depth of about 490–650 ft (150–200 m). *See also* CONTINENTAL MARGIN.
Continental slope The slope between the CONTINENTAL SHELF and CONTINENTAL RISE
Convergent plate boundary The boundary where two PLATES move toward each other. In the OCEAN, it is associated with a TRENCH. *See also* DIVERGENT PLATE BOUNDARY.
Current A flow of water in an OCEAN, SEA, or river, usually in a fixed pattern.
Coral atoll A ring of CORAL REEF growing on a submerged ISLAND. It encloses a coral LAGOON.
Coral polyp An anemonelike animal. Some coral polyps grow in colonies on the chalky skeletons of their ancestors. Millions of polyps form the main structure of a CORAL REEF.
Coral reef A submerged, rocky structure formed mainly from the chalky skeletons of hard CORAL POLYPS.
Coriolis effect The effect of the Earth's rotation on moving bodies. It causes GYRES to turn clockwise in the Northern Hemisphere and counterclockwise in the Southern Hemisphere.
Cosmic radiation High-energy, radioactive particles that enter the Earth's ATMOSPHERE from space.
Crust The Earth's outer rocky layer on which the land and sea lie. The crust ranges in thickness, from oceanic crust, as little as 2.5 mi (4 km) thick, to continental crust, up to about 30 mi (50 km) thick.
Crustacean An INVERTEBRATE with a hard outer skeleton and jointed limbs. COPEPODS, crabs, lobsters, and shrimps are examples of crustaceans.
Delta A (usually) triangular-shaped, low-lying deposit of SEDIMENT at the mouth of a river.
Deposit feeder An animal that feeds by

consuming particles in or on the SEDIMENT of the seabed.

Deposition The settling of SEDIMENT in layers.

Divergent plate boundary The boundary where two PLATES are moving away from each other. This happens, for example, at a MID-OCEAN RIDGE. *See also* CONVERGENT PLATE BOUNDARY.

Earthquake A shaking of the Earth's CRUST caused by the release of energy from a VOLCANO or by the movement along a fracture in the Earth's crust.

Echo sounding *See* SONAR.

Ecosystem The COMMUNITY of living ORGANISMS and their environment. It also includes non-living components of the environment, such as chemicals, and physical factors, such as heat and light.

El Niño current A southeast movement of warm, surface water that occurs off the coast of Peru and Ecuador, usually around Christmas time. It cuts off UPWELLING and marks the end of the productive fishing season.

Epifauna Organisms that live on the surface of soft SEDIMENT deposits (sand and mud). *See also* INFAUNA.

Estuary A place where a river flows into the sea, and where fresh water mixes with salt water.

Erosion The processes by which material is worn away from the Earth's surface and moved. Winds, WAVES, and CURRENTS are major forces of erosion. *See also* WEATHERING.

Eukaryote A single-celled or multi-celled ORGANISM with complex cellular organization. The cells have a nucleus. *See also* PROKARYOTES.

Extinction The dying out of a species or a range of species.

Food chain The sequence in which one ORGANISM become food for another, which in turn is eaten by another organism. *See also* FOOD WEB.

Food web The interconnections between FOOD CHAINS in a COMMUNITY.

Fossilization The process in which the remains of a dead organism are preserved naturally in rocks.

Fringing reef A CORAL REEF that forms alongside the SHORE and parallel to it.

Geology The study of the structure of the Earth and the processes that shape it.

Geophysics The use of physical properties, such as sound, electricity, magnetism, or gravity, in the study of GEOLOGY.

Glacier A slow-moving "river" of ice on land.

Global positioning system (GPS) A system of 24 satellites that send out time and identification radio signals to enable people with GPS devices to locate their precise position on the Earth's surface.

Global warming A sustained rise in average temperatures across the world.

Greenhouse effect The warming effect caused when some of the infrared radiation emitted from the Earth's surface is trapped by the gases in the ATMOSPHERE (notably carbon dioxide).

Groin An artificial barrier that extends from the SHORE into the SEA. It interrupts LONGSHORE DRIFT, thereby keeping particles such as sand and pebbles on the BEACH and reducing local EROSION.

Groundwater Water that saturates rock below the ground.

Gulf An alternative word for SEA. It is part of an OCEAN or a body of water connected to an ocean; for example, the Arabian Gulf or the Gulf of Mexico.

Guyot A flat-topped SEAMOUNT; also called a TABLEMOUNT.

Gyre A circular system of surface CURRENTS in major OCEAN BASINS. Typically, major gyres flow clockwise in the Northern Hemisphere and counterclockwise in the Southern, under the influence of the CORIOLIS EFFECT.

Habitat The place where a particular ORGANISM lives.

Hot spot A place in the Earth's MANTLE, distant from a PLATE boundary, where hot MAGMA burns through the CRUST, producing VOLCANOES.

Hurricane A tropical storm with wind speeds greater than 74 mi per hour (120 km/h). Hurricanes are called typhoons in the western Pacific Ocean and cyclones in the Indian Ocean.

Hydrographer A person who studies the chemistry, physical properties, and movement of water in seas, rivers, and lakes.

Hydrothermal vent A crack in the ocean floor that discharges hot water. Such cracks are found at MID-OCEAN RIDGES and support unusual COMMUNITIES of ORGANISMS.

Ice age A very cold period lasting for thousands of years; also called a glacial period. During an ice age ICE SHEETS are larger, GLACIERS are more extensive, and sea levels are lower. The last ice age ended about 15,000 years ago.

Iceberg A floating block of ice that has been shed (calved) from a GLACIER or has broken off an ICE SHELF.

Ice sheet A large, flat expanse of ice and snow that covers land and sometimes sea. Ice sheets cover most of Greenland and Antarctica. *See also* ICE SHELF.

Ice shelf The extension of an ICE SHEET that floats on the sea surface.

Igneous rock Rock formed from the cooling and solidifying of molten rock (MAGMA) that has come from the lower layers of the Earth's CRUST.

Infauna Animals that live in soft sediment (sand or mud deposits). *See also* EPIFAUNA.

Intertidal zone The region of the shore normally covered by the highest tides and exposed by the lowest tides. It also includes tide (rock) pools.

Invertebrate An animal without a backbone (vertebral column). CRUSTACEANS and MOLLUSKS are examples. *See also* VERTEBRATE.

Island A landmass, smaller than a CONTINENT, that is completely surrounded by water.

Krill The common name for a type of shrimplike CRUSTACEAN. Krill are important food for baleen whales in polar waters.

Lagoon A shallow region of sea water separated from the open sea by a CORAL REEF, BARRIER ISLAND, or other strip of land.

Latitude The angular distance north or south from the equator of a point on the

surface of the Earth, measured in degrees.
Lava Molten rock (MAGMA) that erupts onto the Earth's surface.
Limestone SEDIMENTARY rock formed mostly of calcium carbonate. Chalk is an example.
Lithosphere The outer layer of the Earth, which includes the CRUST with the upper part of the MANTLE attached to it. The lithosphere is broken into PLATES.
Longitude The angular distance east or west of the Greenwich Meridian (an imaginary line that passes north-south through Greenwich, London, which is 0° longitude). The International Date Line is 180° longitude.
Longshore drift (longshore transport) The movement of SEDIMENT along a BEACH produced by the action of WAVES approaching and breaking at an angle to the SHORE.
Magma Molten rock beneath the Earth's surface.
Mangroves Trees and shrubs that tolerate salty water and dominate some tropical and subtropical SHORES.
Mantle The thick layer between the CRUST and the Earth's central core.
Mariculture The rearing of fish and SHELLFISH in sea waters for human consumption.
Marine protected area (MPA) An area of the ocean under special legal protection in order to preserve the environment and its inhabitants.
Meiofauna Tiny animals that live among sediment particles.
Metamorphic rock Recrystallized rock formed by the altering of existing solid rock through temperature, pressure, or by chemical reaction.
Meteorite A piece of rock that falls to Earth's surface from space. *See also* ASTEROID and COMET.
Mid-ocean ridge A raised region of the OCEAN floor where PLATES are moving apart. In the rift of a MID-OCEAN RIDGE, new sea floor is created as MAGMA rises from below. A mid-ocean ridge is a type of DIVERGENT PLATE BOUNDARY. *See also* SEA-FLOOR SPREADING.
Migration The mass movement of animals from one location to another, usually to reach food or to find a place to breed.
Mineralogy The study of minerals (the chemical formations in rock).
Mollusk A type of soft-bodied INVERTEBRATE usually enclosed in a hard, external shell. Types of mollusk include clams, octopuses, snails, and squid.
Mutualism A close association (SYMBIOSIS) between two SPECIES in which both benefit.
Navigation The method by which people or animals find their location and their intended direction.
Neap tides TIDES with the smallest tidal ranges. They occur at the first and third quarter of the lunar month, when the Moon and Sun are at right angles relative to the Earth.
Nekton PELAGIC animals that swim powerfully and do not rely on CURRENTS. Nekton include fish, squid, and marine mammals. *See also* PLANKTON.
Nitrogen narcosis A drunklike state caused by excess nitrogen dissolving in the blood when SCUBA divers dive deep while breathing air at high pressure.
Nutrients Chemicals required by PHOTOSYNTHETIC organisms in small amounts to make vital substances for the body. Nitrates and phosphates are important nutrients.
Observatory A place where scientists observe and take measurements of objects in space.
Ocean The continuous body of salt water covering about 71 percent of the Earth's surface. Ocean also refers to one of the five major world oceans: the Arctic, Atlantic, Indian, Pacific, and Southern Oceans.
Ocean basin A hollow in the Earth's crust that fills with sea water to form an OCEAN.
Oceanography The scientific study of the OCEAN and its inhabitants.
Ooze Deep-sea SEDIMENT that is rich in the skeletons of marine ORGANISMS.
Organism Any living thing—microbe, plant, or animal.
Overfishing Harvesting a fish population at a level beyond which its numbers can be replaced by reproduction.
Pack ice A complete, or nearly complete, floating layer of ice that forms when sea water freezes and the floating ice packs together.
Parasitism A type of SYMBIOSIS in which one ORGANISM (the parasite) benefits at the expense of another (the host), which is harmed.
Pelagic Referring to ORGANISMS that live in the surface waters or mid-waters of the OCEAN. *See also* BENTHIC.
Photosynthesis The process by which plants and some PROTISTS and BACTERIA trap sunlight and use it to manufacture carbon-rich (organic) substances such as carbohydrates.
Physiographic map A map that shows the shape of the Earth's surface as viewed from above. It also shows the geological nature of different regions.
Phytoplankton PLANKTON that photosynthesize; also called plant plankton. They are mostly single celled.
Plankton ORGANISMS that float in the WATER COLUMN and drift on OCEAN CURRENTS. They swim weakly, if at all. *See also* NEKTON, PHYTOPLANKTON, and ZOOPLANKTON.
Plate (lithospheric or tectonic plate) A region of the Earth's surface consisting of a large segment of CRUST and attached upper MANTLE. The Earth's surface is a patchwork of about 20 slowly moving plates.
Plate tectonics The study of the processes by which PLATES form, move, and are destroyed, and the effects of these processes. It also refers to the modern theory that explains this.
Prokaryotes Single-celled organisms with simple cellular organization. They lack a nucleus. BACTERIA, cyanobacteria, and archeobacteria are prokaryotes. *See also* EUKARYOTES.
Protist A single-celled organism with complex cellular organization (EUKARYOTE). Some protists, such as diatoms and dinoflagellates, are plantlike. Others, such as parasitic amebas, are animal-like.

Radar (radio detection and ranging) The use of pulsed radio waves to measure the direction and distance of an object.
Radioactive fallout Radioactive substances that fall to Earth after a nuclear explosion but settle well away from the site of the explosion.
Remotely operated vehicle (ROV) A robotic underwater vehicle operated by remote control from a surface vessel or a SUBMERSIBLE. *See also* AUTONOMOUS UNDERWATER VEHICLE.
Respiration The process by which organisms break down food molecules within cells to release energy.
Salinity A measure of the saltiness of water. Normal sea water has a salinity of about 35—in other words, 35 parts of dissolved salts per 1,000 parts of sea water by mass.
Saltmarsh A muddy shore where salt-tolerant grasses and rushes grow.
Satellite remote sensing The use of satellites to detect features of the Earth's surface. Oceanographic satellites can detect color, temperature, height, and roughness of the SEA's surface.
Scuba (self-contained underwater breathing apparatus) A portable device for breathing air under water. The best-known form is the aqualung.
Sea A named part of an OCEAN—for example, Sargasso Sea; also a general term for the world's ocean.
Sea-floor spreading A process producing new oceanic CRUST by UPWELLING of MAGMA along a MID-OCEAN RIDGE.
Seamount An underwater VOLCANO that rises 3,300 ft (1,000 m) or more above the surrounding OCEAN floor. *See also* GUYOT.
Seaweed A multicelled marine ALGA.
Sediment Particles suspended in the water that settle on the BEACH or sea floor. *See also* SEDIMENTARY ROCK.
Sedimentary rock Rock formed from SEDIMENT, or sometimes, from chemicals deposited by water. *See also* DEPOSITION.
Shellfish CRUSTACEANS or MOLLUSKS that have hard outer shells. The term applies especially to ones eaten by humans.
Shore The boundary between land and SEA. It is the region extending from the level of the lowest TIDES to the level reached by the highest waves during storms.
Solar System The Sun and the other bodies, such as planets and ASTEROIDS, that orbit it.
Sonar (sound navigation and ranging) The use of sound to detect the distance and direction of an object under water; also called ECHO SOUNDING.
Species A population of ORGANISMS whose members interbreed to produce fertile offspring.
Spreading ridge *See* MID-OCEAN RIDGE.
Spring tides TIDES with the largest tidal ranges. They occur every two weeks when the Moon, Sun, and Earth are in a straight line.
Stratosphere The upper part of the ATMOSPHERE. It lies just above the lower atmosphere (troposphere), which contains the Earth's weather systems. At the equator the stratosphere extends from about 11 mi (17 km) to 31 mi (50 km) above the Earth.
Subduction The process by which one PLATE descends beneath another.
Submarine A piloted, cigar-shaped underwater craft. Most submarines are used for military purposes, but some are employed for recreation or research. *See also* SUBMERSIBLE.
Submarine canyon A steep, V-shaped valley cut into the CONTINENTAL SHELF or CONTINENTAL SLOPE.
Submersible A small, piloted, underwater craft with a viewing cabin. Most submersibles are used for exploration and research. *See also* SUBMARINE.
Suspension feeder An animal that feeds by filtering particles out of the WATER COLUMN.
Symbiosis A close relationship between two SPECIES. COMMENSALISM, MUTUALISM, and PARASITISM are types of symbiosis.
Tablemount *See* GUYOT.
Temperate region A region of the Earth, between latitudes 40° and 60°, where the climate is highly seasonal, or variable.
Thermocline A layer in the WATER COLUMN across which the temperature drops rapidly, from top to bottom.
Tide The periodic rise and fall of surface water in response to the gravitational attraction of the Moon and Sun. *See also* NEAP TIDES and SPRING TIDES.
Trade winds Steady winds that blow from east to west toward the equator.
Transform fault (transform plate boundary) A boundary between two PLATES where one plate slides horizontally past the other.
Trench A deep trough in the OCEAN floor associated with SUBDUCTION. Most trenches lie at depths between about 20,000 ft (6,000 m) and 33,000 ft (10,000 m).
Tsunami (seismic sea wave) A gigantic wave triggered by a VOLCANO, EARTHQUAKE, or landslide.
Turbidity current An underwater avalanche of suspended SEDIMENT down a SUBMARINE CANYON.
Upwelling A term that describes the rising of cold, NUTRIENT-rich, deep water to the surface of the sea.
Vertebrate An animal with a backbone (vertebral column). Reptiles, birds, and mammals are examples. *See also* INVERTEBRATE.
Volcano An opening in the Earth's CRUST through which LAVA escapes.
Water column The column of water between the sea surface and the seabed.
Wave (sea wave) An up-and-down movement of water that travels along the sea surface.
Weathering Physical, chemical, and biological processes that break down rocks on the Earth's surface. *See also* EROSION.
Wetland Low-lying, flat areas of land that are covered by water or that have water-saturated soils for at least part of the year.
Zooplankton Animal PLANKTON. They include copepods, KRILL, arrow worms, and the larvae (young forms) of fish and many kinds of INVERTEBRATE that live on the seabed.

FURTHER READING

General Reading

Brower, Kenneth, and David Doubilet. *Realms of the Sea.* Washington, D.C.: National Geographic Society, 1991.

Byatt, Andrew, Alastair Fothergill, and Martha Holmes. *Blue Planet: A Natural History of the Oceans.* New York: DK, 2001.

Cousteau, Jacques Yves. *The Ocean World.* 1979. Reprint, New York: Abradale, 1985.

Ellis, Richard. *Encyclopedia of the Sea.* New York: Knopf, 2000.

Groves, Donald. *The Oceans: A Book of Questions and Answers.* New York: Wiley, 1989.

Hall, Michelle, Howard Hall, and Peter Benchley. *Secrets of the Ocean Realm.* New York: Carroll & Graf, 1997.

Kunzig, Robert. *The Restless Sea: Exploring the World Beneath the Waves.* New York: W.W. Norton, 1999.

The Physical Ocean

Broad, William J., and Dimitry Schidlovsky. *The Universe Below: Discovering the Secrets of the Deep.* New York: Simon & Schuster, 1997.

Earle, Sylvia, and Eric Lindstrom. *Atlas of the Ocean: The Deep Frontier.* Washington, D.C.: Nat. Geographic Society, 2001.

Hansen, Gunnar. *Islands at the Edge of Time: A Journey to America's Barrier Islands.* Washington, D.C.: Island, 1993.

Kunzig, Robert. *Mapping the Deep: The Extraordinary Story of Ocean Science.* New York: W.W. Norton, 2000.

Monahan, Dave. *World Atlas of the Ocean.* Toronto: Firefly, 2001.

Prager, Ellen J., and Sylvia A. Earle. *The Oceans.* New York: McGraw-Hill, 2000.

Rice, A. L., and Tony Rice. *The Deep Ocean.* Washington, D.C.: Smithsonian Institution Press, 2000.

Spalding, Mark D., Corinna Ravilious, and Edmund P. Green. *World Atlas of Coral Reefs.* Berkeley: University of California Press, 2001.

Life in the Ocean

Boschung, Herbert T., and Daniel W. Gotshall, eds. *National Audubon Society Field Guide to North American Fishes, Whales, and Dolphins.* New York: Knopf, 1983.

Bright, Michael. *Sharks.* Washington, D.C.: Smithsonian Institution Press, 2002.

Cohat, Yves, and Anne Collet. *Whales: Giants of the Seas and Oceans.* New York: Abrams, 2001.

Connor, Judith. *Seashore Life on Rocky Coasts.* Monterey, Calif.: Monterey Bay Aquarium, 1993.

Dipper, Frances A. *Extraordinary Fish.* New York: DK, 2002.

Ellis, Richard, John E. McCosker, and Al Giddings. *Great White Shark.* 1991. Reprint. Stanford, Calif.: Stanford Univ. Press, 1995.

———. *The Search for the Giant Squid.* New York: Penguin, 1999.

Folkens, Pieter, Randall R. Reeves, Brent S. Stewart, Phillip J. Clapham, and James A. Powell. *National Audubon Society Guide to Marine Mammals of the World.* New York: Knopf, 2000.

Gibson, Ray, Benedict Hextall, and Alex Rogers. *Photographic Guide to Sea and Shore Life of Britain and North-West Europe.* New York: Oxford University Press, 2001.

Hoyt, Erich. *Creatures of the Deep: In Search of the Sea's Monsters and the World They Live In.* Toronto: Firefly, 2001.

Mead, James G., Joy P. Gold, and Flip Nicklin. *Whales and Dolphins in Question: The Smithsonian Answer Book.* Washington, D.C.: Smithsonian Institution Press, 2002.

Meinkoth, Norman A. *The National Audubon Society Field Guide to North American Seashore Creatures.* New York: Knopf, 1981.

Miller, David. *Seals and Sea Lions.* Stillwater, Minn.: Voyageur, 1998.

Morton, Alexandra. *Listening to Whales: What the Orcas Have Taught Us.* New York: Ballantine, 2002.

Owen, Welden. *Whales, Dolphins, and Porpoises.* New York: Checkmark, 1999.

Perrin, W. F., Bernd Wursig, and J. G. M. Thewissen, eds. *Encyclopedia of Marine Mammals.* New York: Academic Press, 2002.

Musick, John A., and Beverly McMillan. *The Shark Chronicles.* New York: Times Books, 2002.

Safina, Carl. *Eye of the Albatross.* New York: Henry Holt, 2002.

Sprung, Julian. *Corals: A Quick Reference Guide.* Coconut Grove, Fla.: Ricordea, 1999.

Tackett, Denise Nielsen, and Larry Tackett. *Reef Life: Natural History and Behaviors of Marine Fishes and Invertebrates.* Charlotte, Vt.: Microcosm, 2002.

Thomas, David N. *Seaweeds.* Washington, D.C.: Smithsonian Institution Press, 2002.

Weinberg, Samantha. *A Fish Caught in Time: The Search for the Coelacanth.* New York: HarperCollins, 2000.

Uses of the Ocean

Ballard, Robert D., and Malcolm McConnell. *Adventures in Ocean Exploration: From the Discovery of the Titanic to the Search for Noah's Flood.* Washington, D.C.: National Geographic Society, 2001.

Ballard, Robert D., Rick Archbold, and Ken Marschall. *The Discovery of the Titanic.* New York: Warner, 1995.

Collingridge, Vanessa. *Captain Cook: A Legacy Under Fire.* New York: Lyons, 2002.

Ellis, Richard. *Men and Whales.* 1991. Reprint, New York: Lyons, 1999.

———. *Monsters of the Sea: The Truth about the Loch Ness Monster, the Giant Squid, Sea Serpents, Mermaids, and Other Fantastic Creatures of the Deep.* New York: Lyons, 2000.

Fluckey, Eugene B. *Thunder Below!: The USS* Barb *Revolutionizes Submarine Warfare in World War II.* Champaign: University of Illinois Press, 1992.

Gordon, John Steele. *A Thread Across the Ocean: The Heroic Story of the Trans-Atlantic Cable.* New York: Walker, 2002.

Greenlaw, Linda. *The Hungry Ocean: A Swordboat Captain's Journey.* New York: Hyperion, 1999.

Hearn, Chester G. *Tracks in the Sea: Matthew Fontaine Maury and the Mapping of the Oceans.* New York: International Marine, 2002.

Helvarg, David. *Blue Frontier: Saving America's Living Seas.* New York: W. H. Freeman, 2001.

Hicks, Brian, and Schuyler Kropf. *Raising the* Hunley*: The Remarkable History and Recovery of the Lost Confederate Submarine.* New York: Ballantine, 2002.

Horowtiz, Tony. *Blue Latitudes: Boldly Going Where Captain Cook Has Gone Before.* New York: Henry Holt, 2002.

Junger, Sebastian. *The Perfect Storm: A True Story of Men Against the Sea.* New York: W.W. Norton, 1997.

Kurlansky, Mark. *Cod: A Biography of the Fish That Changed the World.* New York: Walker, 1997.

Lenihan, Daniel. *Submerged: Adventures of America's Most Elite Underwater Archaeology Team.* New York: Newmarket, 2002.

McDougall, Walter A. *Let the Sea Make a Noise . . . : A History of the North Pacific from Magellan to MacArthur.* New York: Basic, 1993.

Philbrick, Nathaniel. *Revenge of the Whale: The True Story of the Whaleship* Essex. New York: Putnam, 2002.

———. *In the Heart of the Sea: The Tragedy of the Whaleship* Essex. New York: Viking, 2000.

Preston, Diana. Lusitania: *An Epic Tragedy.* New York: Walker, 2002.

Pryor, Karen, and Konrad Lorenz. *Lads Before the Wind: Diary of a Dolphin Trainer.* 1975. Rev. ed. Waltham, Mass.: Sunshine, 2000.

Safina, Carl. *Song for the Blue Ocean: Encounters Along the World's Coasts and Beneath the Seas.* New York: Henry Holt, 1998.

Sobel, Dava. *Longitude: The True Story of a Lone Genius Who Solved the Greatest Scientific Problem of His Time.* New York: Walker, 1995.

Stanton, Doug. *In Harm's Way: The Sinking of the USS* Indianapolis *and the Extraordinary Story of Its Survivors.* New York: Henry Holt, 2001.

Trupp, Philip Z. *Sea of Dreamers: Travels with Famous Ocean Explorers.* Golden, Colo.: Fulcrum, 1998.

Woodard, Colin. *Ocean's End: Travels Through Endangered Seas.* New York: Basic, 2001.

WEBSITES

American Museum of Natural History
http://www.amnh.org/nationalcenter/expeditions/blacksmokers/
This section of the Museum of Natural History site is affiliated with the NOVA Abyss project (website information about this project is provided below). It gives information about hydrothermal vents, or black smokers, in an engaging, interactive format.

CIA World Fact Book
http://www.cia.gov/cia/publications/factbook/
This site provides the full contents of the CIA World Fact Book online. Using the search feature, visitors can find statistics and other data—such as significant geographic features, natural resources, natural hazards, and more—about each of the five oceans.

Missouri Botanical Garden
http://mbgnet.mobot.org
This site includes descriptions of the world's biomes (a major type of ecological community such as rainforest or desert), including the three marine biomes—shorelines, temperate oceans, and tropical oceans.

Monterey Bay Aquarium
www.mbayaq.org
This site offers an online visit to the aquarium with live-cams in various exhibits. The site also provides information about marine plants and animals and preservation efforts, as well as an education center for both students and teachers.

Museum of Science, Boston
www.mos.org/oceans
This site provides information in such categories as "the water planet," "oceans in motion," "life in the sea," and "scientists at sea."

National Aeronautics and Space Administration (NASA)
http://daac.gsfc.nasa.gov/CAMPAIGN_DOCS/OCDST/shuttle_oceanography_web/oss_cover.html
This NASA site displays images of the world's oceans taken from the space shuttle. More than pretty pictures, these images can help scientists learn about changes in the oceans such as pollution or destruction of coral reefs.

Neptune
www.pao.cnmoc.navy.mil/Educate/Neptune/Neptune.htm
This site maintained by the Naval Meteorology and Oceanography Command offers students and teachers educational resources for learning about the ocean and oceanography.

NOVA, Abyss
www.pbs.org/wgbh/nova/abyss
Visitors to this site take a journey into the abyss off the Pacific Northwest coast of the United States. Scientists on this mission operate submersibles in the abyss using remote control, offering unique access to this dark and dangerous realm of the ocean. The site offers detailed information about the mission, life in the abyss, and what scientists have learned about this last frontier, as well as dispatches that have come directly from the scientists.

Odyssey
www.pbs.org/odyssey
Visitors to this site learn about the activities aboard the research ship *Odyssey* on its five-year scientific voyage to study the health of the world's oceans. Visitors can track the voyage of the *Odyssey*, attend a class from the sea, hear the songs of humpback whales recorded by scientists on board the *Odyssey*, among other exciting features.

Planet Ocean
www.school.discovery.com/schooladventures/planetocean
Part of the educational site of the Discovery Channel, Planet Ocean offers in-depth looks at some of the creatures found in the ocean, and resources for further research.

Save the Whales
www.savethewhales.org
This organization educates about marine mammals and their preservation. The site offers detailed descriptions of a variety of whales, dolphins, porpoises, and other marine mammals.

Scripps Institution of Oceanography
www.sio.ucsd.edu
The Scripps Institution in La Jolla, California, trains marine biologists and oceanographers. The site includes a lively "education corner" that covers such topics as the seahorse, plate tectonics, and giant kelp. Visitors can also learn about cutting-edge science through news releases posted to the site.

Sustainable Seas Expedition (SSE)
http://sustainableseas.noaa.gov
SSE is a project run by the National Oceanographic and Atmospheric Administration. The mission of this five-year project is to explore and research the national marine sanctuaries of the United States. The project's site offers information about educational opportunities for students and teachers that are available through SSE. Visitors can follow the current SSE mission or learn about past missions. Also found on the site are live events, a video gallery, a photo gallery, and details about the marine sanctuaries.

Woods Hole Oceanographic Institution
www.whoi.edu
Among the world's premier centers for the study of the ocean, Woods Hole (WHOI) is dedicated to research and higher education in the field of oceanography. At the WHOI site, news releases inform visitors about the latest scientific advances in oceanography. Dive and Discover—an interactive section of the site—allows visitors to join WHOI scientists on their dives, with daily updates, photos, videos, and e-mails.

World Wildlife Federation
www.panda.org
This site offers news releases concerning environmental issues, including those affecting the oceans; information about conservation education; and opportunities to get involved in conservation.

Places to Visit

Below is information about a number of the major aquariums in the United States, Canada, and Bermuda, as well as museums and parks relating to the study of the ocean. There are many more aquariums throughout the world.

American Museum of Natural History
Central Park West at 79th Street
New York, NY 10024-5192
Tel. 212-569-5100
www.amnh.org
Included in the museum's vast collection is the Hall of Ocean Life. The hall includes a 94-ft (29-m) long model of a blue whale as well as displays of aquatic creatures and ocean environments from around the world.

Aquarium of the Pacific
100 Aquarium Way
Long Beach, CA 90802
Tel. 562-590-3100
www.aquariumofpacific.org
Visitors to the aquarium journey through the Pacific Ocean's three regions: southern California/Baha, the Tropical Pacific, and the North Pacific. The aquarium houses 550 species in 17 major habitats, including a shark tank.

Audubon Aquarium of the Americas
Canal Street at the River
1 Canal Street
New Orleans, Louisiana 70130
Tel. 800-774-7394
www.audubaninstitute.org/aoa
Part of the Audubon Nature Institute, the aquarium exhibits a Caribbean coral reef in a 30-ft (9-m) aquatic tunnel. Some of the aquarium's featured creatures include seahorses, jellyfish, sharks, and sea otters. The sharks are found in the Gulf of Mexico exhibit featuring a replica of an offshore oil rig. Residents of this exhibit include endangered sea turtles, stingrays, and 10-ft (3-m) sharks. The aquarium houses one of the world's largest and most diverse shark collections, including several species not found anywhere else in the United States. The aquarium also offers educational and conservation programs.

Bermuda Aquarium, Museum, and Zoo
40 North Shore Road
Flatts Village, Bermuda
Tel. 441-293-2727
www.bamz.org
This aquarium focuses on species from oceanic islands and offers conservation and education programs. Exhibits include the North Rock Exhibit providing a view of Bermuda's coral reef, local trails, and Discovery Cove.

Cabrillo Marine Aquarium
3720 Stephen White Drive
San Pedro, CA 90731
Tel. 310-548-7562
Fax 310-548-2649
www.cabrilloaq.org
Dedicated to promoting knowledge and conservation of marine life in southern California, this aquarium is located on Cabrillo Beach near Los Angeles. Exhibits include whales and dolphins, sharks, sea lions and seals, kelp forest, and rocky shores. Visitors can explore tide pools at the Point Fermin Marine Life Refuge, within walking distance of the aquarium.

Calvert Cliffs State Park
Point Lookout State Park
P.O. Box 48, Scotland, MD 20687
Tel. 301-872-5688
www.dnr.md.us/publiclands/southern/calvertcliffs
These cliffs, which dominate the shoreline of the Chesapeake Bay for 30 mi (48 km), were formed over 15 million years ago when southern Maryland was covered by a warm, shallow sea. More than 600 species of fossils have been found in the cliffs; the teeth of various sharks are the most abundant fossils.

Cold Spring Harbor Whaling Museum
Main Street, P.O. Box 25
Cold Spring Harbor, NY 11724
Tel. 631-367-3418
Fax 631-692-7037
www.cshwhalingmuseum.org
Located on Long Island, New York, this museum preserves and displays artifacts from the whaling history of Long Island. Highlights of the collection include a fully-equipped 19th-century whaleboat, scrimshaw (elaborately carved whalebone), whaling tools, navigational aids, model ships, and maritime art. Documents in the collection include ships' logs, correspondence of the Cold Spring Whaling Company, and records from the Cold Spring Harbor Custom House.

The Dallas World Aquarium
1801 North Griffin Street
Dallas, TX 75202
Tel. 214-720-2224
www.dwazoo.com/aquarium.asp
The exhibits at this aquarium include a tunnel through a 22,000-gallon (83,280-liter) tank, offering a panoramic view of a coral reef; life at the top of the food chain in the Predators exhibit; and ten tanks featuring marine life from around the world—Palau, Southern Australia, Lord Howe Island, Solomon Islands, Fiji, the Bahamas, British Columbia, Sri Lanka, Indonesia, and Japan.

Everglades National Park
40001 State Road 9336
Homestead, FL 33034-6733
Tel. 305-242-7700
Fax 305-242-7728
www.nps.gov/ever
The only subtropical preserve in the United States, this park—which spans the southern tip of Florida and Florida Bay—includes mangrove and cypress swamps, and marine and estuarine environments. Ranger-led tours, boat tours, canoe rentals, and nature trails are some of the activities available to park visitors.

The Florida Aquarium
701 Channelside Drive
Tampa, FL 33602
Tel. 813-273-4000
www.flaquarium.net
This aquarium focus on Florida's diverse marine life. The Wetlands exhibit features a limestone cavern, cypress swamps, mangrove forests, terrapins, and sting rays. At the Bays and Beaches exhibit visitors move from brackish water to sea grass meadows. The No Bone Zone is dedicated to marine invertebrates.

Monterey Bay Aquarium
886 Cannery Row
Monterey, CA 93940
Tel. 831-648-4800
www.mbayaq.org
This aquarium focuses on the marine life of Monterey Bay in northern California, and the oceans beyond. The Habitats Path takes visitors through numerous marine environments, including a kelp forest, reefs and pilings, the sandy seashore, estuaries, rocky shores, beaches and dunes, and open waters. The aquarium has special exhibits, such as Mysteries of the Deep, focusing on Monterey Bay's 2-mi (3.2-km) deep canyon.

Nantucket Whaling Museum
13 Broad Street
Nantucket, MA 02554-3502
Tel.508-228-1894
www.nha.org/whalingmuseum
Housed in a former sperm whale oil candle factory on Nantucket Island off Cape Cod, Massachusetts, the museum contains whaling memorabilia, paintings, and scrimshaw (elaborately carved whalebone). Exhibits include an illustration of how Nantucketers harpooned and processed whales while at sea, the skeleton of a 45-ft (14-m) finback whale, and artifacts from the *Essex,* a whale ship that was destroyed at sea by a sperm whale.

National Aquarium in Baltimore
Pier 3, 501 East Pratt Street,
Baltimore, MD 21202
Tel. 410-576-3800
www.aqua.org
This five-level aquarium located on the Baltimore harbor has exhibits including sting rays, the water cycle, survival through adaptation featuring electric eels and a giant Pacific octopus, puffins, an Atlantic coral reef, and sharks in the open ocean. The aquarium also includes the Marine Mammal Pavilion with dolphin shows, a life-size humpback whale model, and a discovery cove where visitors can handle animals and learn about human stewardship of marine life.

National Museum of Natural History
10th Street and Constitution Avenue,
NW Washington, DC 20560
Tel. 202-357-2700
www.mnh.si.edu
Located on the Mall in Washington, D.C., the Museum of Natural History includes two exhibits related to the ocean. Life in the Ancient Seas focuses on marine life that lived prior to the two mass extinctions and exhibits fossils that tell the story of evolution in the seas. Marine Ecosystems explores a tropical coral reef and a rocky shore. Also on display are a giant squid, a life-size model of a blue whale, and a giant marlin.

New Bedford Whaling Museum
18 Johnny Cake Hill
New Bedford, MA 02740-6398
Tel. 508-997-0046
Fax 508-997-0018
www.whalingmuseum.org
New Bedford, Massachusetts, was the whaling capital of the United States, and during the height of the whaling industry from 1825 to 1860 New Bedford was the most prosperous city in the country. This museum houses artifacts from the city's whaling past, including paintings, scrimshaw (elaborately carved whalebone), model ships, harpoons, figureheads, full-size whale boats, and whale skeletons. The museum also includes a research library with ships' logs, diaries, maps, charts, and other documents.

New Bedford Whaling National Historic Park
33 William Street
New Bedford, MA 02740
Tel. 508-996-4095
www.whalingmuseum.org
Created to preserve the historic whaling port of New Bedford, Massachusetts, this park includes historic buildings, some of which are open to the public, the schooner *Ernestina,* and the waterfront.

New England Aquarium
Central Wharf
Boston, MA 02110
Tel. 617-973-5200
www.neaq.org
Featuring one of the largest cylindrical saltwater tanks in the world, this aquarium

offers multi-level views of the sea turtles, moray eels, sharks, and tropical fish that live inside. The tank is a scientific replica of a Caribbean coral reef including the 35 major species of corals and sponges found in the Caribbean. The aquarium's research department studies the migratory patterns of the right whale, and the giant skeleton of a right whale is on display.

New Jersey State Aquarium
1 Riverside Drive
Camden, NJ 08103
Tel. 856-365-3300
www.njaquarium.org
In addition to exhibits featuring a range of marine life, this aquarium offers unique programs for students including aquarium sleepovers and sea camp.

New York Aquarium
Surf Avenue & West 8th Street
Brooklyn, NY 11224
Tel. 718-265-3474
http://wcs.org/home/zoos/nyaquarium
Located along the Coney Island boardwalk in Brooklyn, New York, this aquarium houses creatures from beluga whales to walruses, from shark to seahorses. Major exhibit areas include Sea Cliffs, a 300-ft (91-m) long re-creation of the rocky Pacific coast, and Explore the Shore, where visitors can touch starfish and experience the energy of a 400-gallon (1,540 liter) wave crashing on a rocky shore. The aquarium includes two Touch Tanks, where starfish and horseshoe crabs can be handled.

Oregon Coast Aquarium
2820 S.E. Ferry Slip Road
Newport, OR 97365
Tel. 541-867-3474
Fax 541-867-6846
www.aquarium.org
The former home of Keiko (star of *Free Willy*) before the whale was released into the open ocean, this aquarium provides a detailed look at the Oregon coast in four indoor galleries. The galleries include Sandy Shores, a replica of the unique habitat found around harbor piers and pilings; Ocean Exploration Station, an exhibit dedicated to jellyfish; and Rocky Shores, featuring a large touch pool where visitors can handle residents of a tide pool such as starfish and anemones. The aquarium also includes a discovery lab and the Whale Theater, which shows Oregon's sharks and gray whales.

Pittsburgh Zoo and PPG Aquarium
One Wild Place
Pittsburgh, PA 15206
Tel. 412-665-3640 or 800-474-4966
www.pittsburghzoo.com/aquarium.asp
Featuring fresh and saltwater exhibits, this aquarium includes a two-story open ocean tank that includes sharks and other fish in a complete ecosystem. A clear tunnel allows young visitors to crawl along as sting rays swim above and alongside them.

Point Defiance Zoo and Aquarium
5400 North Pearl Street
Tacoma, WA 98407-3218
Tel. 253-591-5337
www.pdza.org
The centerpiece of this aquarium is North Pacific Community exhibit, which displays native marine life. Other smaller exhibits highlight the habitats and creatures of the Puget Sound. The Marine Discovery Center and Simpson Lab allows visitors an up-close view of hermit crabs, sea urchins, and other intertidal creatures; a peek at plankton through a microscope; and a look at "alien" creatures that have invaded the Puget Sound.

John G. Shedd Aquarium
1200 South Lake Shore Drive
Chicago, IL 60605
Tel. 312-939-2435
Fax 312-939-8069
www.sheddnet.org
The world's largest indoor aquarium, Shedd houses such exhibits as Pacific Northwest, Seahorse Symphony, Caribbean Reef, Fishes from around the World, and Wild Reef. Each exhibit is a detailed re-creation of a complex marine community. For example, in the Pacific Northwest exhibit visitors walk through pine trees and past streams to the shore of a pine forest where they will find beluga whales and Pacific white-sided dolphins swimming in salt water.

Vancouver Aquarium
845 Avison Way
Vancouver, BC V6B 3X8, Canada
Tel. 604-659-3474
Fax 604-659-3515
www.vanaqua.org
The aquarium teaches conservation in a variety of settings from kayaking trips to behind-the-scenes tours of the aquarium. Special programs include Trainer Tours, which provide a glimpse into the life of a marine mammal trainer and allow visitors to get near to the animals including the beluga whales, and sleepovers.

Virginia Marine Science Museum
717 General Booth Boulevard
Virginia Beach, VA 23451
Tel. 757-437-6000
www.vmsm.com
With more than 300 hands-on exhibits, an outdoor aviary, a marsh habitat, and a nature trail, this aquarium offers visitors a variety of opportunities to learn about Virginia's marine life. The aquarium's IMAX theater presents "larger-than-life" marine and nature movies. Also available through the museum are ocean excursions including dolphin watching, whale watching, and salt marsh boat trips.

Waikiki Aquarium
2777 Kalakaua Avenue
Honolulu, HI 96815
Tel. 808-923-9741
Fax 808-923-1771
http://waquarium.otted.hawaii.edu
Part of the University of Hawaii, this aquarium is located next to a living coral reef. The exhibits and programs of the aquarium focus on aquatic life of Hawaii and the tropical Pacific. Special programs include a night reef walk.

Series Index

Bold type indicates volume numbers; page numbers in *italics* refer to illustrations

abalones **2:** 23
Aborigines **3:** 6
abyss **2:** *18*, 19
abyssal plains **1:** *16*, 17, **2:** *18*
abyssal ridges **1:** *14*, 17
acid rain **1:** 38, **2:** 10, **3:** 48, 59
aerial reconnaissance **1:** 45
aerosols **4:** 24
Africa **1:** 10, **4:** 14
  sailing round **3:** 9
African Plate **1:** *12*
Africans, slaves **3:** 7
Agassiz, Alexander **3:** *26*, 27, **4:** 34
Agassiz, Louis **3:** 27, **4:** 34
aggregrates **1:** 41
Agulhas Current **1:** 57
air temperature **1:** 4, 46
aircraft carriers **3:** 4, 19, *19*
Alaska **1:** 45, **2:** *41*, **3:** 6, 7, 12, 13, *21*, 55
  coast **1:** 26
  gold **3:** 47, **4:** 19
albatrosses **2:** 35, **4:** 22
  black-browed **2:** *35*
  wandering **2:** 35
Albert I, Prince, of Monaco **3:** 27, **4:** 34
algae **1:** 22, 43, **2:** *7*, 14, 15, 44, 45, *45*, 49, 51, 55, **3:** 51
  brown **2:** 14
  of coral reef **2:** 52, 60
  coralline **2:** 51
  evolution **2:** 32
  green **2:** 51
  red **2:** 51
  symbiotic relationship **2:** 60, *61*
algal bloom **1:** *39*, **2:** 13, 56, **3:** 56
Almagest **4:** 43
Alvin (submersible) **2:** 57, **3:** *33*, 34, 37, **4:** 33, 35
Amazon, tidal bore **1:** 54
amber **3:** 9, **4:** 43
amphibians **1:** 4, **2:** 32
amphipods
  depths reached **4:** 28, *29*
  in food web **2:** *17*
Amundsen, Roald **3:** 12, **4:** 31, 34–5, 44
anchors, damage done by **2:** 55
anchovies **1:** 51, *51*, 52, 53, **2:** *5*, 16, **3:** 41, 42, *42*, 53, **4:** 10, 19
  in food web **2:** *17*
Andaman Sea **4:** 17
anemones *see* sea anemones
angelfish, regal **2:** *53*
anglerfish **2:** 19, *50*, *58*, 59, **4:** 28, *29*
animals
  of dark zone **2:** 19
  depth endured **4:** 28
  in food chains **2:** 16–17, *16*, *17*
  largest **2:** 4
  marine, extinctions **2:** 10, 11
  soft-bodied, fossils **2:** 8
  species **2:** 4
  water content **1:** 36
Anomalocaris **2:** 9, *9*
Antarctic Circle **3:** 13
Antarctic Circumpolar Current **1:** 49, **4:** 22
Antarctic Convergence **4:** 22
Antarctic Ocean **1:** 44, **4:** 22
  icebergs **1:** 45, *45*
Antarctic Plate **1:** *12*
Antarctic Treaty **4:** 24
Antarctica **1:** 10, *10*, 45, 49, **3:** 13, 55, **4:** 14, 22
  scientific stations **4:** 22
  temperatures **4:** 22
  tourism **4:** 22
antibiotics, used in fish farming **3:** 45
aqualung **3:** 31, *31*, **4:** 33, 36
aquariums **3:** *55*
Aquarius (underwater laboratory) **3:** 35
Aquarius-2000 **4:** 33
Arabian Gulf **3:** 5
Arabian Plate **1:** *12*
Arabs **3:** 10
archeologists, marine **3:** 36
archeobacteria **2:** *7*
archer fish **2:** *55*
Arctic **3:** 25, **4:** 42–3
  people of **4:** 4
  tourism **4:** 6
Arctic Circle **3:** *13*, **4:** 4, 6
Arctic Ocean **1:** 5, 6, 7, 44, **4:** 4–6, *4*, 44
  basin **4:** 4
  data **4:** 6
  depth **1:** 6
  fishing **4:** 6
  icebergs **1:** 45, *45*, **4:** 4
  map **4:** *6*
  mineral resources **4:** 4
  size **1:** 6, **4:** 6
Argo II (ROV) **3:** 34
Argo project **4:** 31
Aristotle **3:** 24, 30, **4:** 32
arrow worms **2:** 16, 17, *20*, 21
arthropods **2:** 8–9, 32
asthenosphere **1:** 13
astrolabe **3:** 11, *11*, **4:** 30
astronomy **4:** 43
Atlantic Ocean **1:** 5, 6, **2:** 17, **4:** 8–12, *8*
  animal species in **2:** 4, 56
  barrier islands **1:** 29
  birth **1:** 10
  changes **1:** 11, **4:** 8
  coastline **1:** 24, 28
  currents **4:** 8
  data **4:** 8, 10, 12
  fishing **4:** 10
  floor **1:** 16
  map **4:** 10
  mineral resources **4:** 8–10
  North
    data **4:** 10
    map **4:** *10*
  pollution **4:** 12
  seabed mapping **3:** 28
  seas **4:** 8
  size **1:** 7, **4:** 8, 10, 12
  South
    data **4:** 12
    map **4:** *12*
  tides **1:** 55
  tourism **4:** 12
  upwelling **1:** 51
Atlantis **3:** 16, **4:** 8
Atlantis (research ship) **3:** 34, **4:** 41
atmosphere **1:** 8, 46, **2:** 6, **4:** 43
atolls, coral **1:** *22*, 23, *23*, **4:** 37
atomic weapons **3:** 19
  tests **4:** 42
auks **2:** 35
Australia **1:** 10, *10*, **3:** 6, 13, 14 **4:** 14, 15
Austronesians **3:** 6
autonomous underwater vehicles (AUVs) **3:** 33, **4:** 28, *29*
avalanches, underwater **1:** 16
Aysheaia **2:** *9*
Azores **4:** 8
Aztecs **3:** 7

bacteria **1:** 9, **2:** 4, 5, 6–7, *7*, 17, 20, 57
  chemosynthetic **2:** 57
  energy source **2:** 6–7
Baffin, William **3:** 12
Baffin Bay **3:** 22
Baffin Island **3:** 12, **4:** 6, 38
Bahamas **1:** 47, **3:** 11, 20
Bahr el Baqar drain **1:** 33
Baja California **2:** *43*
baleen **2:** 40, *40*, **3:** 38
Ballard, Robert **3:** *37*, **4:** 33, 35
ballast water
  discharging **4:** 12
  organisms in **3:** 53
Baltic Sea **4:** 8
Bangladesh **1:** 31, 59, **3:** 61, **4:** 15
bar **1:** 29
Barents Sea **4:** 6
barges **3:** 20
barnacles **2:** 21, 23, 44, 45, *45*, 51, **3:** 25, **4:** 37
barracudas **2:** 30, 52
barrages, tidal **3:** 48–9, *48*
barrier islands **1:** 29
barrier reefs **1:** 22, *22*, 23, **4:** 37
Barton, Otis **3:** 33, **4:** 32
basalt **1:** 18
basins **4:** 8
bathyscaphe **3:** 33, **4:** 28, *29*, 33, 37, 43
bathysphere **3:** 33, **4:** 28, *29*, 32
battleships **3:** 19
Battuta, Ibn **4:** 35
bay barrier **1:** 29
bays **1:** 5, 26
beach compartments **1:** 26–7
beach hoppers **2:** 46
beaches
  artificial **1:** 25
  below cliffs **1:** 26
  coastal **1:** *17*, 24
  formation **1:** 28–9
  muddy **1:** 29
  pollution **3:** 51
  raised **1:** *34*
  sandy **1:** 25, 29
    life on **2:** 46–7, *46–7*
  starved of sand **1:** 27
  tourism and **3:** *4*, 54
beadlet anemone **2:** *23*
Beagle, HMS **3:** 25, *25*, **4:** 31, 37
bears, polar **1:** 61, **4:** 6
Beaufort Sea **3:** 12, **4:** 4
Beebe, William **3:** 33, **4:** 32
Behm, Alexander **4:** 32
"bends" **3:** 30, 35
Bengal, Bay of **1:** 5, 31, 58, 59
Benguela Current **1:** *49*
benthic environment **2:** *4*, 5
Berann, Heinrich **4:** 45
Bering, Vitus **3:** 12
Bering Sea **2:** *43*, **3:** 7, 38, **4:** 20
Bering Strait **3:** 12
Bermuda Triangle **3:** *16–17*, 17
biodiversity **3:** 56–7
biology, marine **3:** 24, 25
bioluminescence **2:** 18, 19, 59
birds
  beaks **2:** *49*
  droppings **1:** 30
  effect of El Niño on **1:** 52
  in estuaries **2:** *48*, 49
  harm caused to, by oil **3:** 50, **4:** 6
  hunting **4:** 4
  of mudflats **1:** 31
  seabirds **2:** 17, 34–5, **3:** 57, **4:** 22
    feeding methods **2:** *34–5*
    food **2:** 33, 34
    in food web **2:** *17*
    migrations **2:** *42*, 43
    threat to **3:** 5
  shore birds **2:** 45, 46, 49
  wading birds, threat to **3:** 52
  of wetlands **1:** 32
Bismarck, wreck **4:** 35
bivalve mollusks **2:** 23, 46, 47, 48
black hake **4:** 23
Black Sea **3:** 10, 53, 59, **4:** 8, 44
bladderwrack **2:** *14*

blenny **2**: 61
bloom, algal **1**: *39*, **2**: 13, 56, **3**: 56
blubber **3**: 38
boobies **2**: *34*
Borneo **1**: 20, **4**: 20
bottom fishing **3**: 41
boums **3**: 10
Boyle, Robert **3**: 24, **4**: 32, 35
Brahmaputra River **4**: 14–15
  delta **1**: 31
brine cells **1**: 44
British Navy **1**: 59
brittle stars **2**: 14, 23, 51
Brunel, Isambard Kingdom **3**: 15
bubblenet feeding **2**: 41, *41*
buildings, underwater **3**: 60
buoys, marker **3**: 23
Burgess Shale **2**: 8–9
Bushnell, David **3**: 32, **4**: 32
Bushnell, Ezra Lee **4**: 32
butterfly fish **2**: 53, *53*

Cabot, John **3**: 12
calcium carbonate **2**: 52, *52*
Calicut, India **3**: 11, **4**: 38
California, Gulf of **4**: 44
Callithamnion **2**: *15*
Calypso **4**: 37
Cambrian explosion **2**: 9
Canada
  fishing rights **3**: 58
  hunting **4**: 6
  inhabitants **3**: 7
  Rockies **2**: 8–9
Canary Current **1**: *49*
cancer, drugs against **2**: 52
Cancun, Mexico **1**: *33*
Cano, Sebastian del **4**: 30, 41
canoes
  double-hulled **3**: 6
  hollowed-out trunks **3**: *7*
canyons, submarine **1**: 16, 26–7
capelin **2**: 31, **4**: 6
caravels **3**: 14
carbohydrates **1**: 38, 39, **2**: 7, 16
carbon **1**: 38, 40
carbon cycle **1**: 38–9, *38–9*
carbon dioxide **1**: 4, 38–9, *38–9*, 61, **2**: 4, 10, **3**: 48, 61, **4**: 24
carbon sink **1**: 39
cargo, loading and unloading **3**: 20-1
cargo ships **3**: 4, *8–9*, 9, 19
Caribbean **3**: 11, 59
Caribbean Plate **1**: *12*
Caribbean Sea **1**: 5, **3**: 59, **4**: 8, 12
  Hurricane Georges **1**: *58*, *59*
Carpenter, William **3**: 26
Carson, Rachel **4**: 35–6
Cartier, Jacques **4**: 30–1
cartography **4**: 43
cassiterite **3**: 47, **4**: 15
catshark **2**: *56*
causeways **1**: *29*
caves **1**: 26
Celebes **4**: 17
cephalopod mollusks **2**: 23
cetaceans **2**: 36, 38
Chaetopterus **2**: 46
chalk **1**: 24
Challenger, HMS **3**: 26–7, **4**: 31, 32, 34, 45
Challenger Deep **4**: 33
Champlain, Samuel de **3**: 7
char **4**: 6
Charles V of Spain **4**: 41
charts
  depth **4**: 32
  harbor-finding **4**: 30
  marine **3**: 24
  Mercator projection **4**: 31
  wind and current **4**: 42
chemical weathering **1**: 24
chemicals **1**: 37
  medical **3**: 57
  pollution from **2**: 55, **3**: 50, 57
  recycling **1**: 38–9, *38–9*
chemosynthesis **2**: 7
Chernobyl **3**: 48
Chesapeake Bay **1**: 30
Chiengtang River, tidal bore **1**: 54
chimneys **1**: *41*
Chinese
  explorers **3**: 10, **4**: 30
  farming fish **3**: 44
  inventions **3**: 11
  migration to U.S. **3**: 7
chitons **2**: *45*, 51
cholera **1**: 53, **3**: 51
chordates **2**: 9
chronometers **3**: 12, *12*, 22, 24, **4**: 31, 39
cilia **2**: 47, 48
cities, coastal, effect on habitats **3**: 52
clams **1**: 31, 43, **2**: 15, 23, 46, 47, 49, 57
  catching **3**: 41
  depths reached **4**: 28, *29*
  razor **2**: 47, 50
  soft-shelled **2**: 50
  symbiotic relations **2**: *61*
  threats to **3**: 5
cleaning stations **2**: 60–1
cliffs, coastal **1**: 26, *27*
climate **1**: 4
  changes **1**: 60–1, **2**: 10, **3**: 57
  effects of currents on **1**: 51
  El Niño **1**: 52–3, *52–3*, 61, **2**: 55, **3**: *29*
  studying history of **4**: 24
clippers **3**: 14, *14*, 15
clouds **1**: 4, 8, 36, 38
clownfish **2**: *53*
cnidarians **2**: 22
coal **1**: 61, **3**: 61, **4**: 4, 22
  cargo of **3**: 19
coast, resorts on **3**: 52, 54
coastal estuaries **1**: 30
coastal lagoons **1**: 33, *33*
coastal wetlands **1**: 32–3
coastline
  changes to **1**: 24–5, *24*
  erosion **4**: 12
  threat to life on **2**: 54
cobras **2**: 33
cod **2**: 16, **3**: 27, *42*, **4**: 6, 10
  Atlantic, migrations **2**: 42
  dwindling numbers **3**: 43, 61
  North Atlantic **2**: 31, 61
  reproduction **2**: 31
coelacanth **2**: 11, *11*
cold seeps **2**: 57
colors, warning **2**: 53
Columbus, Christopher **1**: 47, **3**: 11, 14, **4**: 8, 30, 36
comb jellies **2**: 16, **3**: 53
comets **1**: 8, *8*
commensalism **2**: 60, *61*
communities
  deep-sea **2**: 57
  under threat **2**: 54–5
compass **3**: 11, **4**: 30
  gyroscopic **3**: 22, *22*, **4**: 31
computers
  controlling thrusters **3**: 34
  use in navigation **3**: 23
  wave studies on **4**: 42
conservation **3**: 56–7
Conshelf I **4**: 33
container ships **3**: 4, 20–1, *20*
continental drift **1**: 12, **3**: 28, **4**: 32, 40, 45
continental margin **1**: 16
continental rise **2**: *18*
continental shelf **1**: 16, *17*, 19, 24
  depth **4**: 22
  islands on **1**: 20
  threat to life on **2**: 54
continental slope **1**: 16, *17*
convection currents **1**: 12–13
convoys **3**: 19
Cook, Captain James **3**: 13, 22, 24, **4**: 31, 36, 39
copepods **2**: 20–1, *20*, 23
  in food web **2**: *17*
Copernicus, Nicolaus **4**: 43
copper **1**: *41*, **3**: 9, **4**: 4, 10
coral **1**: *28*, **2**: 9, 22
  bleaching **1**: 61, **3**: 35
  as food **2**: 30
  poisons produced **2**: 52
  spawning **2**: 52
  species
    antler **2**: *53*
    brain **2**: *53*
    plate **2**: *53*
    sea fan **2**: *22*
  threats to **3**: 5, 52, 53, **4**: 17, 19
coral atolls **1**: *22*, 23, *23*, **3**: 25, **4**: 31
coral polyps **1**: 22, **2**: *52*
coral reefs **1**: 20, 22–3, *22*, 43, **2**: 15, 52–3, **3**: *52–3*, **4**: 19, 37
  animal life on **2**: 52, 53, *53*, **4**: 15
  classification **1**: *22*, 23
  fringing **1**: *22*, 23, **4**: 19
  locations **1**: *23*
  pollution **4**: 17, 19
  research into **3**: 35
  strangled **3**: 51
  threats to **2**: 54, 55, **3**: 52
  tourism and **3**: 54, **4**: 17
Coriolis effect **1**: *46*, *47*
cormorants **2**: 34, 35, *35*
Cousteau, Jacques **3**: 31, **4**: 33, 36–7
crabs **1**: 32, **2**: *4*, 14, 21, 22, 23, 46, 51, 52, 56, **3**: 53
  catching **3**: 41
  feeding **2**: 33, 45
  as food **2**: 45
  life cycle **2**: *21*
  sharing home **2**: 60
  species
    edible **2**: *22*
    soldier **2**: *55*
cranes *see* derricks
crawfish, larva **2**: *20*
Crete **3**: 16
crocodiles
  American **2**: 33
  food **2**: 33
  Indo-Pacific saltwater **2**: 33
  saltwater **2**: *54*
cruise ships **3**: *21*, 55
cruisers **3**: 19
crust
  continental **1**: 15
  oceanic **1**: 15
crustaceans **2**: 8, 9, 14, 15, 23
  migrations **2**: 42
  in mud **2**: 48
  parasitic **2**: 61
CTD sampling device **3**: 28–9
Cuba **4**: 8
curlews **2**: *48*, *49*
currents
  air **1**: 47
  convection **1**: 12–13
  effect on coastline **1**: 24
  equatorial **1**: 48
  identifying source of **3**: 9
  longshore **1**: 27, 29, 30
  ocean **1**: 48–51, *48*, *49*, *50*, 61
    change of direction **2**: 11
    generating power **3**: 49
  subsurface **1**: 50–1, *50*
  surface **1**: 48–9
  temperature and **1**: 42
  winds and **1**: *47*
cuttlefish **2**: 23
cyanobacteria **2**: 4, *5*, 6, *6*, 7, 12–13, 45, 49
cyclones **1**: 58, 59, **4**: 15, 18
cyclostomes **2**: 24

dams **1**: 27
Danube, River **4**: 41
Danubius **4**: 41
dark zone **2**: *18*, 19, 58
  animals of **2**: 58–9
darter, American **2**: *54*
Darwin, Charles **1**: 23, **3**: 25, **4**: 31, 37
days, lunar **1**: 54, 55
DDT **3**: 51
dead man's fingers sponge **2**: *23*
dead zone **2**: 13
DECCA **4**: 31
decomposers **2**: 17
decompression sickness **3**: 30, 35
deep scattering layer **2**: 19
deep-sea life **2**: 56–7
Dei Gratia **3**: 17
deltas **1**: 30–1, *31*
Demainbray, Stephen **4**: 39
deposit feeders **2**: 49, 51
deposition **1**: 24, 29
depth charges **3**: 19
depth zones **2**: 18–19, *18*
Depths of the Sea **4**: 45
derricks **3**: 21

deserts **1**: 47
destroyers **3**: 19, *23*
Devil's Slide, Britain **1**: *18*
diamonds **3**: 47, **4**: 10
Dias, Bartholomeu **4**: 30, 43
diatoms **2**: *12–13*, 13, 49
in food chain **2**: *16*
in food web **2**: *17*
diesel engines **3**: 15
Dietz, Robert **4**: 33, 37
dimethyl mercury **3**: 51
dimethyl sulfide (DMS) **1**: 38
dinoflagellates **2**: 12, 13, 49
dinosaurs **2**: 10, 11
Dinosaurs, Age of **1**: 10
Discovery **4**: 38
Distribution of Marine Life **4**: 32
diving **3**: 30–1, *30–1*
depths reached **4**: 28, *29*, 37
tourism and **3**: *52–3*, 54–5
diving bell **3**: 30, **4**: 32
diving suits **3**: 30–1, *30–1*, **4**: 33
DNA **2**: 7, 12
dockers **3**: 21
dog whelks **2**: 45, 51
dogfish **2**: 16, *29*, **3**: 43
doldrums **1**: *47*
dolphins **2**: 36, 38
communication **1**: 43
echolocation **2**: 38, *38*
in fishing nets **3**: 41
in food chain **2**: *16*
in food web **2**: *17*
intelligence **2**: 39
schools **2**: 38
social ties **2**: 38
downwelling **1**: 51
Dreadnought, HMS **3**: 19
dreadnoughts **3**: 18–19
Drebbel, Cornelis **3**: 32, **4**: 32
dredging **2**: 55, **3**: 25, 26, 52
drift nets **3**: 41, *41*, **4**: 19
drilling ships **3**: 28
drugs
from corals **2**: 52
in marine organisms **3**: 57
ducks **1**: 31, 32
dugong **2**: 36, *36*
dulse **2**: *15*
Dumas, Frédéric **4**: 37
Dutch East India Company **3**: 14, **4**: 40
dysentery **1**: 53

Earle, Sylvia **4**: 37–8
Earth
area of water on **1**: 4
asthenosphere **1**: 13
core **1**: 5
crust **1**: 12
formation **1**: 8, *9*
lithosphere **1**: 12
magnetic field **1**: 15, **4**: 40, 44
plates **1**: 12–13, *12*
timeline **4**: 26–7
upper mantle **1**: 12
earthquakes **1**: 13, 34–5, **4**: 45
causing tsunamis **1**: 57, **4**: 18, 19
East China Sea **4**: 20
East Pacific Rise **4**: 19
East Wind Drift **4**: 22
Easter Island **3**: 6, **4**: 36
ebb tide **1**: 54
echinoderms **2**: 23, 46, 47, 51
echo sounding *see* sonar
echo soundings **2**: 19
echolocation **2**: 38, *38*
ecosystems, marine **3**: 56–7
ectoparasites **2**: 61
Ediacara Hills, Australia **2**: 8
eels **2**: 24, 49
migrations of **2**: 42
species
gulper **2**: *58*
moray **2**: 52, *60*
egrets **2**: 49
Egypt **3**: 54
farming **1**: 31
water cleaning **1**: 33
Egyptians
exploration **4**: 30, 40
ocean-going ships **3**: 8, *8*
Einstein, Albert **4**: 43
El Niño **1**: 52–3, *52–3*, 61, **2**: 55, **3**: 29
El Niño Southern Oscillation (ENSO) **1**: 53
electricity
by ocean currents **3**: 49
by tidal power **3**: 48–9, *48*
by wave power **3**: 49
by wind power **3**: 49, *49*
fish producing **2**: 29
OTEC (ocean thermal energy conversion) **3**: 49
electronic charts **3**: 23
Eleutherobia **2**: 52
Ellesmere Island **4**: 6
Emperor Seamount chain **1**: *21*
Endeavour HMS **3**: 13, *13*
endoparasites **2**: 61
energy
from food **2**: 16, 17
renewable **3**: 48–9, 61
English settlers, in North America **3**: 7
Enteromorpha **2**: *15*
ENVISAT satellite **4**: 31
equatorial currents **1**: 48, *49*
Eratosthenes **4**: 30
Erebus **4**: 44
Erik the Red **3**: 10, **4**: 30
Eriksson, Leif **3**: 10, **4**: 30, 38
erosion **1**: 24, *24*, 26, *26–7*, 28, 37
estuaries **1**: 6, 30–1, *30*
estuary
life in **2**: 48–9
saltiness **2**: 48
threat to **2**: 54
eukaryotes **2**: 7
Eurasia **1**: 10, *10*, 11
Eurasian Plate **1**: *12*
Europa **1**: 9
Europe **1**: 11
continental shelf **4**: 8
Everest, Mt **1**: 7, *7*
evolution **1**: 20, **2**: 4, 7
theory of **3**: 25, **4**: 31, 37
Ewing, Maurice **4**: 32–3, 45
exclusive economic zone (EEZ) **3**: *58*, 59
exploration
beneath ocean
early methods **1**: 4–5
modern methods **1**: 5
surface, timeline **4**: 30–1
undersea, timeline **4**: 32–3
explorers, voyages **3**: 8–13, *10, 11*
extinctions, mass **2**: 10–11
Exxon Valdez **3**: *50*

factory ships **3**: 39
"famine ships" **3**: 7
FAO (Food and Agriculture Organization) **3**: 43
farming **1**: 31
Faroe Islands **1**: *20*
fat innkeeper **2**: 60
fathoms **1**: 7
fats **2**: 7, 16
ferries **3**: 20
fertilizers **1**: 33, 37, **2**: 13, 14
Fiji **3**: 6
fins **2**: *26*, 27, *27*
anal **2**: *26*, 27, *27*
dorsal (back) **2**: *26*, 27, *27*
pectoral **2**: *26*, 27, 31
pelvic **2**: *26*, 27, *27*, 31
tail (caudal) **2**: 26, *26*, *27*
fish **2**: 5, 14, 15
ancestors **2**: 9
attracting mate **2**: 59
bony **2**: 26, 27, *27*, 30–1
breeding on land **2**: 31
cartilaginous **2**: 26, *26*
communication **1**: 43
of coral reefs **2**: 52, 53, *53*
deep-sea **2**: *56*
feeding methods **2**: 30
fins **2**: *26*, 27, *27*, 31
as food **3**: 40
genetic engineering **3**: 45
gills **2**: 24, 26, *27*
grooming **2**: 60–1
harm caused to, by oil **3**: 50
hermaphroditic **2**: 59
internal and external parts **2**: *26, 27*
jawed **2**: 26–7, *26, 27*
jawless **2**: 24–5, *24, 25*
lateral line **2**: *27*, 30
migrations **2**: 42
mouth-breeders **2**: 31
of mudflats **2**: 49
nurseries **2**: 49
nursery grounds **4**: 20
ocean farming **3**: 44–5, *44, 45*
dangers of **3**: 45
plankton-feeding **1**: 51, *51*
pollution and **4**: 12
reproduction **2**: 31
on sandy shore **2**: 47
schooling **2**: 30–1, 53
warning colors **2**: 53
of wetlands **1**: 32–3
fish farming **2**: 54
fish larva, in food web **2**: *17*
fish louse **2**: 60
fishing
catches **3**: 5
commercial **2**: 54, **3**: 40–1, 57
disputes over **3**: 58
dwindling stocks **3**: 42–3, **4**: 6
excessive **3**: 27
habitat damage caused by **3**: 53
locating shoals **3**: 40–1
methods **3**: 40–1, **4**: 15, 19
for pearls **3**: *31*
policing **3**: 59
preservation of stocks **3**: 34, 42
quotas **3**: 59, **4**: 6
rights **4**: 10
in Southern Ocean **4**: 22–3
fishing grounds, managing **3**: 59
fishing nets **3**: *40*, 41, **4**: 15
fishing vessels **3**: 4–5, *59*
Fitzroy, Robert **3**: *25*
fjords **1**: 30
flagella **2**: 13
flags of convenience **3**: 20
flamingos **1**: 31
flatfish **2**: 21
flatworms **2**: 61
FLIP (FLoating Instrument Platform) **3**: 34–5, *35*
flippers **2**: 36
floes **1**: *44*
flood tide **1**: 54
flooding **3**: 61
caused by El Niño **1**: 53, *53*
protection from **1**: 27
Florida Keys **3**: 57
flounders **2**: 51, **4**: 10
flukes (tail) **2**: 36
flying fish **2**: 31, *31*
fogs **1**: 59
Food and Agriculture Organization (FAO) **3**: 43
food chains **2**: 16–17, *16, 17*, 21, **3**: 40, 51, **4**: 22
food webs **2**: 16, 17, *17*
foraminiferans **2**: 20, 56
Forbes, Sir Edward **4**: 32
forests, underwater **2**: 14, *15*
fossil fuels **1**: 40–1, *40*, 61, **3**: 5, 48, 61
formation **1**: *40*
fossils **2**: 6, 8–9, *9*, 10
Foster, Howard **4**: 45
Fram (steam sailing ship) **3**: *13*, **4**: 42
Franklin, Benjamin **1**: 49, **4**: 31
Franklin, Sir John **3**: 12, **4**: 44
friction **1**: 13
frigatebirds **2**: 35, *35*
fringing reef **1**: *22*, 23, **4**: 19, 37
Frobisher, Martin **3**: 12
frogs **2**: 32
fronds **2**: 14, *14*
Fulton, Robert **3**: 15, 32
Fundy, Bay of **1**: *55*
fungi **2**: *7*, 44

Gagnan, Emile **3**: 31, **4**: 33, 36
Galápagos Islands **1**: 20, *21*, **2**: *33*, **4**: 37
galleys **3**: 18
Galveston Bay, Texas **2**: 55
Gama, Vasco da **3**: 11, 14, **4**: 30, 38
Ganges River **1**: 19, **4**: 14–15
delta **1**: 31
gas
drilling for **3**: 46–7, *46*
greenhouse **3**: 48
land-based deposits **3**: 46
location **1**: *40*
natural **1**: 40–1, *40*, 61, **3**: 61, **4**: 4, 8, 19

under sea ice **4**: 22
gas hydrates **1**: 41
gas platforms **3**: *46*
gastropod mollusks **2**: 23, 45
geese **1**: 31, 32, **4**: 4
genetic engineering **3**: 45
Geographia **4**: 43
geologists **1**: 12
Germany, anti-pollution measures **4**: 12
ghost ships **3**: 17
Gibraltar, Straits of **1**: 11
gill nets **3**: 41
gill rakers **2**: 30
gills **2**: 7, 24, 26
Gjöa (steam sailing ship) **3**: 12, **4**: 34
Glacier Bay, Alaska **3**: *21*
Glacier, USS **1**: 45
glaciers **1**: 30, 44–5, **4**: 22
glauconite **4**: 15
Global Coral Reef Monitoring Network **2**: 55
global positioning system (GPS) **3**: *23*, 41, **4**: 31
global warming **1**: 39, 51, 60–1, *60*, **3**: 5, 29, 34, 48, 61, **4**: 6, 24, 42
Globtik Tokyo **3**: *21*
Glomar Challenger **3**: 28, **4**: 33
GLORIA (Geological LOng Range Inclined Asdic) **3**: *34*
goblinshark **2**: *56*
gold **1**: 40, **3**: 47, **4**: 19
  prospecting for **3**: 14
  trade in **3**: 8, 9
Gondwana **1**: 10, **4**: 14
GPS (Global Positioning Systems) **3**: *23*, 41, **4**: 31
Grand Banks **1**: 59
Grand Banks, Newfoundland **3**: 58
Grand Forks, North Dakota **1**: *53*
granite **1**: 18, *18*, 26
grasses **2**: 12
  marsh **1**: 32, 33
gravel
  on beaches **1**: 28, 41
  erosion by **1**: 24
  extracting **1**: 41, **4**: 10, 19
gravity **1**: 54, 55
grease ice **1**: 44
Great Barrier Reef **1**: 22, 23, **2**: 61, **3**: 54
Great Barrier Reef Marine Park **3**: 57, **4**: 19
Great Britain **4**: 8
Great Britain **3**: *14–15*, 15
Great Lakes **2**: 24
Great Rift Valley **1**: *11*
Greece **3**: 16, 20
Greeks **3**: 4, 9, 18, **4**: 43
  exploration **4**: 30
  migration to U.S. **3**: 7
greenhouse gases **1**: 41, *60*, 61, **3**: 48, 57
Greenland **1**: 20, 45, *45*, **3**: 10, *39*, **4**: 6, 30, 42, 45
Greenland Current **4**: 4
Greenpeace **3**: *56–7*
Greenwich Mean Time (GMT) **3**: 12, **4**: 39
Greenwich Meridian **3**: 22, **4**: 31
groins **1**: *24–5*, 25
grooming **2**: 60–1
groundwater **1**: 36, *37*, 41
grunion **2**: 31
Gulf
  barrier islands **1**: 29
  coast **1**: 28
Gulf Stream **1**: 48, 49, 50, 51, 59, **3**: 49, **4**: 8, 41
  navigational chart **4**: 31
gulfs **1**: 5
gulls **2**: 35, **4**: 22
guns, used at sea **3**: 18–19
guyots **4**: 40
gyres **1**: 48, **4**: 15, 19
gyrocompass **3**: 22, *22*

habitats, destruction **2**: 10, **3**: 52–3, 54–5
haddock **2**: 61, **4**: 6, 10
  dwindling numbers **3**: 43
hagfish **2**: 24–5, *25*, 56
Halley, Edmond **3**: 30, **4**: 32
Hallucigenia **2**: 8, *8*
Hanno **4**: 30
Hansa Carrier **1**: 49
Hardy, Alister **4**: 38
harpoons **3**: 39
Harrison, John **4**: 31, 38–9
Harrison, William **4**: 39
hatchetfish **2**: 18–19
Hatteras Plateau **1**: 16
Hawaiian Islands **1**: *17*, *20–1*, **3**: 6, 13, 34
He, Zheng, Admiral **3**: 10, **4**: 30, 39
headlands **1**: 26, 29
heart urchins **2**: 47
heat **1**: 42–3
heavy metals, pollution from **2**: 55, **3**: 51
Heezen, Bruce **3**: 28, **4**: 33, 40, 45
Heligoland **4**: 43
Hell's Gate, Kenya **1**: *11*
Helluland **4**: 38
Henry the Navigator, Prince **3**: 11, **4**: 30, 39
Henry VIII of England **3**: 36
Hensen, Victor **4**: 39–40
herbivores **2**: 16
herons **2**: 49
herring **1**: 51, **2**: 16, 17, **3**: 27, 41, *41*, 42, *42*, **4**: 10
  feeding **2**: 30
Hess, Harry **1**: 14, 15, **4**: 33, 37, 40, 45
heterostracan **2**: 24
Heyerdahl, Thor **3**: 8, **4**: 31, 40
high seas **3**: 58, *58*
high-pressure zones **1**: 47
highways, oceans as **3**: *5*, 6–7, *6*
Himalayas **1**: 13, *13*
Histoire Physique de la Mer **4**: 41
holdfasts **2**: 14, *14*, *15*
Hong Kong **3**: *11*, 20, 60
Honshu **4**: 20
horse latitudes **1**: *46*, 47
hot-water (hydrothermal) vents **1**: 5, 15, *41*, **2**: 6, 57, **4**: 33, 35
hotspots **4**: 14
  living **2**: 5
Hudson, Henry **3**: 12, **4**: 31, 40–1
Hudson Bay **1**: 61, **3**: 12, **4**: 8, 31, 40
Hudson River **4**: 40
Hudson Strait **4**: 40
humans, migrations **3**: 6–7, *6*, *7*
hunting
  in Arctic regions **4**: 4, 6
  of sea mammals **3**: 38–9, *38*, *39*
Hurghada, Egypt **3**: 54
Hurricane Georges **1**: *58*, *59*
hurricanes **1**: *58*, **4**: 8, 18
Hvalfjorden, Iceland **3**: *38*
hydrocarbons **1**: 40
hydroelectricity **1**: 27
hydrogen **1**: 40
hydrogen sulfide **2**: 7, 57
hydrological cycle **1**: 36
hydrothermal (hot-water) vents **1**: 5, 15, 41, **2**: 6, 57, **3**: 33, **4**: 33, 35

ibis, scarlet **2**: *55*
Ibn Battuta **4**: 35
ice **1**: 9, 36, 44–5, *44*, **4**: 4, 6, 22
ice ages **1**: 29, 34, 35, **3**: 6, 27
ice shelves **1**: 44, **4**: 22
ice-breakers **4**: 4
icebergs **1**: 44–5, *45*, **4**: 4, 22
Iceland **4**: 8, 43
  fishing rights **3**: 58
ICES (International Council for the Exploration of the Sea) **3**: 27
iceworms **2**: 57
igneous rock **1**: 18, 19, *19*
IGOS (Integrated Global Observing Strategy) **4**: 31
iguana, marine **2**: *33*
IIP (International Ice Patrol) **1**: 45, **4**: 4
IMO (International Marine Organization) **3**: 53
impermeable rock **1**: 41
Incas **3**: 7
incense **3**: 8
India **1**: 10, 11, **3**: 10, 11, **4**: 14, 15
Indian Ocean **1**: 5, 6, **4**: 14–17, *15*
  changes **1**: 11
  data **4**: 17
  fish and fishing **2**: 11, **4**: 15
  map **4**: *17*
  mineral resources **4**: 15
  pollution **4**: 15–17
  sea snakes **2**: 33
  seas **4**: 17
  size **1**: 7
  voyages across **3**: 10, 11
  water temperatures **2**: 55
Indo-Australian Plate **1**: *12*
Indo-West Pacific **4**: 15, 19
Indonesia **3**: 6, 10, **4**: 15
  forest fires **1**: *52–3*
Indus River **4**: 14–15
industrial waste **2**: 13
  pollution from **3**: *50–1*, 51, **4**: 12
industry, on estuaries **1**: *30*
infrared radiation **1**: 60–1, *60*
Ingstad, Anne Stine **4**: 38
Ingstad, Helge **4**: 38
insecticides **3**: 57
insects **2**: 32
Integrated Global Observing Strategy (IGOS) **4**: 31
Intergovernmental Panel on Climate Change (IPCC) **1**: 60
International Council for the Exploration of the Sea (ICES) **3**: 27
International Ice Patrol (IIP) **1**: 45, **4**: 4, 41
International Marine Organization (IMO) **3**: 53
International Seabed Authority **3**: 59
International Shark Attack File (ISAF) **2**: 28
International Whaling Commission (IWC) **3**: 38, 39
intertidal zone **2**: 44–5
intruson **1**: 19
Inuit **3**: 7, *25*, 39, *39*, **4**: 44
invertebrates **2**: 22–3, *22*, **4**: 45
IPCC (Intergovernmental Panel on Climate Change) **1**: 60
Ireland **4**: 8
Irish, migrants **3**: 7
iron **1**: 61, **4**: 4, 22
  cargo of **3**: 19
ISAF (International Shark Attack File) **2**: 28
Iselin, Columbus O'Donnell **4**: 41
islands **1**: 17, *17*, 20–1
  barrier **1**: 29
  inshore **1**: 20–1
    connection to mainland **1**: *29*
  oceanic **1**: 20, 21, *21*
  offshore **1**: 20, 21
  Pacific, people of **3**: 6–7
  volcanic **1**: *22*
Italians, migration to U.S. **3**: 7
ivory **3**: 8
IWC (International Whaling Commission) **3**: 38, 39

jaegers **2**: *34*, 35
Jahre Viking **3**: 21
Jamestown, Virginia **3**: 7
Japan **3**: 10
Japan, Sea of **4**: 20
Japanese, migration to U.S. **3**: 7
Jarrett, Jim **4**: 33
Jason (ROV) **3**: 34
Jason Junior (ROV) **4**: 33
Jason Project **4**: 35
Java **4**: 17
jawless fish **2**: 24–5, *24*, *25*
jaws, of fish **2**: 26
jellyfish **2**: 5, *5*, 8, 9, 16, 21, 22, *59*
  common **2**: *22*
jet-streams **1**: 53
jewelry **3**: 9, 14
Jews, migration to U.S. **3**: 7
JIM suit **4**: 37
Joides Resolution (drilling ship) **4**: 33
Juan de Fuca Plate **1**: *12*
junks **3**: 10, *11*

Jupiter (planet) **1**: 9

Kaiko (ROV) **4**: 33
Kaneohe Bay, Hawaii **3**: 51
kelps **2**: 14, *15*, 51
  sugar **2**: *14*
Kerguelen Islands **4**: 14
kinetic energy **3**: 49
knotted wrack **2**: *14*
Kon-tiki (raft) **4**: 31, 40
kraken **3**: 16
krill **1**: 43, **2**: 20, 23, **3**: 16, **4**: 23
  in food chain **2**: *16*
  in food web **2**: *17*
  reflecting sound waves **2**: 19
Kuroshio Current **3**: 49

La Jolla, California **1**: 26, **3**: 27, 34
Labrador **1**: 49, **3**: 10, **4**: 36
Labrador Current **1**: 49, 59, **4**: 41
Labrador Sea **4**: 4
lagoons **1**: *22*, 23, *23*
  coastal **1**: 33, *33*
  threat to life on **2**: 54
lakes **1**: 36
Lamont (Lamont Doherty) Geological Observatory **4**: 33
lampreys **2**: 24, *24*, 25, *25*
land
  reclamation **1**: *30*, **3**: 60
  signs of, at sea **3**: 6, 9
  temperature **1**: 4
land bridges **3**: 6, 7
landmasses
  collision **2**: 11
  rise and fall **1**: 34–5
landslides, causing tsunamis **1**: 57, **4**: 18
lanternfish **2**: 19
Lanzarote, Canary Islands **1**: *37*
larvaceans **2**: 21
larvae **2**: 21
lateral line system **2**: *27*, 30
latitudes **1**: 46–7, **3**: 11, **4**: 30
Laurasia **1**: 10
lava **1**: *9*, 13, *15*, **4**: 19
  cooling **1**: 18
Law of the Sea **3**: 58–9, **4**: 31, 41
lead **1**: 33, **3**: 51, 56, **4**: 4
  pollution **4**: 24
Lebanon **3**: 8
levees **1**: *31*
Liberia **3**: 20
lice, fish **3**: 45, *45*
lichens **2**: 44–5
light **1**: *42*, 43
  bioluminescent **2**: 18, 19, 59
  on water **1**: *42*
lighthouses **3**: 23
Lightning, HMS **4**: 32
lightships **3**: 23
limestone **1**: 19, 22, 24, 26
limpets **2**: 44, 45, 51
lionfish **2**: 53
lithosphere **1**: 12
Little Hercules (ROV) **4**: 33
living hotspots **2**: 5
lobsters **2**: 15, 21, 23
  farming **3**: 44
lodestone **3**: 11
Loiho (volcano) **1**: *21*
London, England **1**: 49, **3**: 20
longitude **3**: 11, 12, 22, 24, **4**: 31, 38–9
longships **3**: 10
longshore currents **1**: 27, 29, 30
longshore drift **1**: 25, 57
LORAN **4**: 31
Los Angeles, earthquakes **1**: 13
louse **2**: 60
Lovelock, James **1**: 38
luciferase **2**: 59
luciferin **2**: 59
lugworms **2**: 46, *49*
  casts **2**: *46–7*, *49*
lunar day **1**: 54, 55
lunar month **1**: 55
Lusitania, wreck **4**: 35
Luxor **3**: 8
Luzon **4**: 20

mackerel **2**: *4*, **3**: 41
Madagascar **4**: 17
Magellan, Ferdinand **3**: 11, 12, 14, **4**: 18, 30, 41
magma **1**: 13, 14, 15, **4**: 19
  cooling **1**: 18
  magnesium **1**: 15
magnetic field **1**: 15, **4**: 40, 4.44
magnetic north **4**: 44
magnetic reversals **4**: 40
magnetite **3**: 11
Maldives **3**: 61, **4**: 15
mammals
  harm caused to, by oil **3**: 50, **4**: 6
  hunting **3**: 38–9, **4**: 4, 6
  marine **2**: 5, 17, 21, 36–7
  threat to **3**: 5
manatees **2**: 36, **3**: 55
manganese **3**: 47
manganese nodules **1**: 19, *41*
mangrove swamps **2**: 49
  clearing **2**: 54, **4**: 12, 17, 20
  life in **2**: *54*
  threat to **2**: 54
mangroves **1**: 32, *32*, **3**: 52
manta ray **2**: 28, *28*
Manzala, Lake **1**: 33
mapmaking **4**: 43, 45
  of ocean floor **3**: 27, **4**: 45
marble **1**: 19
Marianas Trench **1**: 7, **3**: 27, 33, **4**: 18, 32, 33, 37, 43
mariculture **3**: 44–5, *44*, *45*
marinas **3**: 5
marine life **3**: *24*
  effects of tides on **1**: 54
  studying **2**: 4–5, **3**: 24–5
marine organisms **1**: 4
marine protected areas (MPAs) **3**: 54, 55, 57
Markland **4**: 38
marlin **2**: 16, *30*
Marquesas Islands **4**: 36, 40
marsh grasses **1**: 32, 33
Marsigli, Count L.F. **4**: 41
Mary Celeste **3**: 17
Mary Rose **3**: 36, *36*, **4**: 33
mass extinctions **2**: 10–11
Massachusetts **3**: 7
Maud **4**: 34, 44
Mauna Kea **1**: *7*
Mauritania **4**: 10
Maury, Matthew Fontaine **3**: 24, **4**: 31, 32, 41–2
Mayans **3**: 7
meadows, seagrass **2**: 15, 55, *55*
medicines **3**: 57, **4**: 41
Mediterranean Sea **1**: 5, 11, **3**: 8, 9, 10, 15, 59, **4**: 8, 12
  tidal range **1**: *55*
megalops stage **2**: *21*
megamouth **3**: 16
Melanesia **4**: 20
Melbourne, Australia **3**: *14*
Mercator, Gerhardus **4**: 31
merchant vessels **3**: 20
mercury **3**: 51
merlin **2**: *5*
mermaids **2**: 36
message in a bottle **1**: 49
metallic nodules **1**: *41*
metals **3**: 47
  deposits **1**: 19
  heavy, pollution from **2**: 55, **3**: 51
  pollution from **2**: 55
metamorphic rock **1**: 19, *19*
Meteor (research vessel) **3**: 27, **4**: 32
meteorites **1**: 8, **2**: 11, **4**: 37
  crater **2**: *10–11*
methane **1**: 41, 61, **2**: 7, 57
methane hydrate (methane ice) **2**: 57
Mexico, Gulf of **1**: 5, 48, 55, **2**: 13, 57, **3**: 5, 46, 51, **4**: 8, 12
Miami Beach, Florida **1**: 29
microbes **1**: 8, 30
  species **2**: 4
Micronesia **4**: 20
Mid-Atlantic Ridge **1**: 16, 21, **3**: 28, **4**: 8, 33, 35, 40, 45
mid-ocean ridges **1**: 14, *14*, *15*, 16, *16*, **4**: 33
migrations **2**: 42–3
  human **3**: 6–7, *6*, *7*
Milky Way **1**: 9
Minamata, Japan **3**: 51
Mindanao **4**: 20
minerals **3**: 47
  placer deposits **3**: 47
  in sea **1**: 40-1, **4**: 4, 8–10, 15, 19
  under ice **4**: 22
mining, undersea **3**: 46–7, *46*
Minoan civilization **3**: 16
Mississippi River **3**: 20
  delta **1**: *31*
mist **1**: *42*
mites **2**: 9
mollusks **2**: 15, 23, 45
  bivalve **2**: 23, 46, 47, 48
  cephalopod **2**: 23
  gastropod **2**: 23, 45
monkeys, proboscis **2**: *54*
monogenean **2**: 61
monsoons **4**: 15
monsters, sea **3**: 16–17
month, lunar **1**: 55
Moon
  orbiting Earth **1**: 54, 55
  tides and **1**: 54, *54–5*, 55
Moorish idol **2**: 53, *53*
moray eels **2**: 52, *60*
Morley, Lawrence **1**: 14–15
mosses **2**: 32
mother-of-pearl shells **3**: 30
mountains
  growth **1**: 13
  submarine **1**: 5, 7, *7*, *15*, 16
Mozambique **4**: 38
MPAs *see* marine protected areas
mud, on beaches **1**: 28
mudflats **1**: 31, **2**: 48, *48*, 49
mudhoppers **2**: 49
mudskipper **2**: *54*
mullet **2**: 49
multicelled organisms **2**: 7, *7*, 8
Munk, Walter **4**: 42
Muscovy Company **4**: 40
mussels **2**: *4*, 21, 23, 45, 57
  farming **2**: 54, **3**: 44
  species
    common **2**: *23*
    zebra **3**: *53*
mutualism **2**: 60, *61*
Mykines Island **1**: *20*
myths **3**: 16–17
Myxine **2**: *25*

Namibia
  diamonds **3**: 47
  fishing **4**: 10
Nansen, Fridtjof **4**: 4, 31, 42–3
Nansen Basin **4**: 4
narwhal **2**: 38
NASA **1**: 8, 9
National Oceanic and Atmospheric Administration (NOAA) **1**: 53, **4**: 38
National (steamer) **4**: 39–40
Native Americans **3**: 7
natural gas *see* gas, natural
natural selection **3**: 25, **4**: 31, 37
Nautilus (early submarine) **3**: 32
Nautilus, USS (nuclear submarine) **3**: 19, 32, **4**: 33
navies **3**: 4, 19
navigation **2**: 39, **3**: 9, 22–3, *23*
  instruments of **3**: 11, 22–3, *23*, **4**: 31
  for migrations **2**: 42, 43
Navstar satellite **4**: 31
Nazca Plate **1**: *12*
neap tide **1**: 55
Necho **3**: 9
nekton **2**: 5
New Caledonia **4**: 36
New Guinea **3**: 6, **4**: 20
New York City **3**: 20
New Zealand **3**: 7, 13, **4**: 20
Newfoundland **1**: 59, **3**: 10, **4**: 8, 36, 38
nickel **1**: *41*, **4**: 10
Niger River **3**: *40*
nilas **1**: 44
Nile, River **3**: 8
  delta **1**: *31*
Nimitz, USS **3**: 19
Niña **4**: 36
Ninety East Ridge **4**: 14

nitrates **1**: 33, **3**: 45, 51, 56
nitrogen
  sources **2**: 13
  trapping **2**: 12–13
nitrogen narcosis **3**: 30
NOAA (National Oceanic and Atmospheric Administration) **1**: 53
Nobile, Umberto **4**: 35
North America **1**: 10, 11
  peoples of **3**: 7
North American Plate **1**: *12*
North Atlantic Drift **1**: 49, 51, **4**: 4, 8
North Atlantic gyre **1**: *49*, 51
North Graham Land **4**: 44
North Pole **1**: 47, **4**: 4, 42–3
North Sea **1**: *35*, **4**: 8
North Star **3**: 9
Northeast Passage **4**: 40
Northern Territory, Australia, meteorite crater **2**: *10–11*
Northwest Passage **3**: 12, **4**: 6, 30–1, 34, 36, 40, 44
Norway, salmon farming **3**: *45*
Norwegian Current **4**: 4
Norwegians **3**: 16
notochord **2**: 24
Nova Scotia, Canada, tidal power **3**: 49
nuclear power **3**: 48
nuclear weapons **3**: 19
nucleic acids **2**: 7
nucleus **2**: 7
nutrients **1**: 43, 50, 51
  nitrogen-based **2**: 12
  recycling **1**: 30

oarweed **2**: *14*
Obelia **2**: 22, *22*
Ocean Beach, Tasmania **2**: *39*
Ocean City, New Jersey **1**: 25
ocean thermal energy conversion schenes OTEC) **3**: 49
ocean trenches **2**: *18*, 19
oceanographers **1**: 5
oceanography **3**: 26–7
Oceanography **4**: 44
Oceans: Their Physics, Chemistry and General Biology **4**: 44
oceans
  see also named oceans
  anatomy **1**: 16–17, *16–17*
  changes in **1**: 10–11, *11*
  circulation **1**: 36, **4**: 45
  color **1**: 6
  depth **1**: 6–7, 16, 17
  measuring **3**: 26
  depth zones **2**: 18–19, *18*
  effects **1**: 4
  floor (seabed) **1**: 5, 16
    changes **1**: 12, 13, 14–15, *14*, **4**: 8
    deep **2**: 56
    formation **1**: 13, **3**: 28, **4**: 8, 19
    law concerning **3**: 59, **4**: 31, 41
    mapping **1**: 5, 7, **3**: 27
    shallow **2**: 50–1
    surveying **3**: *34*
  formation **1**: 8–9, *11*
  ice cover **4**: 22
  layers **1**: 43
  managing **3**: 58–9
  names **1**: 5
  physical structure **1**: 4
  protecting **3**: 56–7
  seas in **1**: 5, **4**: 8
  size **1**: 6, 7
  temperatures, measuring **3**: 28–9
  water color **3**: 29
octopuses **2**: 23, 51, *51*
  giant **3**: 16
Odaraia **2**: *9*
Odontogryphus **2**: *9*
Oikopleura **2**: *20*
oil **3**: 61
  crude, transporting **3**: 21, *21*
  deposits **1**: *40*, **4**: 4, 15, 19
    searching for **3**: 47
  drilling for **3**: 46–7, *46*, 57
  extracting **3**: 5, **4**: 4, 8
  ingredients of **1**: 40–1
  land-based deposits **3**: 46
  location **1**: 40, *61*
  pollution from **2**: 55, **3**: 5, 50–1, *50*, **4**: 12, 17
    in cold conditions **4**: 6
    spraying **3**: *51*
  under sea ice **4**: 22
oil rigs **3**: *46*, 51
Okhotsk, Sea of **4**: 20
Olenoides **2**: *9*
olive oil **3**: 9
Oman **4**: 15
On the Origin of Species by Means of Natural Selection or the Preservation of Favoured Races in the Struggle for Life **4**: 37
Open Sea **4**: 38
orcas (killer whales) **2**: *17*, 38, 39
  in food web **2**: *17*
Oregon, gold **3**: 47
organisms
  earliest **2**: 6–7
  groups **2**: 5
  multicelled **2**: 7, *7*, 8
  single-celled **1**: 4, **2**: 4, *5*, 22, 47, 49, 61
  smallest **2**: 4
OTEC (ocean thermal energy conversion) schemes **3**: 49
otter trawls **3**: *40*
otters, sea **2**: 15, *37*, **3**: 5
overfishing **3**: 27, 42–3, *59*, **4**: 6, 15
overharvesting **2**: 10, **4**: 19
oxygen
  in atmosphere **2**: 7
  levels **2**: 6
  organisms surviving without **2**: 48
  production **2**: 4, 7, 12
oystercatchers **2**: 45
oysters **2**: 23
  farming **2**: 54, **3**: 44, *44*
ozone layer **4**: 24

Pacific Ocean **1**: 5, 6, *6*, 7 **4**: 18–20, *19*
  birth **1**: 10–11
  breaking waves **1**: 56
  changes **1**: 11, **4**: 19
  coastline **1**: 24
  countries bordering **4**: 19
  currents **4**: 19
  data **4**: 20
  depth **1**: 6, 7
  island inhabitants **3**: 6–7
  islands **4**: 19, 20
  map **4**: *20*
  mineral resources **4**: 19
  pollution **4**: 19–21
  ports **4**: 19
  rising coast **1**: 26
  sea snakes **2**: 33
  seas **4**: 20
  size **1**: 6, 7, **4**: 18, 20
  trough in **1**: 7
  upwelling **1**: 51
  waves **1**: 56–7, *56*
Pacific Plate **1**: *12*
Pacific Tsunami Warning Center **1**: 57
pack ice **1**: 44, *44*, **4**: 4, 6
Painted Desert, Utah **1**: *18*
Pakistan **4**: 15
Palos Verdes hills, California **1**: 26
Panama **3**: 20
Panama Canal **3**: *5*
Panama Isthmus **1**: 11, **4**: 34
pancake ice **1**: 44
Pangaea **1**: 10
Panthalassa **1**: 10
Papua New Guinea **1**: 20, *41*
parasitism **2**: 60, 61
Paris, Oceanographic Institute **4**: 34
parrotfish **2**: 30
Parry, Sir William **4**: 44
pearl fishermen **3**: *31*
pearls **3**: 44
Peary, Robert **4**: 34
pelagic environment **2**: *4*, 5
pelicans **2**: *34*, 35
penguins **2**: 34, *34*, **3**: 51, **4**: 22
  Adélie **2**: 34
  emperor **2**: 34
  Galápagos **1**: *51*
perch, sand- **2**: *51*
periwinkles **2**: 45
Permian period **2**: 10
Persian Gulf **3**: 10, **4**: 15, 17
Persians **3**: 18
pesticides, dumping **4**: 12
petrels **2**: 35
  diving **2**: *34*
  storm- **2**: *34*, 35
petroleum oil *see* oil
Pfiesteria **2**: 12
pheromones **2**: 59
Philippine Plate **1**: *12*
Philippines **3**: 11
Phoenicians **3**: 9, **4**: 30
photography
  satellite **1**: 5, *5*
  underwater **1**: 5
photosynthesis **1**: 4, 22, 38–9, *38*, 43, 61, **2**: 4, 7, 11, 12, *13*, 18, 55, 60
phyllosoma larva **2**: *20*
Physical Geography of the Sea **3**: 24, **4**: 31, 42
phytoplankton **1**: 4, 16, 20, 38, 39, *39*, 51, 52, 61, **2**: 4, *4*, 5, *12–13*, 18, 46, 49, 50, 56, **4**: 15
  benefits of **2**: 4, **3**: 56
  bloom **1**: *39*, **2**: 13, 56, **3**: 56
  in carbon cycle **1**: *38*, 39
  controlling growth of **1**: 61
  damage to **4**: 24
  distribution, on satellite image **3**: 29, *29*
  effect of nitrates on **3**: 51
  El Niño and **1**: 52
  in estuaries **2**: 49
  in food chain **2**: 12, 16, 17, **4**: 5, 22
  harm caused to, by oil **3**: 50
  in marine communities **2**: *4*
  monitoring **3**: 34
  photosynthesis **1**: 4, *38*, 39, **2**: 4, 12, 18, **4**: 5
  producing DMS **1**: 38
  size range **2**: 5
  types **2**: 12–13
  uncontrolled growth **3**: 50, 56
  weight of **2**: 4
Piccard, Auguste **3**: 33, **4**: 33, 43
Piccard, Jacques **1**: 7, **3**: 33, **4**: 33, 37, 43
Piccard, Jean **4**: 43
pilchard **3**: *42*
Pilgrims **3**: 7
pillow lava **1**: *15*
pilot fish **2**: *61*
Pimlico Sound, North Carolina **1**: 30
pinnacles **1**: 26
pinnipeds **2**: 36–7
Pinta **4**: 36
pipelines, underwater **3**: 5
pirates **3**: 14
placer deposits **3**: 47
plaice **2**: 51, **4**: 6, 10
planets
  formation **1**: 8
  life on **1**: 8–9
  water on **1**: 8–9
plankton **1**: 39, 40, **2**: 5, 12, 23, 43, 46, **4**: 10, 39
  see also phytoplankton; zooplankton
  bloom **1**: *39*, **2**: 13, 56, **3**: 56
  colors **2**: 13
  creating dead zone **2**: 13
Plankton Expedition **4**: 39–40
plants
  in food chains **2**: 16
  marine **1**: 20
  respiration **1**: *38*, 39
  species **2**: 4
  water content **1**: 36
plastic, pollution from **3**: 50
plate tectonics **1**: 15, **3**: 28, **4**: 19, 33, 37
plates **1**: 12–13, *12*, 19
  effects on coast **1**: 24
  sea levels and **1**: 34–5
platinum **1**: *41*
plover, ringed **2**: *49*
poisons
  produced by corals **2**: 52
  in red tides **2**: 12
polar bears **1**: 61, **4**: 6
polar seas **1**: 43
Poles, migration to U.S. **3**: 7
pollution **1**: 37, **3**: 50–1, **4**: 4, 12, 19–20

chemical 2: 55, 3: 50, 57
in cold conditions 4: 6
damaging ozone layer 4: 24
eliminating 3: 59
of estuaries 1: *30*
from fish farms 3: 45
industrial waste 3: *50–1*, 51, 4: 12
lead 4: 24
metal 2: 55, 3: 51
oil 2: 55, 3: 5, 50–1, *50*, *51*, 4: 6, 12, 17
spraying 3: *51*
plastic 3: 50
sewage 2: 13, 55, 3: 51, 4: 12
of wetlands 1: 33
polymetallic nodules 4: 19
Polynesia 4: 20
Polynesians 3: 6–7
polyps 2: 47, 50–1
coral 2: 52, *52*, 55, 60
Porcupine, HMS 4: 32
Porcupine Abyssal Plain 1: 16
pororoca 1: 54
porpoises 2: 36, 38
schools 2: 38
social ties 2: 38
threats to 3: 5
vaquita 3: 56
ports 3: 5, 20
Portsmouth, England 3: 36
Portuguese sailors 3: 11, 4: 30
Portuguese man-of-war 2: 22
portulans 4: 30
potash 4: 15
prawns 2: 56
precipitation 1: 36, *37*
primary consumers 2: 16
Prince of Wales Island 4: 44
proteins 1: 38, 2: 7, 12, 16
Ptolemy (Ptolemaeus), Claudius 4: 30, 43
puffins 2: *34*, 35
Punt 3: 8
purse seine nets 3: *40*, 41, *41* 4: 15
Pytheas 4: 30, 43

Queen Mary 3: *15*

Ra (reed boat) 3: 8, *8*, 4: 40
radar 3: 22–3, *23*, *29*
radar altimetry 1: 7, 34, 3: 29, 4: 33
radar reconnaissance 1: 45
radiation 1: 4
infrared 1: 60–1, *60*
solar 1: *60*
ultraviolet (U.V.) 2: 6
radioactive fallout 4: 24
radioactive waste 3: 5
radiolarians 2: 20, 3: *26*
rafts, log 3: 8
ragworm 2: *22*
rain 1: 8
acid 1: 38, 2: 10, 3: 48, 59
rainforest, tropical 2: 52, 4: 12
La Rance, France, tidal barrier 3: *48*, 49
ranching, ocean 3: 45
rattail fish 2: 56
rays 2: 27, 28–9, 51
fins 2: 29
food 2: 28, 29
reproduction 2: 29
species
electric 2: 29
manta 2: 28, *28*
sting- 2: 29
razor clams 2: 47, 50
razorbills 2: 35
Red River, N. Dakota 1: *53*
Red Sea 1: 11, *11*, 13, *13*, 3: 8, 10, 15, 52, 4: 17
red tides 2: 12
redshanks 2: *49*
reed boats 3: 8, *8*, 4: 40
Reef Check 2: 55
remoras 2: 60
remote-sensing 3: 29
remotely operated vehicles *see* ROVs
reptiles 2: 32–3
breathing 2: 32
reproduction 2: 32
research, marine 3: 34–5
resorts, coastal 3: 52, 54
respiration 1: *38*, 39
Revolution 3: 13
ribbon worms 2: 22
ridges 1: 14, *14*, *15*, 16, 17
Rift Valley 1: *11*
rifts 1: 14, 15, 16, *16*
Rihlah 4: 35
Ring of Fire 1: 57, 4: 19
Ritter, William 3: 27
rivers 1: 6, 19, 36, *37*
damming 1: 27
as highways 3: 8
sediment from 1: 27, 28
La Rochelle, France, mariculture 3: *44*
rock
arch of 1: *27*
on coastline 1: 26–7, *27*
cycle 1: 18–19, *19*
igneous 1: 18, 19, *19*
impermeable 1: 41
metamorphic 1: 19, *19*
molten (liquid) 1: 8, *9*, 13
sedimentary 1: 18–19, *18*, *19*, 26, 38
types 1: 18
weathering 1: 24
rockpools, creatures in 2: 22
rocks
acids eating away 2: 7
formation 2: 8
rocky shores 2: 44–5, *44*, *45*
Rodinia 2: 9
Romans 3: 4
rorquals 2: 40
Ross, James Clark 3: 25, *25*, 4: 32, 44
Ross, John 3: 25, *25*, 4: 32, 44
Ross Island 4: 44
Ross Sea 4: 34, 44
Ross Shelf 4: 44
roundworms 2: 56, 61
ROVs (remotely operated vehicles) 3: 33, *35*, 45, 4: 28, *29*, 33
rowing ships 3: 18
Royal Observatory, Greenwich 4: 31
Royal Society (of London) 3: 24
Russia 3: 7
Russians, migration to U.S. 3: 7

sailing vessels 3: 14–15
reed 3: 8, *8*
St Lawrence, Gulf of 4: 30
St Lawrence River 4: 30, 36
St Louis, Missouri 3: 20
Salamis 3: 18
salmon 2: 49
farming 2: 54, 3: 44, 45, *45*
in food chain 2: *16*
in food web 2: *17*
migrations 2: 42
salps 2: *20*, 21
salt 1: 36, 44
concentrations of 1: 33
effect on currents 1: 50
in estuary 1: 30–1
in pack ice 1: 44
plants tolerant of 1: 32
sources 1: 37
salt marshes 1: 32, 33, 2: 49, 3: 56
saltiness, measuring 3: 28–9
saltpans 1: *37*
Samoa 3: 6
San Andreas Fault 1: *12*, 13, 30
San Francisco, earthquakes 1: 13
San Francisco Bay 1: 30
sand
on beaches 1: 25, 26–7, 28–9, *28*, *29*, 41
creatures in 2: 22
erosion by 1: 24
mining 1: 41, 4: 10, 19
sand dollars 2: 47
sand eels 2: 17
sand mason worms 2: *46*, 47
sand-perch 2: *51*
sandstone 1: *19*
Sandwell, David 4: 33
Santa Maria 4: 36
Santorini 3: 16
sardines 1: 51, 3: 41, 42, 4: 10, 19
Sargasso Sea 1: 5, 2: 42
satellite altimetry 1: 7, 34, 3: 29, *29*, 4: 33
satellite images, "false-color" 3: *29*
satellite photography 1: 5, *5*
satellites, signals from 3: 23, 29, *29*, 4: 31
Saudi Arabia 4: 15
scallops 2: *4*
scavengers, deep-sea 2: 56
schooling 2: 30–1, 53
Scott, Robert F. 4: 34
Scripps Institution for Biological Research 3: 27, 34, 4: 31, 34, 37, 42, 44
scuba diving 3: 30, 31, *31*, 4: 33, 37
depths reached 4: 28, *29*
sea anemones 2: 22, 45, *53*, 3: 53
beadlet 2: *23*
jewel 2: *23*
sea bass, antarctic 4: 23
Sea Change: A Message of the Oceans 4: 38
sea cows 2: 15, 36
Stellar's 3: 38
threat to 2: 55
sea cucumbers 2: 47, 56, 3: 53
leopard 2: *57*
sea defenses 1: 25, *25*
sea floor *see* oceans
sea floor spreading 1: 14, 15, 4: 33, 37, 40, 45
sea fogs 1: 59
sea ice *see* ice
sea lettuce 2: *15*
sea levels
changes 1: 34–5
longterm 1: 34
measuring 1: 34
rises 1: 61
sea lions 2: 28, 36, 37
Californian 2: 37, *37*
sea otters 2: 15, *37*, 3: 5
sea pens 2: 50–1
sea slugs 2: 14
sea snakes *see* snakes
sea spiders 2: *57*
sea squirts 2: 51
larva 2: 21
sea urchins 2: *4*, 8, 14, 15, 21, 23, *23*, 45, *45*, 51, 3: 35
seabed *see* oceans: floor
seabirds *see* birds
seagrasses 2: *4*, 14–15, *15*, *36*
meadows 2: 55, *55*
seahorses 2: *31*, 55, 3: 56
seals 2: 14, 28, 36, 37, 4: 22
in fishing nets 3: 41
in food web 2: *17*
harm caused to, by oil 3: 50
hunting 3: 39, *39*, 4: 4, 6, 24
species
crabeater 2: 37, *37*
elephant 2: 37
fur 2: 37
harp 4: 6
ringed 4: 4
seamounts 1: *17*, 3: 53, 4: 19
seamouse 2: 51
seas
color 1: 6
definition 1: 5, 4: 8
exclusive economic zone 3: *58*, 59
frozen 1: 44–5
law of 3: 58–9, 4: 31, 41
territorial 3: 58, *58*
Seasat satellite 3: *29*, 4: 31
seawalls 1: *25*
seaweeds 1: 16, 2: 5, *5*, 14, *14–15*, 44, *44*, 45, 46, 49, 50
brown 2: *14*, 45
colors 2: *15*
effect of nitrates on 3: 51
farming 3: 44
grazers among 2: 51
green 2: 45
red 2: 45
secondary consumers 2: 16
sediment 1: 17, 18–19, 25, 26, 2: 10
on beaches 1: 28
levees 1: *31*
river-borne 1: 19, 27
sediment fan 1: 19
sedimentary rock 1: 18–19, *18*, *19*, 26, 38
seine nets 3: 41, *41*, 4: 15, 19
seismic sea waves 1: 57

Seven Miles Down **4**: 37
Severnaya Zemlya **4**: 44
sewage **2**: 13, 55, **3**: 5, 51, **4**: 12
Seychelles **1**: *28*, **4**: 14, 15
Shanghai **4**: 19
Shark Bay, Western Australia **2**: 39
sharks **2**: *4*, 14, 26, 27, 28, 52, 56
  attacks on humans **2**: 28
  fish living on **2**: 60
  food **2**: 28, 33
  in food web **2**: *17*
  legislation to protect **2**: 28
  reproduction **2**: 29
  senses **2**: 28
  species
    basking **2**: 28, *29*
    bull **2**: 28
    catshark **2**: *56*
    dogfish **2**: 16, *29*, **3**: 43
    goblin **2**: *56*
    great white **2**: 28, *29*, *61*
    megamouth **3**: 16
    saw **2**: *29*
    tiger **2**: 28
    whale **2**: 28, **3**: 54–5
shearwaters **2**: 35
shelf break **1**: 16
shellfish
  catching **3**: 41
  farming **3**: 44, *44*
  pollution and **4**: 12
shingle **1**: 29
ships **3**: 4–5, 10, 11, 14–15, 20
  bridge of **3**: 22, *22*
  transporting water as ballast **3**: 53, **4**: 12
shipwrecks, investigation **3**: 36–7, *36*, *37*
shoes, floating across ocean **1**: 49
Shokalsky, Yuly Mikhaylovich **4**: 44
shores **1**: 24
  rocky **1**: 26–7
    life on **2**: 44–5, *44*, *45*,
  sandy **1**: 28–9, *28*, *29*
    life on **2**: 46–7, *46–7*
  waves breaking on **1**: 57
shrimp **1**: 43, **2**: 14, 23, 47, 51, 52, 59, **4**: 15
  farming **2**: 54, **3**: 44
  reflecting sound waves **2**: 19
Siberia **2**: 10, **3**: 6
Siebe, Augustus **3**: 30–1, **4**: 32
Silent World **4**: 37
silica **2**: 20
silicate **2**: 13
silks **3**: 14
silver **3**: 9
Singapore **3**: *52*, **4**: 19
single-celled organisms **1**: 4, **2**: 4, *5*, 22, 47, 49, 61
Sinus Borealis **1**: 10, **4**: 4
sirenians **2**: 36
skates **2**: 27, 28–9, 51
  fins **2**: 29
  food **2**: 28, 29
  reproduction **2**: 29
skuas **2**: 35, **4**: 22
sky, color **1**: 6
slate **1**: 19
slaves, transported from Africa **3**: 7
Slocum, Joshua **4**: 31
Smith, Walter **4**: 33
snails **1**: 32, **2**: 14, 23, 56
snakes, sea **2**: 33, 52
  banded **2**: *33*
  yellow-bellied **2**: 33
snorkels **4**: 36
snowballs **1**: 8
solar power **3**: 61
solar radiation **1**: *60*
sole **2**: 49
Somali Plate **1**: *12*
sonar (echo sounding) **3**: 27, 28, *28*, 34, *34*, 36, 37, **4**: 32, 41
  use by fishing vessels **3**: 40–1
sonar surveys **1**: 5, 7, 43
sound **1**: 42, 43, **4**: 42
  made by whales **1**: 43
soundings **1**: 7, **3**: 26
South Africa **4**: 15
South America **1**: 10, 11
South American Plate **1**: *12*
South Atlantic gyre **1**: *49*
South China Sea **4**: 20
South Georgia **3**: 39
South Pole **1**: 47, **4**: 22, 34
Southern Cross **3**: 9
Southern Ocean **1**: 5, 6, 44, **3**: 39, **4**: 22–4, *23*
  data **4**: 24
  fishing **4**: 22–3
  hunting **4**: 24
  map **4**: *25*
  size **1**: 7
Spanish settlers **3**: 7
species
  mass extinction **2**: 10
  numbers **2**: 4
spermaceti **2**: 38, **3**: 38–9
Sperry, Elmer **3**: 22, **4**: 31
Spice Islands **4**: 41
spices **3**: 14
spiders **2**: 9
spineless creatures *see* invertebrates
spiracles **2**: 27
spit **1**: 29
splash zone **2**: 44–5
sponges **2**: 9, 16, 22, *23*, 51, **3**: 53
  dead man's fingers **2**: *23*
Spray **4**: 31
spreading ridge **1**: 13
spring tide **1**: 55
squid **2**: 5, *5*, 16, 22, 23, 58
  in food web **2**: *17*
  giant **2**: 22, **3**: 16, *17*
  ink produced by **2**: 18
  migrations **2**: 42
stacks **1**: 26, *27*
starches **1**: 39, **2**: 16
starfish **2**: *4*, 22, 23, 45, 46, 47, *47*, 51, 52, **4**: 44
  blue **2**: *22*
  feeding **2**: 45
stars, navigation by **3**: 9
steam **1**: 8
steam turbines **3**: 15
steamships **3**: 4, 15, *15*, 18–19
stinging cells **2**: 22
stingrays **2**: 29
storm waves **1**: 56, 57
storm-petrel **2**: *34*, 35
storms **1**: 58
  effects on sea levels **1**: 34
  worst **1**: 59
stromatolites **2**: 6, *6*
subduction **1**: 13
submariners **3**: 32
submarines **3**: 4, 15, 19, 32, **4**: 32
  depths reached **4**: 28, *29*
  military **3**: 33
  nuclear-powered **3**: 19, *19*, **4**: 28, *29*
  tourist **3**: 54
submersibles **1**: 5, **3**: *27*, 32, *32–3*, 33, 37, **4**: 38
  depths reached **4**: 28, *29*
Suez Canal **3**: 15, 21
sugar kelp **2**: *14*
sugars **2**: 16
Sulawesi **4**: 17
sulfates **1**: 38
sulfur **1**: 15, 38
sulfur cycle **1**: 38
sulfur dioxide **1**: 37, 38, **2**: 10, **3**: 48
sulfur springs **2**: 6
sulfuric acid **1**: 38
Sumatra **4**: 17
Sumerians **3**: 8
Sun
  eclipse **4**: 36
  heat from **1**: 46
  navigation by **3**: 9
  tides and **1**: 54, *54–5*, 55
sunlight **1**: 4, 6
  deprivation of **2**: 11
  trapping **2**: 7
sunlit zone **2**: 18, *18*
superocean **1**: 6
supertankers **3**: 4, 5, 21, *21*
  oil spillage from **3**: 50
supertrawlers **3**: 41
surfing **1**: *56*
surgeon fish **2**: *53*
Surtsey **4**: 8
Sverdrup, Harald **4**: 42, 44
swells **1**: 56
swim bladder **2**: 27, *27*
Sydneyia **2**: *9*
symbionts **2**: 60
symbiosis **2**: 60–1

Tahiti **3**: 6, 13
Tampa Bay, Florida **3**: 52
tankers **3**: 21, *21*, *47*, **4**: 15
  oil spillage from **3**: 50, *50*, **4**: 17
Taormina, Sicily **1**: *28–9*
Tarr Inlet, Glacier Bay, Alaska **3**: *21*
Tasmania **3**: 13
tea **3**: 14
tectonic forces **1**: 24
tectonic plates **1**: 15, **4**: 19, 33
telegraph cables **4**: 32
temperate regions **1**: 43
temperature:
  changes **1**: 36–7
  and marine life **1**: 43
  ocean's effects on **1**: 4
terebellids **2**: 51
terns **2**: 35
  arctic, migration **2**: *42*, 43
Terra Australis **4**: 36
territorial waters **3**: 58, *58*
tertiary consumers **2**: 16–17
Tethys Sea **1**: 10, 11
Tharp, Marie **3**: 28, **4**: 33, 40, 45
Thera **3**: 16
thermocline **1**: 43, 51
thermohaline circulation **1**: 50, 51
Thomson, Charles Wyville **3**: 26, **4**: 31, 32, 34, 45
Thule **4**: 43
tidal barrages **3**: 48–9, *48*
tidal bores **1**: 54
tidal bulges **1**: 54
tidal power **3**: 48–9, *48*
tidal surges **1**: 54
tide gauge **1**: 34
tides **1**: 34
  in Bay of Fundy **1**: *55*
  cause of **1**: 54–5
  effect on coastline **1**: 24
  effect on marine organisms **1**: 54
  frequency **1**: 54–5
  neap **1**: 55
  ranges **1**: *55*
  red **2**: 12
  rhythm of **1**: 54–5
  spring **1**: 55
Tigris (reed boat) **4**: 40
timelines
  Earth **4**: 26–7
  surface exploration **4**: 30–1
  underwater exploration **4**: 32–3
Timor **3**: 6
tin **3**: 9, 47, **4**: 15, 19
Titanic **1**: 45, **4**: 4
  wreck of **3**: 37, *37*, **4**: 33, 35
toads **2**: 32
tombolo **1**: *29*
Tonga **3**: 6
tool users **2**: *37*
toothed wrack **2**: *14*
toothfish **4**: 23
topshells **2**: 45
torpedoes **3**: 19, 32, **4**: 42
tortoises **2**: 32
tourism **3**: 5, 20, 54–5, **4**: 6, 12, 19
  in Antarctica **4**: 22
  damaging habitats **3**: 52–3, **4**: 17
toxic substances **3**: 5
trade **3**: 4, 8–9, 10, 11, 14, 18, 20
trade winds **1**: 47, *47*, 48, *52*
transform faults **1**: *14*
trash fish **3**: *43*
trawl nets **3**: *40*, 41, 52–3, 57
trawlers **3**: 41
trees **2**: 12
trenches **1**: 7, 13, 14, *14*, 15, 16–17, *17*
  ocean **2**: *18*, 19, **4**: 8
Trieste (submersible) **1**: 7, **3**: *32–3*, **4**: 33, 37, 43
trilobites **2**: 8–9, *9*
triremes **3**: 4, 18
Triton, USS **4**: 33
trophic levels **2**: 16
tsunamis **1**: 57, *57*, **4**: 18, 19, 42
Tuamotu Islands **4**: 40
tube feet **2**: 23
tubenoses **2**: 35

tubeworms **2**: 51, 57, **4**: 28, *29*
tuna **1**: 53, **2**: 16, 31, **4**: 19
catching **3**: 41, **4**: 15
species
bluefin **3**: *43*
skipjack **1**: 53
yellowfin **1**: 53
trapped in net **3**: *43*
turbines **3**: 15
turbot **3**: 58
Turtle (submarine) **3**: 32, *32*, **4**: 32
turtles **2**: *5*, 32–3, **3**: 57
in fishing nets **3**: 41
migrations **2**: 42
reproduction **2**: 32–3
sex determination **2**: 32
species
green **2**: 15, 32, 42, *42*
loggerhead **2**: *32*
threats to **3**: 5
twilight zone **2**: 18–19, *18*, 58
typhoons **4**: 18

U-boats **3**: 19
ultraviolet (U.V.) radiation **2**: 6
Undersea Research Group **4**: 36–7
underwater craft, depth reached **4**: 28, *29*
United Nations,
Environment Program (UNEP) **3**: 59
Food and Agriculture Organization (FAO) **3**: 43
United States **3**: 15
upwelling **1**: 50–1, **2**: 5, **4**: 10, 15, 22
uranium **4**: 4
U.S., immigration **3**: 7, *7*
U.V. (ultraviolet) radiation **2**: 6

valleys **1**: 16
Vampyroteuthis **2**: 58
Van Dieman's Land (Tasmania) **3**: 13
vaquita **3**: 56
Venice **1**: 35, *35*
vents, hydrothermal (hot-water) **1**: 5, 15, *41*, **2**: 6, 57, **4**: 33, 35
vertebrates, endangered **3**: 56
Victoria Island **4**: 6
Victoria Land **4**: 44
Vikings **3**: 10
Vine, Fred **1**: 14–15
Vinland **4**: 38
viper fish **2**: *58*
Virgin Islands **4**: 37
volcanoes **1**: 8, *9*, 13, 15, 37
activity **1**: 14, *17*
causing tsunamis **1**: 57, **4**: 18
eruptions **2**: 10, 11
extinct **1**: *17*, **4**: 19
island **1**: 17
oceanic **1**: *14*
Ring of Fire **4**: 19
underwater **3**: 34, **4**: 14, 40
Voyage of Discovery and Research to Southern and Antarctic Regions **4**: 44
voyages **3**: 4
of ancient peoples **3**: 6–7
of discovery **3**: 8–13, *10*, *11*
of scientific exploration **3**: 24

waders **1**: 31
Wallace, Alfred R. **4**: 37
walruses **2**: 36, 37, **4**: 4
Walsh, Don **1**: 7, **3**: 33, **4**: 33, 43
warships **3**: 4, 10, 15, 18–19, *18*, *19*
submarine attacks on **3**: 32
used for research **3**: 28
water
area covered **1**: 4
in atmosphere **1**: 61
in body fluids **2**: 6
chemical makeup **1**: 15
chemical recycling **1**: 38–9
circulation **1**: 36, 42
cleaning **1**: 33
evaporation **1**: 4, 36, *37*
heat storage **1**: 36–7
light on **1**: *42*
on planets **1**: 8–9
properties **1**: 36
source of **1**: 8
water vapor **1**: 8
waterspout **1**: *59*
waves **1**: 56–7
absorbing impact **1**: 25, *25*, 26
analyzing **4**: 42
biggest **1**: 57
conflicting **1**: 56
effect on coastline **1**: 24, *27*
effects of wind on **1**: 56
height **1**: 56
hitting beach **1**: 25
movement of water **1**: *56*
power of **1**: 57, **3**: 49
rogue **1**: 57
sea waves **1**: 56, *56*
storm waves **1**: 56, 57
winds and **1**: 56
weapons, modern **3**: 19
weather **1**: 4
ocean **1**: 58–9
weathering **1**: 24, 37
Wegener, Alfred **1**: 12, **4**: 32, 40, 45
West Indies **3**: 11
wetlands **3**: 52
coastal **1**: 32–3
whale sharks **3**: 54–5
whalebone **2**: 40
whales **2**: 36
baleen **1**: *43*, **2**: 21, 40–1, *40*, *41*, **3**: *38*
blowholes **2**: 40
breathing **2**: 36
communication **1**: 43, *43*, **2**: 41
depths reached **4**: 28, *29*
feeding **2**: 40, *40*, 41
in fishing nets **3**: 41
hunting **3**: 38–9, **4**: 6, 24
intelligence **2**: 39
migrations **2**: 43, *43*
navigation **2**: 39
parasites **2**: 61
schools **2**: 38–9, **3**: 55
social ties **2**: 38–9
song of **2**: 41
species
blue **2**: 4, *4*, 40, 41, *41*
bow-head **3**: 39
fin **2**: *41*
gray **2**: 39, 40, *40*, *43*, **3**: 39, *54*, **4**: 4
humpback **1**: *43*, **2**: 40, 41, *41*, **4**: 22
killer **2**: *17*, 38, 39, *39*
in food web **2**: *17*
minke **3**: 39, **4**: 24
narwhal **2**: 38
pilot **2**: 39, *39*
right **2**: 40, *41*, **3**: 39, 56
rorquals **2**: 40
sei **2**: *41*, **3**: 39
sperm **1**: *43*, **2**: *19*, 38, *41*, **3**: *17*, 39, **4**: 28, *29*
toothed **1**: *43*, **2**: 38–9
stranding **2**: 39, *39*
threats to **3**: 5
uses **3**: 38–9
Whales, Bay of **4**: 34
whaling **2**: 43, **3**: 39
whaling stations **3**: 39
whelks, dog **2**: 45, 51
whiting **2**: 61
WHOI *see* Woods Hole Oceanographic Institution
wildlife
of estuary **1**: 30–1
of oceanic islands **1**: 20
sea temperature and **1**: 42–3
use of light **1**: 43
of wetlands **1**: 32
wind farms **3**: 49, *49*
winds
change of direction **2**: 11
Coriolis effect **1**: *46*, *47*
currents and **1**: *47*
deflection **1**: *46*
effects on sea levels **1**: 34
global **1**: 46–7, *46*, *47*
offshore **1**: 34
onshore **1**: 34
waves and **1**: 56
wine **3**: 9
winkle, mangrove **2**: *54*
Woods Hole Oceanographic Institution (WHOI) **3**: 27, *27*, 34, **4**: 31, 35, 41
wool **3**: 15
World Conservation Union (WCU) **3**: 43, 56
World Ocean Floor **4**: 45
worms **1**: 5, 22, **2**: 8, 9, 14, 15, 17, 22, 45, 46–7, 52, **4**: 44
in mud **2**: 48
parasitic **2**: 61
predators **2**: 46–7
segmented **2**: 46
species
arrow **2**: 16, 17, *20*, 21
flat **2**: 61
iceworms **2**: 57
ragworm **2**: *22*
ribbon **2**: 22
roundworms **2**: 56, 61
sand mason **2**: *46*, 47
scale **2**: 60
seamouse **2**: 51
terebellids **2**: 51
tubeworms **2**: 51, 57, **4**: 28, *29*
Worth, Lake **3**: 52
wrack
bladderwrack **2**: *14*
knotted **2**: *14*
toothed **2**: *14*
wrasse, cleaner **2**: 61

Xenophthalmichthys **2**: *56*

Yellow Sea **4**: 20
Yemen, fishing rights **3**: 58

zebra mussels **3**: *53*
Zeebrugge, Belgium, wind farm **3**: *49*
zircon **4**: 15
zoea larva **2**: *21*
zones, depth **2**: 18–19, *18*
zooplankton **2**: *4*, 20–1, *20*, *21*, 30, 40, 46, 50, 56, **4**: 5
depth found **2**: 18
and alien organisms **3**: 53
in food chain **2**: 16, 21
harm caused by oil **3**: 50
sieving from water **2**: 30
size range **2**: 5

Picture Credits

Photographs

Abbreviations

A Ardea; C Corbis; FLPA Frank Lane Picture Agency; NHPA Natural History Photographic Agency; SPL Science Photo Library

1 W Wisniewski/Frank Lane Picture Agency; 3 Adam Woolfitt/Corbis; cover photograph: Powerstock.

Illustrations

Andromeda Oxford Ltd 5, 7, 9, 10–11, 12–13, 14, 16–17, 18, 20–21, 23, 25; Tony Truscott Designs 26–27; Alan Hancocks 29.

EXPLORING THE OCEAN

VOLUME 3

USES OF THE OCEAN

Trevor Day

OXFORD
UNIVERSITY PRESS

Consulting Editor Professor Dorrik Stow
University of Southampton Oceanography Centre, UK

Published in the United States of America by
Oxford University Press, Inc.
198 Madison Avenue
New York, NY 10016
www.oup.com

Oxford is a registered trademark of
Oxford University Press

Planned and produced by
Andromeda Oxford Limited
11-13 The Vineyard
Abingdon
Oxon OX14 3PX
UK
www.andromeda.co.uk

Library of Congress Cataloging-in-Publication Data
Day, Trevor.
Exploring the ocean / Trevor Day.
p . cm.
Summary: Explores various aspects of the world's oceans including currents, tides, weather patterns, the sea floor, marine life, uses by humans, and ecological issues.
Contents: v. 1. The physical ocean -- v. 2. Life in the ocean -- v. 3. Uses of the ocean -- v. 4. Reference and set index.
ISBN 0-19-515738-9 (set : alk. paper)
1. Oceanography--Juvenile literature. 2. Marine organisms--Juvenile literature. 3. Marine biology--Juvenile literature. 4. Ocean [1. Oceanography. 2. Marine biology. 3. Ocean.] I. Title.

GC21.5.D385 2003
551.46--dc21 2002042583

Project Director Susan Kennedy
Project Manager Derek Hall
Design Tony Truscott
Editors Marion Dent, Deborah Fox, Suzanne Juby
Cartographers Tim Williams, Richard Watts
Picture Manager Claire Turner
Picture Research Alison Floyd
Production Clive Sparling

Printed in the U.A.E. by Oriental Press Limited, Dubai

USING THIS SET

This four-volume set describes the fascinating world of the ocean through informative text and colorful illustrations and maps. Each volume explores a different theme. **Volume 1** is concerned with the physical ocean: its origin and structure, and associated features such as tides, weather systems, islands, and coral reefs; **volume 2** describes the varied natural life in the ocean: evolution, marine animals and plants, food webs, ocean zones, communities of creatures, and conservation; **volume 3** looks at the way humans use the ocean: exploration from earliest times to the present day, exploitation of ocean resources, pollution, and ocean management.

In each volume, feature boxes highlight items of special interest. Cross-references at the bottom of each left-hand page point the reader to further information on related topics in other sections and volumes: the relevant volume number is given in bold type, followed by the page numbers. Each volume also has its own subject index.

Volume 4 contains a comprehensive series index, together with further reference resources, including maps, datafiles, full descriptions of the five oceans, biographies of men and women who have contributed to our knowledge of the oceans, timelines, places to visit, websites, further reading, and a glossary.

Throughout the set the following abbreviations are used for measurements:–
Length in (inches), cm (centimeters), ft (feet), m (meters), mi (miles), km (kilometers)
Area sq in (square inches), sq cm (square centimeters), sq ft (square feet), sq m (square meters), sq mi (square miles), sq km (square kilometers)
Volume cu mi (cubic miles), cu km (cubic kilometers)
Temperature F (Fahrenheit), C (Celsius)
Weight Given in U.S. tons, followed by metric tonnes

Previous page: A tanker carrying crude oil to a refinery.

Right: Fishermen inspecting a farmed salmon at Eastport, Maine, for the presence of fish lice.

Contents

Uses of the Ocean	4
Ocean Highways	6
Early Seafarers	8
The Age of Discovery	10
To the Ends of the Earth	12
From Sail to Steam	14
Myths and Monsters	16
Ships of War	18
Giant Ships	20
Modern Navigation	22
Studying the Ocean	24
Oceanography is Born	26
Sonar and Satellites	28
Diving	30
Underwater Vessels	32
Research at Sea	34
Investigating Shipwrecks	36
Hunting Sea Mammals	38
Fish from the Sea	40
Dwindling Fish Stocks	42
Ocean Farming	44
Undersea Mining	46
Renewable Energy	48
Marine Pollution	50
Habitat Destruction	52
Marine Tourism	54
Protecting the Oceans	56
Managing the Oceans	58
Prospects for the Future	60
Index	62
Picture Credits	64

Uses of the Ocean

The ocean covers nearly three-quarters of our planet. Humans have used it for thousands of years as a source of food and transportation. Today we extract gas and oil from the seabed and treat the ocean as a vast place for recreation.

Blue seas, rolling surf, and sandy beaches attract millions of people to the coast each year, like here in Australia. Activities such as spear fishing and speedboating impact on marine and coastal environments.

Today, when we travel to distant countries, we usually go by air. But from way back in human history to the relatively recent past, people used the oceans as marine highways to voyage to distant lands—both to find new places to inhabit and to trade goods with people living in other parts of the world.

Until about 180 years ago, seafarers had only oars or sails to power their small wooden ships. They made use of winds and surface currents to move their craft across the ocean. The coming of the first steamships in the 19th century led the way to today's diesel-powered supertankers and container ships—vessels that carry raw materials such as oil and coal, industrial goods, and food such as grain, fruit, and meat in bulk across the ocean. Big ships like this are much more cost effective than airplanes for carrying cargo around the world, but they also require large port facilities for the loading and unloading of goods.

Since early times, people have built warships to fight battles at sea. The ancient Greeks and Romans, for example, built triremes—ships powered by sails and by oarsmen seated in three tiers one above the other. Triremes moved rapidly through the water and had pointed prows to ram enemy ships. Today, many countries have navies equipped with huge aircraft carriers and nuclear-powered submarines to protect their coastal borders and to act as peacekeepers in other parts of the world.

People who live beside the ocean have commonly built crafts to take them out into coastal waters to fish with nets or to hunt marine mammals such as seals and whales. Mostly these fishing vessels stayed close to shore and returned to

See also **vol 1:** 4–5, 6–7

harbor after a few days or weeks at sea. Today's modern fishing vessels are equipped with huge nets that can scoop up hundreds of thousands of fish at a time like giant vacuum cleaners. Large fishing trawlers with refrigerated compartments can remain at sea for months. In the last 50 years, the annual seafood catch worldwide increased by 400 percent.

Out of balance with nature

Too much fishing is reducing fish populations in all the world's oceans. Other sea creatures, from tiny corals and clams to turtles, porpoises, and whales are also directly threatened by human activities such as hunting, species collecting, and tourism.

We depend on fossil fuels such as oil and natural gas for energy to heat our homes, drive our cars and planes, and power our industries. Oil companies construct massive drilling platforms to extract this natural wealth from beneath offshore waters in places such as the Gulf of Mexico and the Arabian Gulf. Underwater pipelines and supertankers carry crude oil to onshore oil refineries around the world. Oil spills from tankers pollute the ocean waters, causing harm to seabirds and mammals such as sea otters that swim in them. Global warming due to overuse of fossil fuels is likely to cause sea levels to rise in the not-so-distant future. This will flood low-lying islands and coastal regions.

People use oceans as dumping grounds for all kinds of toxic substances, including human sewage and radioactive waste from nuclear power stations. Development for new tourist marinas or ports destroys the natural feeding grounds of millions of shorebirds and other creatures.

Looking to the future

The impact of human activities on the oceans affects us all, whether we live in countries or regions bordering the ocean or inland. Oceanographers—scientists who study the ocean environment and the creatures that live in it—have shown what a precious and important resource the ocean is for life on Earth. The world's nations must learn to manage the Earth's oceans more carefully by sharing scientific knowledge that will protect the sea and its living communities.

Oceans are vital highways for international trade and industry. Tugboats guide a commercial ship as it enters the Pacific end of the Panama Canal on its way to the Atlantic.

Ocean Highways

Ocean voyages have shaped human history. The oceans act as highways for the movement of people between landmasses.

Modern humans like ourselves first emerged in Africa around 100,000 years ago. By about 12,000 years ago, people had spread across the globe to inhabit all the continents except Antarctica. During this long passage of time the Earth underwent at least two ice ages, when more water was frozen and stayed on land and sea levels were consequently lower, revealing more of the land. For example, land bridges linked Siberia to Alaska and Britain to Europe, and people walked from one to the other. But as the ice melted, the land bridges disappeared, and then people could only travel between these landmasses by boat.

Island-hopping

Archeologists believe that the ancestors of Australia's native peoples, the Aborigines, reached the continent more than 60,000 years ago. At this time of low sea levels Australia and New Guinea formed a single landmass, known to geographers as Sahul. The stretches of water that separated Sahul from the island chains of Indonesia were much narrower than today, but even so the first people to arrive in Australia must have got there by boat, by "hopping" from island to island. Today, the gap between the last island in the chain, Timor, and Australia is around 300 mi (485 km) wide, but 60,000 years ago it was less than 100 mi (160 km). Seafarers had to sail beyond sight of land to reach Australia, though they would have known by signs such as cloud formations and the flight of birds that land was close.

Across the South Pacific

Nearly 4,000 years ago, people known as Austronesians crossed by boat from Asia to New Guinea. From there, they spread through the island chains off the southeast coast of New Guinea, and by 1500 BCE they had reached the islands of Fiji, Tonga, and Samoa in the South Pacific, more than 1,500 mi (2,400 km) to the east of New Guinea.

Their descendants, the Polynesians, were even more outstanding sea voyagers than the Austronesians. They sailed their large double-hulled canoes across vast stretches of ocean to settle the scattered islands of the eastern South Pacific, including Hawaii and Tahiti, in French Polynesia. One Polynesian group even settled on tiny Easter Island, one of the most remote places on Earth, some 2,300 mi (3,765 km)

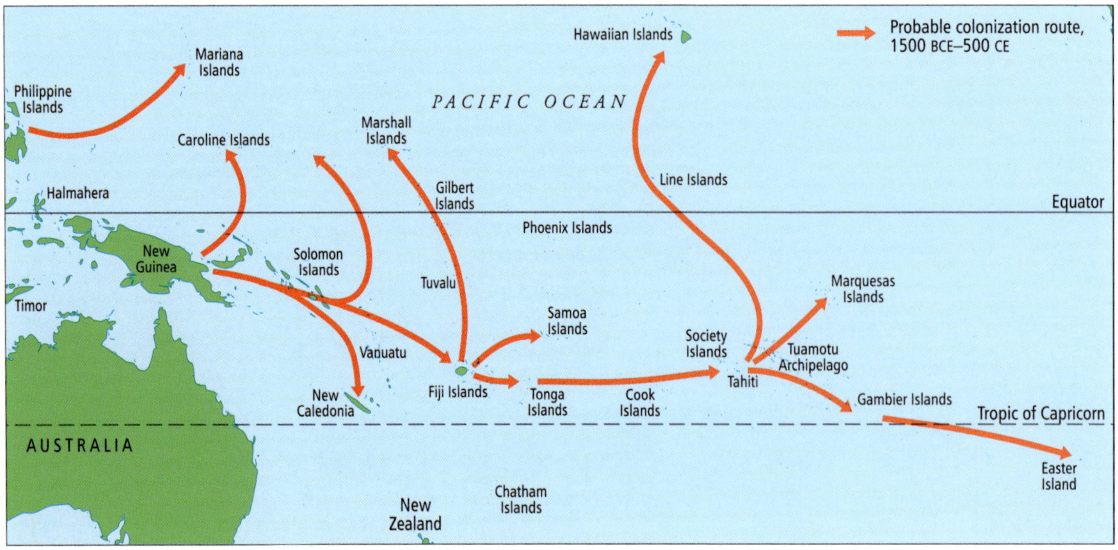

Shown here are the sea journeys made by the people who originally settled the remote islands of the South Pacific.

See also **vol 1:** 4–5, 6–7; **vol 3:** 8–9, 10–11, 12–13

These people, from Okavango, Botswana, are hollowing out a tree trunk to make a canoe. Okavango is a large inland delta in southern Africa, which is often flooded. It is likely that the first seafarers used canoes similar to these on their island-hopping journeys.

from South America. The Polynesians reached New Zealand, the last landmass on Earth to be inhabited by ancient peoples, around 1400 CE.

Migration to North America

The earliest inhabitants of North America were ice age people from eastern Asia who crossed by a land bridge, over what is now the Bering Sea, from eastern Russia to Alaska between 12,000 and 20,000 years ago. From there, groups of hunters followed herds of grazing animals onto the Great Plains, and spread throughout North and South America. These people were the ancestors of the Inuit, Aztecs, Mayans, Incas, and other Native American peoples.

The Native American peoples of North America were undisturbed by outsiders for thousands of years until, some 500 years ago, waves of migrants began to cross the Atlantic Ocean from Europe. The Spanish, among the first to arrive, settled in Florida in 1565. In 1607 the first permanent English settlement was established at Jamestown, Virginia. The following year, the Frenchman Samuel de Champlain founded Quebec in eastern Canada. In 1620 the Pilgrims sailed from England and landed at Plymouth, where they founded the colony of Massachusetts.

European colonists were responsible for a further wave of human immigration to North America: the unwilling arrivals were African slaves, transported on European ships from West Africa to the Caribbean and North America. Around 10 million Africans were transported in this way between the early 1600s and 1800s. The conditions on board the slave ships were so dreadful that between 10 and 20 percent died on the passage before ever reaching North America.

By the 1840s, steamships were regularly crossing the Atlantic between Europe and the United States, bringing a growing stream of European migrants escaping poverty and religious persecution at home. Many U.S. citizens with Irish roots are descended from the thousands who arrived on the "famine ships" of 1846–50 after a plant disease completely destroyed the Irish potato crop, leaving thousands of people dead from starvation.

European immigration to the United States peaked at the end of the 19th century, when thousands of Polish and Russian Jews escaped persecution and hardship by buying cheap passages across the ocean. They traveled in cramped conditions in the bottom of the ships, enduring seasickness and discomfort for the hope of a better life. Thousands of others came from the impoverished countries of southern Europe, especially Italy and Greece. The Pacific Ocean also became an immigration highway, bringing large numbers of Chinese and Japanese to the west coast of the United States to work in the gold fields and on the railroads.

Immigrants to New York wait to cross to Manhattan from Ellis Island. By the early 1900s, when this photograph was taken, most immigrants were from eastern and southern Europe.

EARLY SEAFARERS

Ancient peoples ventured to sea for all sorts of reasons. Their need to discover and conquer new lands was probably as strong as the modern desire to explore outer space.

Reconstructing the past

Many people have built replicas of ancient sailing craft to test if these ships could make long sea journeys. For example, in 1969 the Norwegian anthropologist Thor Heyerdahl built a reed boat based on designs taken from ancient Egyptian paintings and carvings, which he called *Ra* (the name of the Egyptian sun god). *Ra*'s first voyage was unsuccessful, but a year later, Heyerdahl built *Ra II* (shown below) and sailed it from Morocco across the Atlantic Ocean, arriving at Barbados, in the West Indies, 57 days later. However, there is no evidence that the Egyptians ever made such a voyage.

Early sailors probably made their ocean-going craft by hollowing out tree trunks with stone axes, lashing logs together to make a raft, or binding bundles of reeds together to form a hull, the main floating part of a boat. Ancient Egyptian carvings, dating from around 3000 BCE, show people gathering reeds from beside the River Nile and making boats of this kind (see box on left). The Sumerians who lived between the Tigris and Euphrates Rivers in present-day southern Iraq and the people of the Indus Valley civilization in present-day Pakistan also made reed boats.

The Egyptians first built boats to sail down the Nile, which served as their major highway. For example, they transported grain and other harvested crops from the Nile delta by boat to important cities like Luxor in Upper Egypt. By 2500 BCE, Egyptian traders were sailing ships south down the Red Sea to a place they called Punt (probably present-day Somalia), in search of exotic goods such as ebony, gold, incense, and ivory. Other Egyptian merchants sailed north-east across the Mediterranean Sea to Lebanon to bring back cedar wood that was used to make larger sailing vessels.

By the Late Bronze Age (about 1400 BCE), all the countries around the eastern end

An artist's impression shows Egyptian slaves loading a ship with goods, while monkeys run through the rigging. The Egyptians built their ocean-going ships of cedar wood imported from Lebanon. The curved prows were brightly painted and the sails were made of linen.

See also **vol 1:** 4–5, 6–7; **vol 3:** 10–11, 12–13, 22–23, 36–37

of the Mediterranean, from Egypt in the south to Greece in the north, were linked in a circle of trade. Cargo ships sailed from one harbor to the next, carrying goods such as elephant tusks, ostrich eggs, gold and silver jewelry, copper ingots, wine, and olive oil. A thousand years later the Phoenicians (from present-day Lebanon) and Greeks were great traders. They founded trading cities throughout the Mediterranean as far west as Morocco and Spain.

Phoenician and Greek traders even ventured out into the Atlantic Ocean. The Greek historian Herodotus, writing in in the 5th century BCE, tells of goods such as tin and amber "coming from the ends of the Earth." In fact, the tin came from Britain and Ireland, and the amber (yellow lumps of fossilized tree resin, valued as beads in ancient times) came from the Baltic coast of North Germany.

Herodotus records a story that the Egyptian pharaoh Necho, who lived in the 7th century BCE, sent out a fleet with orders to sail southward from the Arabian Gulf down the east coast of Africa. The fleet returned to the Mediterranean Sea some three years later through the Strait of Gibraltar at the sea's western end, having sailed around the southern tip of Africa and back north up the western coast, pushed by winds and currents.

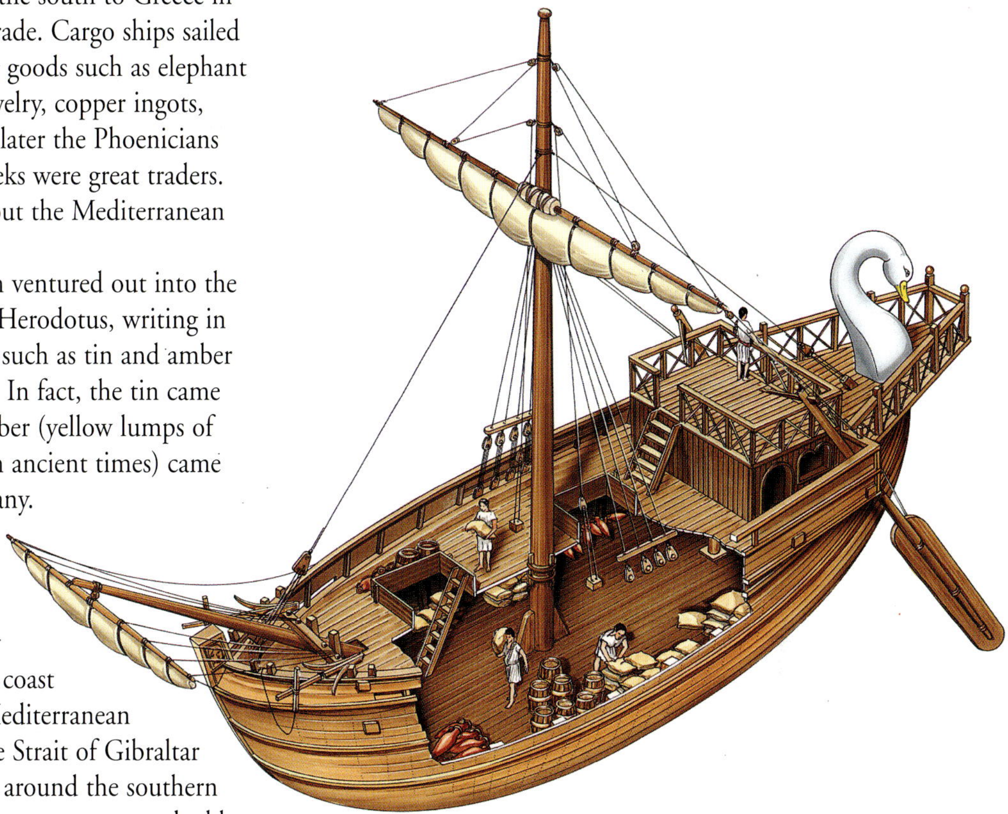

This is a reconstruction of a Roman merchant ship, dating from the 2nd century CE. It would have been around 180 ft (55 m) long, 46 ft (14 m) wide, and 43 ft (13 m) from deck to keel—big enough to carry bulky cargoes such as oil and grain.

Early navigation

Navigation means finding your location and your direction of travel at sea. Early seafarers found their way by sailing close to land and noting the major landmarks. Before the days of written ship's logs, they memorized these landmarks, passing them on by word of mouth to the next generation. Some sailors could tell where an ocean current came from by the color, smell, and taste of its sea water. Birds overhead and clouds in the sky also indicated that land was close by.

Ancient seafarers navigated beyond sight of land by observing the position of the Sun. Knowing that the Sun always travels in the same direction, from east to west across the sky, gives north from south as well. At night sailors used certain stars to check their position. In the Northern Hemisphere the North Star marks north, and in southern skies the Southern Cross constellation shows south.

The Age of Discovery

About 1,000 years ago, most voyages made by European seafarers were still within sight of land. But a few adventurous explorers struck out into unknown waters far from land. By 1522, the most daring had sailed around the world.

In 793 CE Viking pirates from Scandinavia attacked a Christian monastery on Lindisfarne, an island off the coast of England. In the century that followed, Viking ships regularly raided along the coasts of western Europe. But although the Vikings have a reputation for piracy and plunder, they were also explorers who sailed west across the Atlantic and settled as farmers on many islands.

The Vikings were skilled shipbuilders. They constructed their longships from overlapping planks of wood fastened to a central keel with iron nails. At either end was a curved prow. The ships had a single mast mounted with a square linen sail, and were also powered by rowers sitting in line along each side of the ship. Their trading boats were deeper and broader than their warships to make room for cargo.

The Vikings sailed their longboats from Scandinavia down the rivers of Russia to reach the Black Sea. From there they sailed to Constantinople, capital of the Greek Byzantine empire (modern-day Turkey). Other Vikings sailed down the Atlantic coast of Europe and into the Mediterranean Sea, where they attacked cities in southern France and Italy.

Most venturesome of all were the Vikings who sailed west across the Atlantic Ocean. Erik the Red, a Viking chieftain, established a settlement in Greenland, and around 1001 CE his son, Leif Eriksson, made a voyage from Greenland across the Davis Strait to Labrador. Sailing south down the coast he reached a place, which historians believe is Newfoundland, where he spent the winter before returning home.

Middle Eastern and Asian travelers

The Arabs of the 8th to 14th centuries were skilled sailors and navigators, who made long voyages in the Indian Ocean and beyond. Arab ships, called boums, had triangular sails, more than one mast, and were much more maneuverable than European ships with their square sails and single masts.

The Chinese sailed in junks, flat-bottomed ships with squarish sails. One famous Chinese traveler, Admiral Zheng He, sailed widely across the Indian Ocean. In 1405–7 he made a voyage to India with a fleet consisting of 62 large junks and 225 smaller ships. Zheng He also visited Indonesia, East Africa, the Red Sea, the Persian Gulf, and Japan.

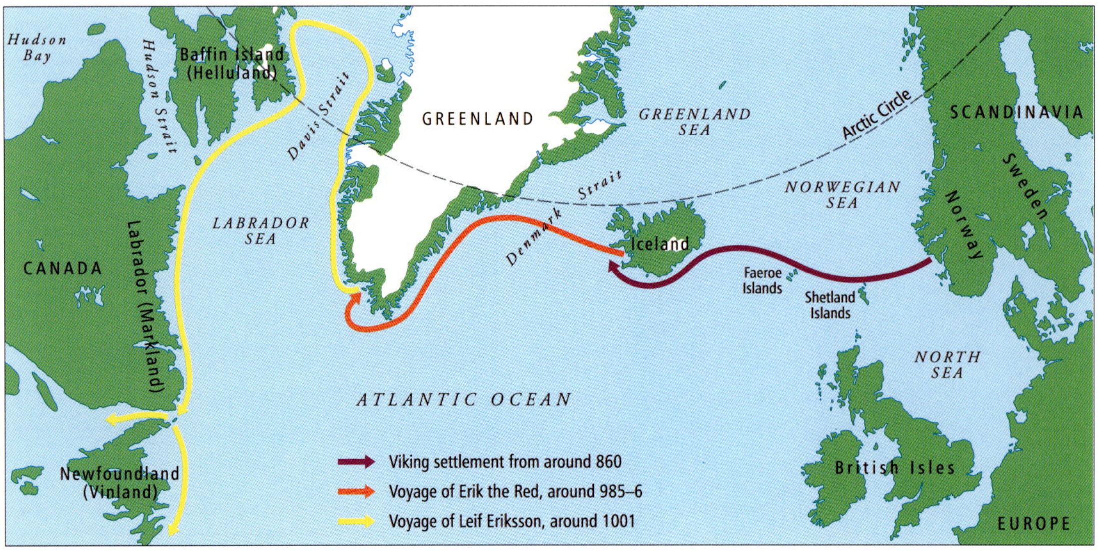

During the 1960s, archeologists found the remains of a Viking settlement in Newfoundland. This may have been the place Leif Eriksson called Vinland because he found so many wild berries (grapes) growing there.

See also **vol 1:** 46–47, 48–49; **vol 3:** 12–13, 14–15, 22–23, 36–37

Navigational instruments

In the 1100s, Europeans began to use the magnetic compass, first invented by the Chinese. With a compass, seafarers could steer their vessel out of sight of land, because the compass needle, made of magnetic metal called lodestone, or magnetite, pointed to the Earth's magnetic North Pole. By the 1400s, seafarers began using the astrolabe (left) to measure the height of the Sun or stars above the horizon. This gave a ship's approximate latitude (its north–south location on the Earth). A ship's longitude (its east–west location) remained largely guesswork, based on a knowledge of currents and an estimate of the vessel's speed in the water.

To the New World

Europeans wanted to find new sea routes to Asia so that they could trade for spices and silks. In 1416 a Portuguese prince, Henry the Navigator, set up a school to train seafarers in the most up-to-date methods of navigation. Portuguese sailors began to sail further and further down the west coast of Africa. Eventually, in 1497, Vasco da Gama sailed around Africa and crossed the Indian Ocean. In 1498 he reached Calicut in India. He had found a new sea route to Asia, then known as the "Indies."

Meanwhile, in 1492, the Italian-born explorer Christopher Columbus, sailing under the Spanish flag, voyaged west across the Atlantic Ocean. He believed he would reach the "Indies" more quickly by this route than by sailing around Africa. Instead, he landed in the Bahamas. But because Columbus thought that the islands he had reached were part of Asia, they came to be known as the West Indies.

Between 1519 and 1522 Portuguese explorer Ferdinand Magellan's expedition sailed from Spain southwest across the Atlantic and around the tip of South America into a "new" ocean he named the Pacific. Magellan died in the Philippines, but 18 members of his crew arrived back in Spain, the first people to sail right around the world.

Today's Chinese junks, like the one shown here in Hong Kong harbor, are very similar to those used in medieval times. The medieval ships were up to 180 ft (55 m) long and 30 ft (9 m) across.

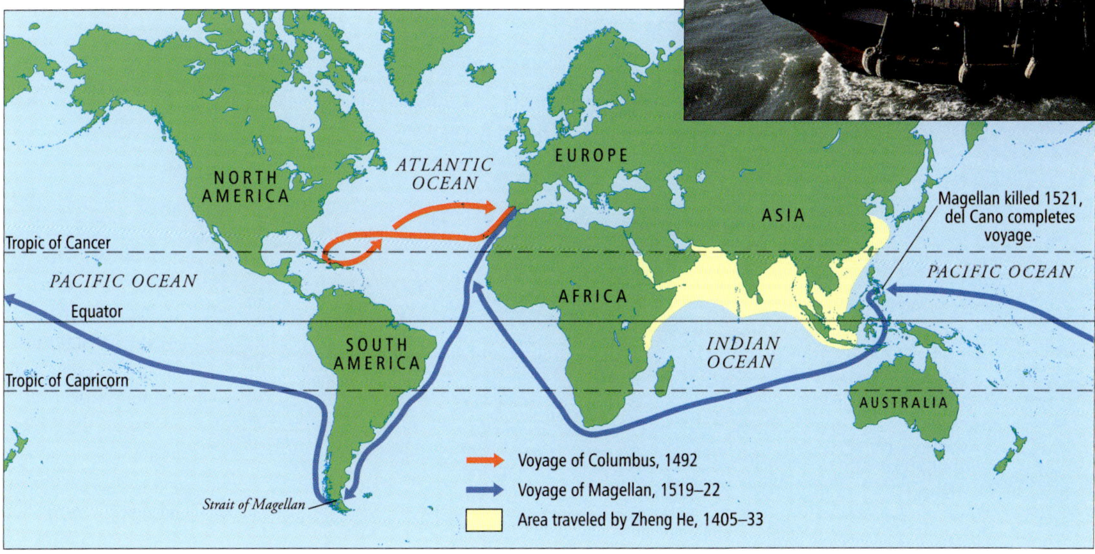

Some of the voyages of Chinese explorer Zheng He, Spanish explorer Christopher Columbus, and Portuguese explorer Ferdinand Magellan, are shown here.

TO THE ENDS OF THE EARTH

After the discovery of the New World, European explorers tried to discover a sea route linking the Atlantic and Pacific Oceans. They also sailed the southern oceans and searched for the "southern continent."

The first European to visit North America since the Vikings was John Cabot who, in 1497, sailed across the Atlantic Ocean from Bristol, England, and landed in either Newfoundland or Cape Breton Island.

Like Columbus, Cabot thought he had reached the rich trading lands of Asia. But Magellan's later expedition showed that the continents of North America and South America lay between Europe and Asia. To reach Asia by sailing west meant traveling all the way around South America and across the Pacific. People wondered if a quicker route existed around the north of what is now called Canada. They named this route the Northwest Passage.

The Northwest Passage

British seafarer Martin Frobisher made three voyages between 1576 and 1578 to search for the Northwest Passage, but was unsuccessful in his quest. Other seafarers such as the Englishmen Henry Hudson and William Baffin and the Dane Vitus Bering charted more of the region, but they also failed to find a route through to the Pacific. They were defeated by the ice in this northerly part of the world. However, their exploits are remembered in places that bear their names—Hudson Bay, Baffin Island, and Bering Strait.

In 1846 the English explorer John Franklin finally discovered a passage through the islands of northern Canada, 500 mi (800 km) north of the Arctic Circle into the Beaufort Sea above Alaska. But his expedition ended in disaster. His two vessels became locked in the arctic ice, where they remained trapped throughout the winter. His crew of 134 tried to walk to safety across the ice but all died. A search party organized by Franklin's widow in 1857 found some of their frozen skeletons, together with written evidence that the party had discovered the Northwest Passage.

In 1903 the Norwegian Roald Amundsen and six companions made the first successful sea voyage through the Northwest Passage in the steam sailing ship *Gjöa.* Today, after nearly 400 years of exploration, the Northwest Passage is navigable to only a few ice-breaking vessels.

Chronometers

Chronometers are very accurate timepieces that enabled sailors to calculate their east–west location, or longitude. The chronometer was set to Greenwich Mean Time (GMT)—the time at zero degrees longitude in Greenwich, London. Each day, sailors compared local noon—when the Sun was directly overhead—with the time shown on the chronometer. The difference between the two times told them how many degrees of longitude the ship had sailed since leaving port. The Earth rotates on its axis through 360° every 24 hours, so each degree of longitude is equivalent to 15 minutes of time. Sailors knew how to work out their latitude using the stars and Sun. Now, if they drew two lines on a map, one indicating their longitude, and the other their latitude, the point where the lines crossed gave their position at sea.

See also **vol 1:** 44–45; **vol 3:** 22–23, 38–39

Captain James Cook's ship HMS *Endeavour* had a crew of 80 seamen and 3 scientists on board.

The voyages of Captain Cook

Since ancient Greek times, European geographers had recognized three continents: Europe, Asia, and Africa. They believed that a "great southern land" lay in the Southern Hemisphere to balance the landmass of Europe and Asia in the Northern Hemisphere. Dutch sailors in the 1600s discovered areas of barren land in the South Pacific to which they gave the name New Holland. One Dutch seaman, Abel Tasman, also discovered an island he called van Dieman's Land (Tasmania today) and was the first European to visit New Zealand.

It was English navigator Captain James Cook who confirmed that New Holland was part of a much larger continent. On his first voyage to the Pacific in HMS *Endeavour* in 1768–71, Cook sailed right around the two islands of New Zealand, charting their coastlines. He then voyaged west until he reached the eastern coast of the country we now know as Australia (which means "southern land"). Cook sailed right the way up the coast, confirming that it was a huge landmass.

Cook was probably one of the greatest explorers of all time. On his second voyage (1772–75), with his vessels *Resolution* and *Adventure*, Cook entered the Antarctic Circle before he was forced to turn back because of ice. He had come within 100 mi (160 km) of Antarctica and sailed right around it without ever seeing it.

On his third voyage (1776–79), Cook set out to see if he could find a way into the Northwest Passage from the Pacific. He sailed from the Polynesian island of Tahiti in the southern Pacific, to the Pacific coast of North America, discovering the Hawaiian islands on the way. He then sailed up the coast to Alaska. He failed to find a way into the passage and returned to Hawaii, where he was killed during a violent dispute with a group of islanders.

The immensely strong Norwegian steam sailing ship, *Fram*, was built for a successful journey across the Arctic Circle in 1893–96. Amundsen later used the *Fram* to reach the South Pole in 1912.

From Sail to Steam

In the mid-1800s, steel-hulled steamships began to replace wooden-hulled sailing ships. Steamships were not reliant on winds, and eventually they became faster and more reliable than sailing ships.

Clippers standing at the Railway Pier at Melbourne Port, Australia, ready to be loaded with wool to carry back to Britain. Even after the coming of steam, fast clippers were favored on long routes such as the Australia–Britain run because they did not need to stop to refuel.

The ships used by the European explorers of the 1400s and 1500s were caravels—small ships with two or more masts from which hung triangular sails. By changing the angle of the sails, seafarers could still travel at speed even if the wind changed direction. The Portuguese perfected the caravel in the early 1400s. Columbus, da Gama, and Magellan all made their epic voyages in caravels.

By the 1700s, European shipbuilders were constructing caravel-like ships, up to 200 ft (60 m) long and 50 ft (15 m) wide. The powerful Dutch East India Trading Company used these vessels to bring back cargoes from the Far East, such as spices, silks, porcelain, and jewels. The ships, fitted with luxury cabins for passengers, were also armed with rows of cannon to fight off pirates and enemy ships.

Age of the clipper

Clippers (so-called because they "clipped" days off a normal journey) were fast, elegant cargo sailing ships used on long-distance trading voyages. The great age of the clipper was from the 1840s to the 1880s, when they raced from one port to another carrying cargo around the world.

Some clippers had as many as seven masts, which carried up to two-thirds of an acre (0.27 ha) of sail. The clippers' sleek hulls were designed to slice through the water, and the ships could reach speeds of over 18 knots (about 21 mi per hour/ 34 km/h).

The California clippers of the 1840s and 1850s carried prospectors from the east coast of America to the gold fields of California (there was no transcontinental railroad until 1869), and to Australia. The British tea clippers of the 1860s raced to be the first back to London with their cargoes of Chinese tea. Speed was everything—the sooner the new crop of tea arrived in England, the more merchants would pay.

See also **vol 1:** 46–47, 48–49; **vol 3:** 18–19, 20–21

The coming of steam

By the early 1800s, steam was coming to replace wind and water power as an energy source to run machines. The American engineer Robert Fulton was one of several inventors who experimented with ways of using steam engines to power ships. In 1807 he designed the *Clermont*, which was propelled by steam-driven paddle wheels and was the first steamship to carry passengers. The *Clermont* cruised on the Hudson River between New York City and Albany, but she was not powerful enough for safe travel at sea.

In 1819 the S.S. *Savannah* was the first steamship to cross the Atlantic. She made the voyage from Savannah, Georgia, to Liverpool, England, in 29 days. Her engines, which operated for only 80 hours of the crossing, were helped out by a full set of masts and sails—as was the case with all steamships for decades to come.

By the 1840s, ship owners were replacing their wooden vessels with ones built of iron, which were bigger, stronger, and posed less of a fire risk. Bigger ships meant more passengers, larger cargoes, and therefore greater profits. In 1845 the *Great Britain*, built by the English engineer Isambard Kingdom Brunel, became the first ocean-going steamship to be built entirely of iron and the first to be driven by a stern-mounted propeller instead of by side paddlewheels.

The opening of the Suez Canal in 1869, linking the Red Sea and the Mediterranean Sea, meant that the days of the tea clipper were almost over. Steamships from Europe could now take a much shorter route to and from the Far East, avoiding the long journey around the south of Africa. In the 1880s, owners of British clippers could still make a living transporting wool from Australia, but as steamships became more efficient and reliable, it was only a matter of time before sail gave way to steam.

The Blue Riband of the Atlantic is a trophy awarded to the ship that makes the fastest crossing of the Atlantic without refueling. First presented in 1935, the trophy was won in 1938 by the British luxury passenger liner, the *Queen Mary*, which is now a tourist attraction in Long Beach, California. She continued to hold the record until another passenger liner, the *United States*, set a new record of 3 days, 10 hours, and 31 minutes in 1952. Since the virtual disappearance of transatlantic travel in luxury liners, later attempts to beat the record have been made by much smaller, custom-built craft.

The *Great Britain* was the first modern passenger liner. Despite her steam engine and advanced six-bladed propeller, she still carried 15,000 sq ft (1,395 sq m) of sail.

By the end of the 1800s, steam turbines (in which steam turns a rotating shaft by pushing on angled blades, like a fan) began replacing steam engines operated by pistons, and oil became an alternative fuel to coal. Today, most merchant ships are powered by large diesel engines, or by steam turbines driven by oil-fired boilers. Most warships and submarines also use diesel engines, but the largest and most powerful are often nuclear-powered. Small sailing vessels are still used to transport cargoes in the Indian Ocean. Elsewhere, however, sailboats are used almost entirely for leisure and recreation.

Myths and Monsters

Sailors' tales of mermaids, sea monsters, and other mysteries are evidence of humankind's long fascination with, and fear of, the ocean.

The oceans are large, deep, and mysterious. For centuries, sailors lived in fear from attacks by sea monsters. Drawings of giant octopuses and imaginary creatures on old maps often marked uncharted parts of the ocean where people believed these monsters lurked. Although such stories may be exaggerated, huge and mysterious marine creatures do exist in the ocean.

In the 1700s, Norwegians believed in a monster, the kraken, which wrapped its suckered tentacles around ships before sinking them. This mythical beast may be based on the giant squid, which grows up to 60 ft (18 m) long. Sometimes dead giant squid are washed up on shore, but no one has ever seen a live specimen under water.

The legend of Atlantis

In 370 BCE the Greek philosopher Plato wrote about a place called Atlantis. This fabled island-state in the Atlantic Ocean was supposed to have contained a city of magnificent buildings and temples decorated with precious metals. According to Plato, the inhabitants of Atlantis were arrogant and dishonest, so the Greek gods punished them by shaking the ground and sinking their island.

People have tried in vain to locate Plato's "lost island." Oceanographers who have mapped the middle of the Atlantic floor have found no trace of an ancient landmass. Many people believe that the myth of Atlantis records ancient memories of a series of huge volcanic eruptions and earthquakes that devastated the Greek island of Thera (present-day Santorini) around 1500 BCE and destroyed the Minoan civilization on nearby Crete.

A new "monster" from the deep

The ocean still abounds with mysteries. The megamouth—or "big mouth"—shark was not discovered until 1976. It is over 16 ft (5 m) in length and weighs 1,500 lb (680 kg). Despite its large size, the megamouth's teeth are small. During the day it feeds off tiny, shrimplike krill near the bottom of the sunlit zone—at a depth of about 500–660 ft (150–200 m). At night, the krill rise toward the surface, and the megamouth follows them to feed. When the krill descend at dawn, the megamouth returns with them.

See also **vol 1:** 58–59; **vol 2:** 22–23, 28–29; **vol 3:** 22–23

Ghost ship

Not all sea mysteries have ancient origins. In 1872 the crew on a British vessel, the *Dei Gratia*, spotted the *Mary Celeste*, an American two-masted ship, drifting in the Atlantic Ocean between the Azores Islands and Portugal. No living, or dead, people were on board. No trace of the *Mary Celeste*'s captain and crew was ever found.

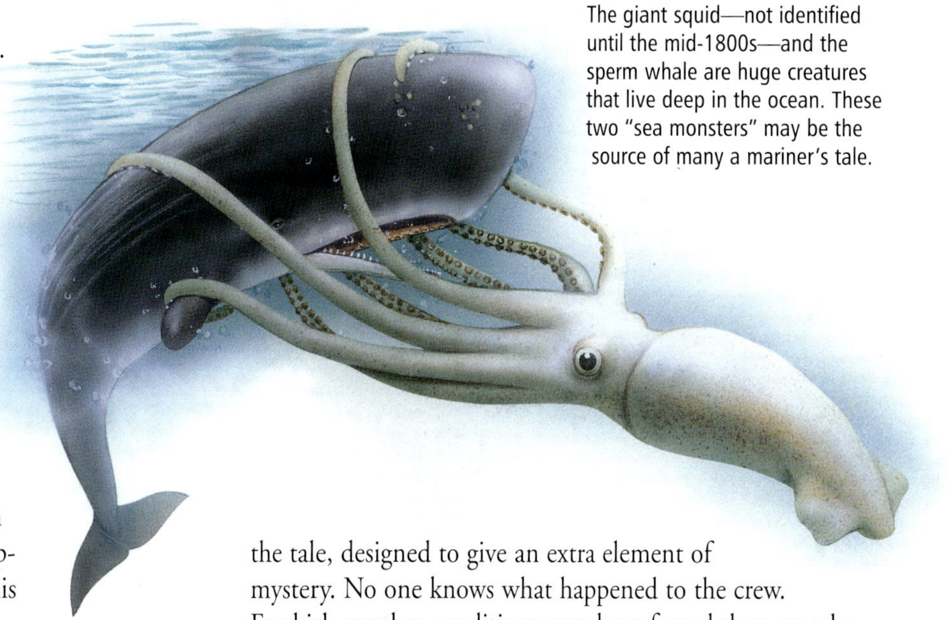

The giant squid—not identified until the mid-1800s—and the sperm whale are huge creatures that live deep in the ocean. These two "sea monsters" may be the source of many a mariner's tale.

When the *Dei Gratia*'s chief mate clambered on board the *Mary Celeste*, he found that the ship's lifeboat and navigational instruments were missing. There were signs that the crew had left in a hurry. Some accounts claim that the captain's table was laid out for a meal, but this was one of many later embellishments to the tale, designed to give an extra element of mystery. No one knows what happened to the crew. Freakish weather conditions may have forced them to take to a lifeboat that later sank. The vessel's cargo of raw alcohol, which is explosive, may have been a reason to abandon the threatened ship. The *Mary Celeste* remains one of the unexplained mysteries of the sea.

The Bermuda Triangle's most mysterious disappearance happened on December 5, 1945, when five U.S. Navy planes vanished during severe weather. Their last radio message said they were running low on fuel and might have to ditch at sea. No wreckage was ever found.

The Bermuda Triangle

The Bermuda Triangle is an area of ocean in the western Atlantic, which is bordered by Bermuda, the Bahamas, and Puerto Rico. Since the 1940s, many ships and more than a dozen aircraft have disappeared without trace here.

People have put forward all sorts of theories to account for these unexplained disappearances. A great many books, films, and television programs have been devoted to the mystery. The wilder explanations include methane-gas explosions, giant whirlpools, and even abductions by extraterrestrial beings.

The Bermuda Triangle lies in a very busy part of the ocean. It is a popular place with yachtsmen, and is used by commercial ships and ferries. Airplanes on their way between the Caribbean islands frequently pass overhead. Tropical storms occasionally sweep the region, and the ocean here has many treacherous currents and swells. Electrical storms that accompany thunderstorms produce static and other interference that disrupts electronic communications. All these hazards are enough to explain why so many ships and airplanes have disappeared here. The truc explanation of the Bermuda Triangle mystery may be that there is no mystery at all.

SHIPS OF WAR

Nations have fought wars at sea for thousands of years. Today's modern warships and submarines patrol the oceans with the aim of maintaining peace.

In 480 BCE a Greek fleet met and defeated a Persian fleet off the Greek island of Salamis. The Greek poet Aeschylus, who took part in the battle, describes how the sea was strewn with broken oars and pieces of timber. His eyewitness account is the earliest we have of a battle at sea.

The Greeks' fighting ships were called triremes. They had about 72 oars, as well as a sail, to move them. The rowers sat in three tiers one above the other. The sail was furled when the trireme went into action, and the rowers powered the ship very rapidly across the water. As the trireme closed on an enemy ship, the rowers propelled the pointed prow, forming a ram, into the opposing ship's side to sink it. The Romans adopted and improved the trireme. Roman warships, called galleys, had five rowers, usually slaves, to work each oar. They also carried a force of armed soldiers, called marines, ready to board and overwhelm the enemy once their ships had been approached and rammed.

Galleys remained the warships of the Mediterranean for hundreds of years. One important development that caused great damage to enemy ships was Greek fire, invented in the 7th century CE. Greek fire was apparently a gasoline-based substance that caught fire spontaneously and could not be put out by water. It was launched at enemy ships through tubes mounted on the prow.

Guns at sea

The next major advance in naval warfare was cannons—metal-barreled guns that fired heavy iron balls to smash the hull and masts of enemy ships. By the 1500s, European warships carried rows of cannons along each side, enabling them to win sea battles without coming into close contact with the enemy. European nations such as England and Spain used their armed navies to conquer territory all around the world and to establish and keep control of international trade.

By the late 1800s, steam engines had replaced sails, and hulls were built of iron instead of wood. The huge battleships of the 1900s came to be called dreadnoughts after the British Navy battleship *Dreadnought*, completed in 1906.

Cannon-carrying warships, like this 17th-century vessel, remained the main naval force until the 20th century. The biggest ships were armed with over 100 cannons and carried over 800 sailors and troops.

See also **vol 1:** 44–45; **vol 3:** 14–15, 16–17, 22–23, 58–59

Submarine warfare

German submarines (U-boats) were a constant menace to Allied convoys in the North Atlantic during World Wars I and II. The crews of surface ships lived in fear of submarine attack, while the crews within the submarines had to endure cramped and claustrophobic conditions and were similarly fearful of being hit by torpedoes or depth charges.

Nuclear submarines (below), armed with atomic weapons, came into use after World War II. In 1954 the U.S. Navy launched USS *Nautilus*, the world's first nuclear-powered submarine. Nuclear-powered submarines can sail nonstop around the world underwater for months on end. They can also sail under polar ice. Britain's HMS *Dreadnought* was the first nuclear-powered submarine to surface at the North Pole.

They carried heavy artillery guns that fired explosive shells to sink other warships or to attack land targets at long range.

In the early 20th century, navies consisted of one or more battleships protected by smaller, faster cruisers and destroyers. One of the navy's roles in times of war was to prevent essential supplies such as iron, coal, and food from reaching ports by attacking enemy shipping. Surface ships and submarines sought out and destroyed cargo-carrying ships. To prevent this happening, cargo ships traveled together in groups called convoys, protected by fighting ships like destroyers.

Modern navies

In today's modern navies, huge aircraft carriers such as USS *Nimitz* have replaced large battleships as the most important surface vessels. These aircraft carriers can carry more than 6,000 crew and up to 90 planes and helicopters, enabling fighter aircraft to strike at enemy targets both on land and at sea in any part of the world. With their supporting group of destroyers and cruisers, carriers carry troops, weapons, and equipment wherever they are needed and can launch attacks on land, sea, or air, using the latest military technology such as laser-guided missile systems.

The biggest modern fleets carry both nuclear and non-nuclear (conventional) weapons. Nuclear weapons are usually carried on submarines that cruise below the ocean surface to keep their whereabouts hidden from other nations. The purpose of these weapons is to deter the enemy from making a first nuclear strike with the possibility of a devastating counterattack.

Submarines keep an "underwater eye" on what is happening on the ocean's surface. Modern navies such as those of the United States, Britain, Russia, China, and France patrol the seas to defend their borders, protect their interests, and prevent wars breaking out in other parts of the world.

A fighter plane takes off from the flight deck of a modern aircraft carrier. Aircraft carriers are the most important non-nuclear weapons in large, modern naval fleets.

GIANT SHIPS

About 90 percent of all the goods transported between continents travel by ship during some part of their journey. It is much cheaper to transport large amounts of heavy goods by sea than it is to carry them by air.

In the 1950s, aircraft became the main way of carrying people across the ocean. Today, most ships' passengers are either tourists on leisure cruises or those using ferries. Huge cargo ships and supertankers that transport goods around the world outnumber all other ocean-going vessels.

A container ship, docked in Felixstowe, England, sails under the Panamanian "flag of convenience." The biggest of these ships can carry more than 5,000 standard-sized containers full of goods on the deck and in the hold.

The great ports

Ports are where ships come to load or unload their passengers and cargo. Most ports are located on coasts, but there are some inland ports on large rivers, such as the port of St. Louis, Missouri, on the Mississippi River. Ports must provide a safe anchorage for ships, as well as good access for transporting goods inland. Originally goods were moved by river and canal—now they are moved by railroad and by road. Port authorities often dredge new, deep-water channels to allow today's giant cargo vessels to reach their docks. If ships are too large to dock they can anchor offshore, and their cargoes are unloaded onto giant barges called lighters.

Most ports have transportation and trading links built up over hundreds of years. This is why so many of today's great commercial cities—New York, London, and Hong Kong—are long-established ports.

Flags of convenience

Trading ships are called merchant vessels. In 2000 the six countries with the biggest merchant fleets (in terms of total tonnage) were Panama, Liberia, the Bahamas, Malta, Greece, and Cyprus—all relatively small countries. The United States, the greatest international trader of all nations, does not appear on the list.

The reason for this is that merchant ships are owned by private companies. Because some countries have safety regulations that are less strict than others and employ crews who receive lower wages, ships registered in countries like Panama or Liberia are cheaper to operate. Lower costs are attractive to shipping companies in the United States, Britain, and elsewhere, and this is why they use foreign-registered ships. Such ships are said to sail under "flags of convenience."

Container shipping

To keep costs down, shipowners want their vessels to carry as much cargo as possible, to make as few stops as necessary, and to unload and reload quickly. Until the 1950s, loading and unloading cargo ships was a slow, expensive business.

See also **vol 3:** 14–15, 18–19, 22–23, 36–37, 54–55

The supertanker *Globtik Tokyo* is 1,240 ft (380 m) long and 165 ft (50 m) wide. Such huge ships are very hard to maneuver, take a long time to stop even from slow speeds, and can only use ports that have deep-water channels up to 100 ft (30 m) deep. Nine-tenths of a supertanker consists of special tanks for holding oil. The crew, the engines, and all the controls are housed in the remaining one-tenth of the ship.

Cranes known as derricks lifted small amounts of goods onto the ships at a time. Men in the ships' holds then stacked these goods by hand or used machinery such as fork-lifts to move the goods about. Dockers were constantly packing and repacking, and loading and unloading.

In the 1960s, shipping engineers developed a metal container built to the same standard size, for transporting goods by sea. The main advantage of using a container is that once the goods are placed inside it—at a factory, warehouse, or farm, often many miles from a port—they remain in the container until they reach their final destination. Trains or trucks carry the full containers to the port, where cranes lift them off and lower them onto ships. At the end of the voyage, cranes lift the containers (still packed) back onto trains or trucks, which carry them on to their final destination for unloading. Some containers are refrigerated to keep perishable produce such as meat, fruit, and vegetables frozen or chilled.

Today, most goods are shipped in standardized containers. Containers do away with the need for large teams of dockers. They are lifted straight on and off by derricks and can be packed tightly together, saving space. Their weight is evenly distributed to balance the ship, so they do not become dislodged in heavy seas.

Supertankers

Tankers (ships carrying crude petroleum oil) were in use by the 1950s. Following a conflict in the Middle East, Egypt closed the Suez Canal in 1956. It was closed again in 1967, and it was not reopened until 1975. Tankers carrying oil from the Middle East to Europe had to travel a much greater distance around southern Africa, rather than take the shorter journey through the Suez Canal. To make the longer trip economical, shipowners built larger tankers, called "supertankers." One supertanker today can carry more than 550,000 tons (500,000 tonnes) of oil at a time. The largest supertanker in the world, the *Jahre Viking,* is 1,590 ft (485 m) long, some 252 ft (77 m) longer than the Empire State Building is high.

Apart from cargo vessels, some of the biggest civilian ships are cruise ships. This one is in Tarr Inlet, Glacier Bay, Alaska. The spectacular Alaska coastline is a favorite destination for cruises.

MODERN NAVIGATION

At the touch of a button, modern seafarers are able to pinpoint their position at sea to within a few yards. Navigation, once an art requiring the use of charts and cumbersome instruments, is now an exact electronic science.

The bridge, or navigation area, on a modern ship has a dazzling display of color monitors and illuminated dials and scales—very different from the chronometer that Cook first used on his second voyage to the Pacific Ocean. On modern vessels, digital readouts and electronic screens replace old-fashioned compasses and paper charts.

Today's navigators still find their position at sea by taking a reading of latitude and longitude, but they now use modern instruments, and their electronic readings are displayed on screens. Since 1884 all ships' navigators have used a common standard for reading longitude. It was decided at an international conference that 0° longitude would run through the Royal Observatory at Greenwich, London, England. This line of longitude is often referred to as the Greenwich Meridian. All longitude readings taken anywhere on Earth relate to this line—for example, a reading of 90° west always means 90° west of Greenwich. The line of longitude on exactly the opposite side of the world to Greenwich is referred to as either 180° west or 180° east (because the earth's circumference totals 360°). The Greenwich Meridian is today marked by a metal strip in the ground, so visitors can straddle the line with one foot in the Western Hemisphere, the other in the Eastern Hemisphere.

Radar

Radar (**ra**dio **d**etection **a**nd **r**anging) is a device that uses radio waves for navigation. It was invented by the Scottish physicist Robert Watson-Watt in 1935. It was used in World War II to warn land-defense forces, airplanes, and ships when enemy planes or ships were approaching. Since then, radar has revolutionized the safety of ships at sea. All large vessels use radar to "see" their surroundings electronically.

The gyroscopic compass

For centuries sailors used magnetic compasses to find North. But compass needles orient themselves to the magnetic North Pole rather than the geographic, or true, North Pole at the northern end of the Earth's axis. The two poles lie some distance apart and navigators had to do complicated calculations to allow for this difference. In 1911 Elmer Sperry developed the nonmagnetic gyroscopic compass, or gyrocompass, which always points to true North. Ships still have gyroscopic compasses as backup in case modern-day electronic systems break down.

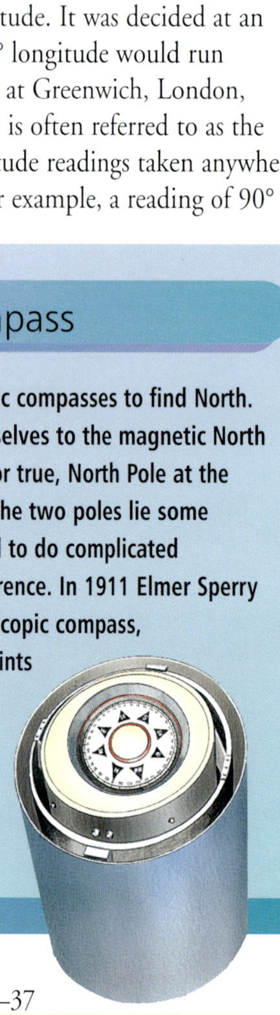

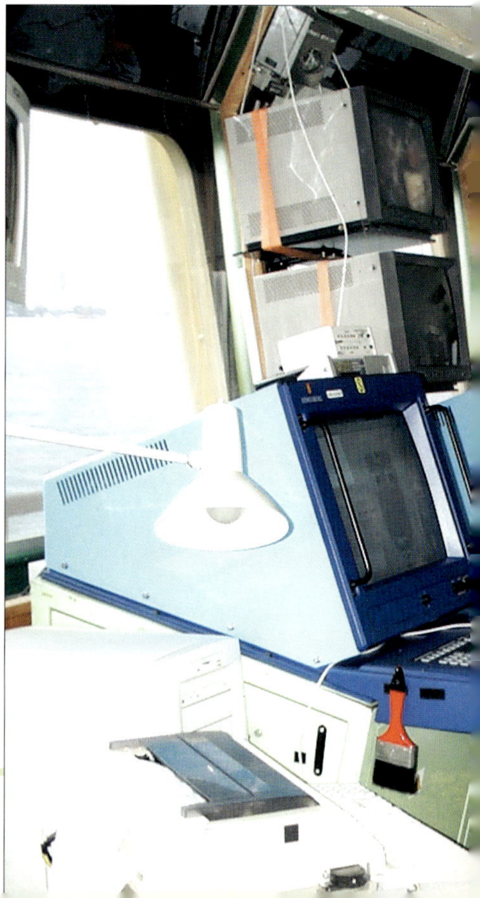

The bridge of a modern ship is equipped with many computer screens and other electronic devices. The information displayed includes the ship's position, its course, the location of other ships in the area, the depth of sea water, and weather forecasts.

See also **vol 3:** 28–29, 34–35, 36–37

A radar scanner or an aerial sends out radio waves in all directions, traveling at the speed of light, and then detects the radio echoes that bounce back from objects in their path. The time interval between the operator transmitting the wave and receiving its echo back measures the distance between the ship and the object.

Modern ships display the returning echoes on a computer screen, and a ship's navigator can detect nearby coasts and islands, as well as approaching vessels. Collisions at sea are now much rarer events, because radar can detect anything in the ship's path, even in fog.

Radar scanning equipment is fixed to the communications mast of this destroyer. On naval vessels, radar not only helps with navigation, but locates other ships and aircraft, too.

Electronic charts

On modern ships, navigators use computers to access the Internet and download the latest weather charts and other information they need to help them guide the ship safely. Electronic navigators, linked to orbiting satellites, provide a continuous display of the ship's position. The captain programs the navigator (known as plotting a course) to steer the ship automatically on a planned route.

Electronic devices occasionally break down, and experienced navigators may have to plot a course on paper charts using traditional aids like compasses and their knowledge of winds and currents—just as navigators did hundreds of years ago. When close to the coast, positional marker buoys, lightships, lighthouses, and even prominent landmarks help to confirm a ship's location.

Sailors can use an electronic portable navigator—little larger than a television remote control—to find out their exact position on the Earth. Satellite technology has made this possible. GPS (Global Positioning System) satellites orbit the Earth, transmitting radio signals. By picking up the signals from three or more of these satellites, the GPS device works out a precise location and provides a readout. Advanced GPS devices can measure a ship's position with pinpoint accuracy.

STUDYING THE OCEAN

The world of the ocean has fascinated people for thousands of years. But the sheer size and depth of the ocean presented early scientists with major problems when they tried to discover its secrets.

In the 4th century BCE the Greek philosopher Aristotle made precise observations of the marine world. For example, he listed and described about 180 different types of marine animal. He also put forward ideas to explain why the sea is salty.

Aristotle's theories about the natural world dominated European thought for nearly 2,000 years. It was not until the 1600s that scholars, aided by new scientific instruments such as depth measurers, began to carry out experiments that gave them new insights into the nature of the physical world. For example, the English chemist Robert Boyle undertook a number of investigations into the properties of sea water. In the next century, learned societies such as the Royal Society of London sponsored voyages of scientific exploration to find out more about the life of oceans.

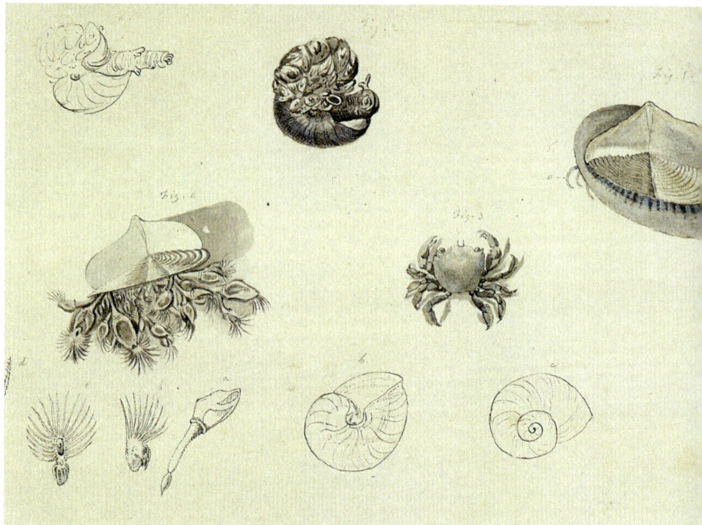

Some of the highly detailed illustrations of marine life painted by William Ellis on Cook's third voyage between 1776–79. They include various mollusks, a crab, and studies of barnacles.

An American pioneer

Matthew Fontaine Maury was a pioneer in early scientific study of the ocean. Maury was a lieutenant in the U.S. Navy and wrote his first book, *Navigation*, on returning from a voyage around the world in 1834. An injury forced him to retire from the navy to work at the Hydrographical Office, and later the Naval Observatory, in Washington, D.C. He used his position to encourage seafarers of all nationalities to send him information about the ocean. Maury assembled the facts in his bestselling book, *Physical Geography of the Sea* (published in 1855), which contained the first map of an ocean basin incorporating depth measurements. His book also had detailed information about currents, water temperatures, wind patterns, the chemical nature of the sea, and marine creatures.

Scientific exploration

Captain James Cook was one of the first people to undertake a modern scientific exploration of the oceans on behalf of the Royal Society. During his Pacific voyages between 1768 and 1779, Captain Cook advanced the frontiers of geographical and scientific exploration, producing new charts and pioneering a scientific approach to taking measurements of sea temperature, wind strength, ocean currents, and water depths. Cook always carried a team of naturalists and artists on board his ships to collect and record the new animals and plants they found.

By the early 1800s, the governments of several seafaring nations were producing highly accurate marine charts. The invention of the chronometer (a device for measuring longitude) enabled navigators to plot their location accurately, and mariners with scientific interests could identify exactly where in the oceans they were making their discoveries.

See also **vol 1:** 22–23; **vol 3:** 12–13, 26–27, 34–35

Challenging ideas

The 1800s saw great advances in the study of marine biology. A leading pioneer in the field was the Scottish explorer John Ross. During an expedition to the Arctic in 1817–18, Ross took deep-sea soundings (depth measurements) and used a bucket-shaped device called a dredge to haul up worms and starfish from 6,300 ft (1,920 m) in Canada's Baffin Bay. Some 20 years later, Ross's nephew, James Clark Ross, hauled up similar species from even greater depths in the Southern Ocean. The Rosses concluded that the same species of deep-sea creatures must live at great depths in many of the world's oceans.

In 1831–36 the English naturalist Charles Darwin traveled on HMS *Beagle* to South America, the Galápagos Islands, New Zealand, and Tasmania, Australia. Darwin's observations of the animals and plants of the different countries he visited led him to conclude that species change very gradually over long periods of time, because offspring inherit from their parents the characteristics that best allow them to survive in their environments. This theory is called evolution by natural selection. Darwin did not publish his controversial ideas about evolution until 1858. But the *Beagle* voyage had already led Darwin to propose a scientific explanation of how coral atolls are formed and to produce a study of barnacles. Scientists still use his findings today.

The marine explorers John Ross, James Clark Ross, and William Thom study charts with two Inuit (Eskimo) helpers (at center of picture). Indigenous people often gave invaluable information to scientists.

Charles Darwin spent more than five years aboard HMS *Beagle*, studying marine life. The *Beagle*'s commander, Robert Fitzroy, had the ship fitted with barometers to assist in detailed studies of the weather.

Oceanography Is Born

Oceanography became a truly international, fully-fledged science between the 1870s and 1920s. Scientists joined marine expeditions to study many aspects of the world's oceans.

In 1872 an expedition led by William Benjamin Carpenter and Charles Wyville Thomson set out from England aboard HMS *Challenger* with the aim of "exploring all aspects of the deep sea." HMS *Challenger* was a converted warship that had been fitted out as a floating research laboratory with all the equipment available for modern scientific study of the ocean. Its supplies included 135 mi (215 km) of rope for taking soundings (measuring the depth of the ocean) and around 12 mi (20 km) of cable to haul the dredges and sampling nets that brought up specimens from the ocean depths.

Challenger's crew of 240 sailors and 9 scientists were to spend the next four years crossing the Atlantic, Indian, and Pacific Oceans in a 79,000-mi (127,140-km) voyage of exploration. This single voyage yielded more information about the world beneath the sea's surface than all the expeditions that had gone before.

Swiss-born American zoologist Alexander Agassiz supervises the care of a collection of marine specimens. Agassiz founded a zoological station at Newport, Rhode Island, and improved deep-sea research techniques.

It took many years for the results of the *Challenger* expedition to be analyzed. This illustration of microscopic, silica-shelled creatures called *Radiolaria* appeared in a book published in 1896.

See also **vol 1:** 16–17; **vol 2:** 20–21; **vol 3:** 32–33, 34–35

Mapping the ocean floor

Prince Albert of Monaco, monarch of a tiny state on the Mediterranean coast, had a strong interest in oceanography and led his own oceanographic expeditions. In 1905 he provided funds so that existing depth data could be made into charts to create the first world map of the ocean floor. Two decades later, a German expedition (1925–27) in the research vessel *Meteor* criss-crossed the southern Atlantic Ocean. The scientists on board studied the currents and chemistry of the sea and used echo sounders to measure the depth and shape of the ocean floor. This expedition was the first to use the electronics that revolutionized oceanography.

An underwater research vessel, or submersible, at Woods Hole Oceanographic Institution. Through the work of early pioneers such as Alexander Agassiz, Woods Hole has become world famous.

The scientists on board *Challenger* examined every item brought to the surface, and identified 4,000 new species. They established beyond doubt that marine organisms live even in the deepest parts of the oceans, feeding on the rain of organic particles filtering down from the upper parts of the ocean and on the carcasses of large animals such as whales. *Challenger*'s crew measured the Mariana Trench—the deepest known feature on the surface of the Earth—to a depth of about 26,850 ft (8,185 m). Since then the trench has been shown to extend to 35,840 ft (10,924 m) deep.

It took dozens of scientists more than 20 years to examine all the samples and analyze all the information that *Challenger* brought back to England. Their scientific reports filled more than 50 thick volumes, which laid the foundations of modern oceanography.

Marine laboratories

The first of many laboratories devoted to carrying out research into marine life was established in Naples, Italy, in 1872, the year that *Challenger* left on its mission. In 1873 zoologist Alexander Agassiz founded a marine laboratory on America's east coast. Agassiz was the son of the Swiss geologist Louis Agassiz who established the place of the ice ages in Earth's history and later moved to the United States.

Alexander Agassiz studied engineering and chemistry at Harvard, and made a fortune in copper mining, which he used to pursue his interest in oceanography. In 1888 Agassiz moved his marine laboratory to Woods Hole, Massachusetts, later renamed the Woods Hole Oceanographic Institution. In 1903, William Ritter set up a similar marine research laboratory on the west coast of America. It became the Scripps Institution for Biological Research in 1912 and is now based in La Jolla, California. Although the marine laboratories put their exhibits on public show, their main function was to act as centers for research expeditions, where scientists could examine and analyze their findings. Woods Hole and Scripps remain leading U.S. centers for oceanographic research.

As early as the 1870s, scientists were coming to recognize the problems posed by overfishing (catching so many fish that they cannot replace their numbers by natural breeding). Scandinavian scientists in particular were concerned that cod and herring were being overfished in the North Sea. They realized the need to set up a scientific investigation into the life cycles of these fish in order to learn more about their feeding and breeding patterns. In 1902 the countries bordering the North Sea set up the International Council for the Exploration of the Sea (ICES)—the first international fisheries research organization anywhere in the world—to carry out this important work.

Sonar and Satellites

New technologies have changed the way scientists study the oceans and the seabed. Satellites monitor the ocean surface, sonar and other electronic devices probe the hidden watery depths, and drills bore deep into the ocean floor.

When World War II ended in 1945, the United States and European nations sold many of their warships. It was a chance for marine research institutions to purchase vessels that already had sophisticated electronic equipment on board. Marine scientists were able to use this equipment to map much of the deeper parts of the ocean floor for the first time. They used methods such as echo sounding, or sonar (**so**und **n**avigation **a**nd **r**anging), to bounce sound waves off the sea floor, which told them the depth and shape of the ocean floor.

In 1956 the American scientists Bruce Heezen and Marie Tharp used this information to produce a map of the North Atlantic seabed, showing its rises (mounds) and hollows (dips). Their map revealed the presence of the Mid-Atlantic Ridge, a mountain range that snakes from north to south the length of the ocean. Scientists found that similar ridges exist in other oceans. Because new sea floor is created at mid-ocean ridges, the discovery gave scientists information about how plate tectonics—the theory that Earth's crust is broken up into large plates that are carried along on a deep layer of slowly moving rock—was taking place.

In 1968 the deep-sea drilling ship *Glomar Challenger* bored down into the seabed and took samples of the sea floor. Samples taken close to mid-ocean ridges showed that the rocks here were much younger, or more recently formed, than samples of rock taken farther away from the ridges. This finding confirmed the theory that new sea floor is continually created at the ridges. The new rock pushes older rock away from the ridge, and the ocean floor spreads.

Recording and sampling

Taking readings from the sea to measure conditions such as temperature and saltiness was once very time-consuming. It involved lowering and raising a series of instruments. Today, scientists use a single piece of automatic equipment fitted with multiple-sampling devices so that many readings and samples can be taken at the same time. The equipment is called a CTD because it measures conductivity (saltiness),

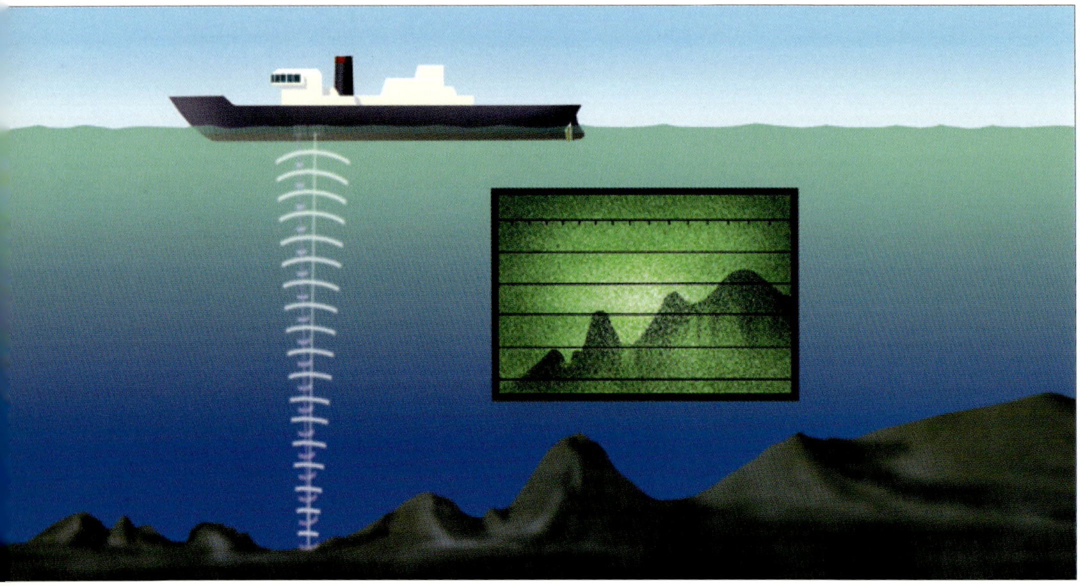

This ship is using sonar, or echo sounding, to check the depth of the water. The technique involves bouncing sound waves off the bottom of the sea and timing how long it takes for the sound waves to return. The longer the time it takes, the greater the depth. The soundings are recorded and displayed on a screen, giving a picture of the seabed's depth and its shape.

See also **vol 1:** 6–7, 42–43; **vol 2:** 12–13; **vol 3:** 34–35

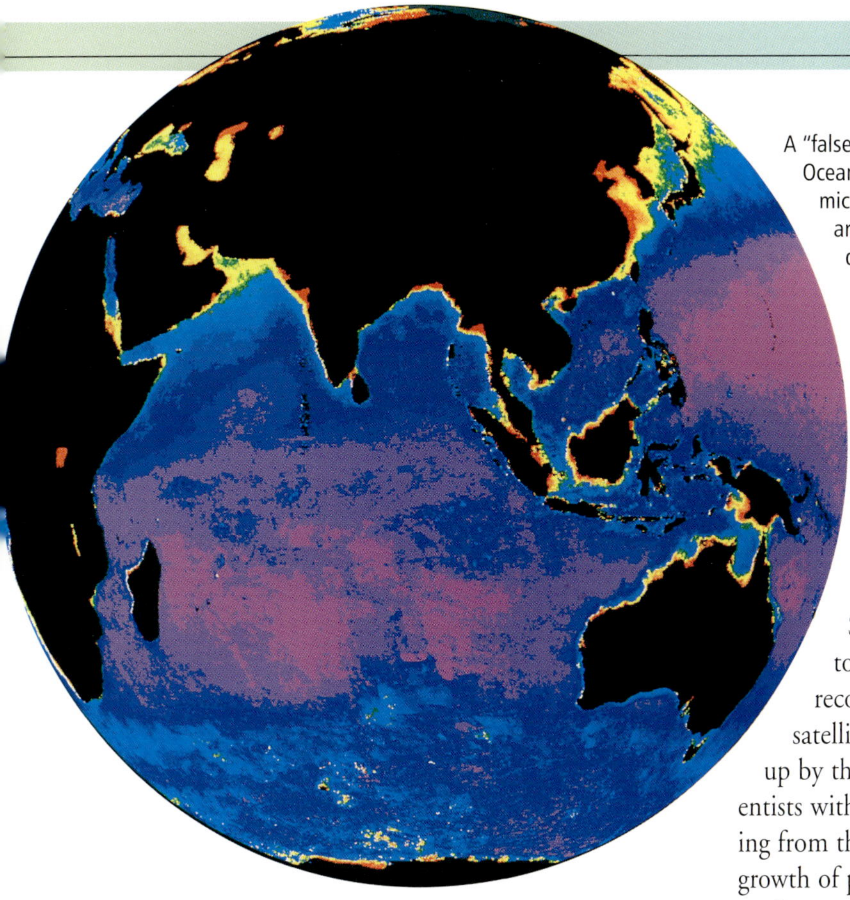

A "false-color" satellite image of the Indian and western Pacific Oceans, showing the distribution of phytoplankton, masses of microscopic floating plants. Phytoplankton is densest in the red areas. Yellow, green, blue, and pink indicate areas of decreasing phytoplankton density; pink is the least dense.

temperature, and depth. It measures water temperature, saltiness, and depth electronically and transmits these readings back to the ship. At the same time, canisters collect water samples at specific depths. The water is analyzed later for plankton and chemical substances.

Eyes in the sky

Satellites orbiting the Earth are one of the newest tools available to the ocean scientist. Satellite recording is also called remote-sensing, because the satellites operate from high in the sky. The data picked up by the satellites and beamed back to Earth provides scientists with much new information about the oceans, ranging from the height of the seabed to changing patterns in the growth of phytoplankton (microscopic floating plants).

Oceanographic satellites carry two types of sensors: active and passive. Active sensors beam down radio waves that bounce back from the sea's surface—this is called radar altimetry. The information gathered in this way helps scientists to plot ocean currents and the shape of the ocean floor, as well as calculate wave height and wind speeds. Passive sensors detect various wavelengths of light that reach them from the sea's surface. The sensors measure surface temperature and the color of the sea near the surface.

Maps of sea temperature allow scientists to plot the movements of ocean currents. Changes in their pattern may warn of climatic events, such as El Niño, that affect the feeding and spawning of fish. Sea-temperature patterns also provide scientists with information about trends such as global warming. Sea water color is affected by various biological, chemical, and physical features of the ocean. For example, massive patches of green are likely to reveal the existence of phytoplankton. These microscopic marine plants provide food for other sea creatures. By seeing how much phytoplankton there is, scientists can then work out how much food is being produced for other marine creatures to feed upon.

The Seasat satellite carries out ocean research using radar. For example, it measures the height of the seabed by bouncing radio waves off the sea surface. The sea surface above a raised portion of the seabed is slightly higher than the surrounding surface, and Seasat can detect this. The readings are relayed to a ground station.

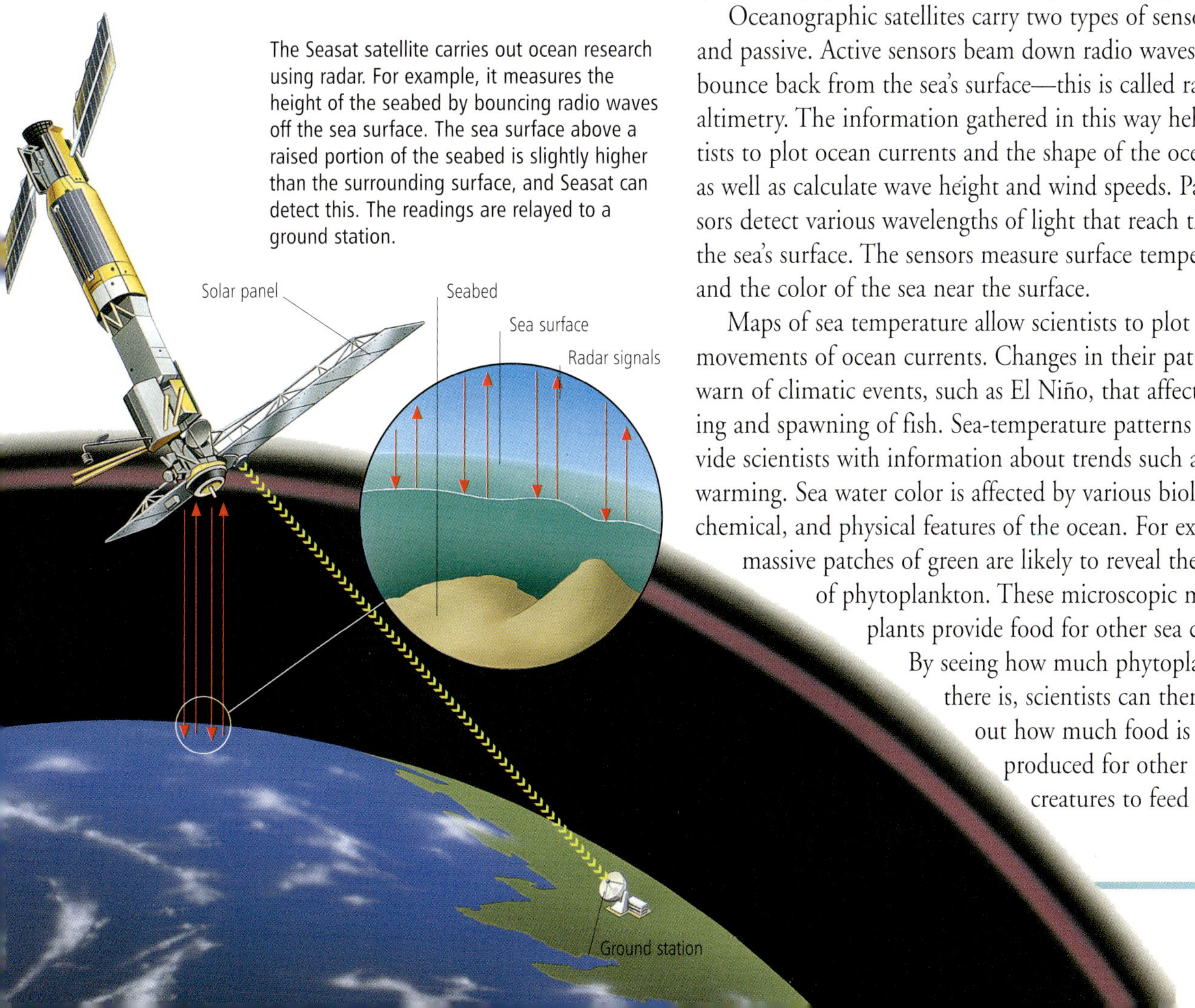

Diving

People have been swimming beneath the sea—diving—for thousands of years. Mother-of-pearl shells, more than 6,000 years old, have been discovered in Southwest Asia (Mesopotamia)—only divers could have collected such shells.

Around 360 BCE, the Greek philosopher Aristotle described how boatmen sent containers of air down to sponge divers on the seabed so they could breathe under water. English scientist Edmond Halley invented the first "modern" diving apparatus, the diving bell, in the early 1700s. Operators lowered a wooden, cone-shaped bell into the water, trapping a pocket of air inside, and anchored it to the seabed with weights. The bell was the source of air for divers working close by. They would take a gulp of air from the diving bell, work in the water for a few minutes, then return to the bell for more air before going off again. Stale air was replenished by barrels containing fresh air, lowered into the water and released into the bell. The bell allowed divers to repair harbor walls and bridge supports under water.

Diving suits

By the 1800s, diving bells were made of metal and supplied with air pumped through tubes. They restricted the divers to working in a small area around the bell. The invention of the first diving suits gave divers great freedom to move about.

The greatest challenges facing a diver under water are the cold, the dark, and the effects of pressure. In the 1830s, the German inventor Augustus Siebe developed the "Standard"

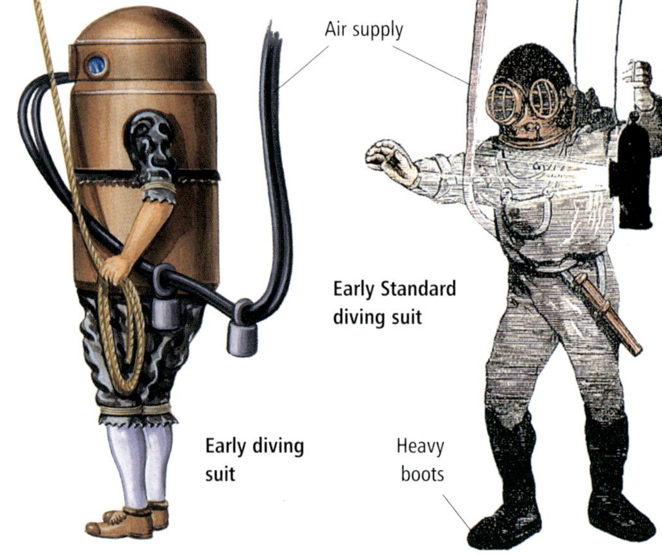

Pressure hazards

Today, most divers who swim in shallow water use scuba gear, but the effects of pressure place limits on how deep they can dive. Few scuba divers venture below depths of 130 ft (40 m), where the air pressure is four times that at the surface. At this depth, scuba divers rapidly use up their air supply, which normally limits a dive to 30 minutes or less.

The high pressure also has other effects. Under pressure, nitrogen (which makes up 80 percent of air) dissolves readily in the blood. This can cause a condition known as nitrogen narcosis, or "rapture of the deep." As a consequence, the diver may behave strangely and take dangerous risks.

Another hazard is decompression sickness. If a diver returns to the surface too rapidly at the end of a deep dive, bubbles of nitrogen can "fizz" out of the blood—an effect similar to releasing the cap on a bottle of fizzy drink. Tiny nitrogen bubbles can block blood vessels and starve vital parts of the body of oxygen, or press on tiny gaps in joints, causing excruciating pain. Divers who get severe joint pain bend over to relieve the agony, which is why decompression sickness is described as "the bends." In severe cases, divers are treated in a decompression chamber. There, the air pressure is raised high enough to force the nitrogen bubbles to dissolve back into the blood.

See also **vol 1:** 42–43; **vol 3:** 32–33, 34–35, 36–37, 54–55

Pearl fishermen in Asia dive without any breathing equipment. They reach depths of up to 100 ft (30 m) and can stay under water for several minutes. But if you really want to explore the underwater world, you need either to take an air supply with you or be connected to an air supply on the surface.

diving suit, which remained the basic design of diving suit for more than 100 years. The Standard suit was made of layers of canvas and rubber to keep the diver dry and protected from the cold. A copper and brass helmet, with a small window in the front strong enough to resist the water pressure, was bolted to the body of the suit. Air entered the helmet through a tube from a ship on the surface. The diving suit was very heavy and had built-in heavy lead and leather boots to aid stability. Divers could not swim in these clumsy suits, but were lowered from boats to the seabed, to depths of up to 200 ft (61 m). Helpers on board the diving ship made sure that the diver's supply of air did not fail. Divers in these suits were able to stay under water for an hour or more.

Standard suits were in common use until the mid-1900s. Even today, divers sometimes still work in suits in which air is supplied from the surface. Divers working at depths of up to 2,000 ft (609 m) use atmospheric diving suits (containing air at normal atmospheric pressure). Made of strong metal, the suits can withstand the pressures found in deep water.

The aqualung

By the late 1800s, inventors were trying to develop ways of enabling divers to be more mobile and not reliant on an air supply from the surface. Early devices relied on recycling the air used by the diver and removing the carbon dioxide that the diver breathed out.

A major breakthrough came in 1943 when the French diving pioneer Jacques Cousteau and engineer Émile Gagnan designed the aqualung, or scuba (**s**elf-**c**ontained **u**nderwater **b**reathing **a**pparatus). The aqualung had a special valve that allowed divers to breathe air under water with much greater ease. Using a high-pressure air supply in tanks on their backs, divers could breathe air at the same pressure as their surroundings, and the air they breathed out rose to the surface as bubbles. The tanks had enough air to last an hour or more. Coupled with a rubber diving suit, mask, and flippers, this equipment gave the diver new freedom under water.

Some of the major developments in diving suits are shown here.

Scuba diver with aqualung

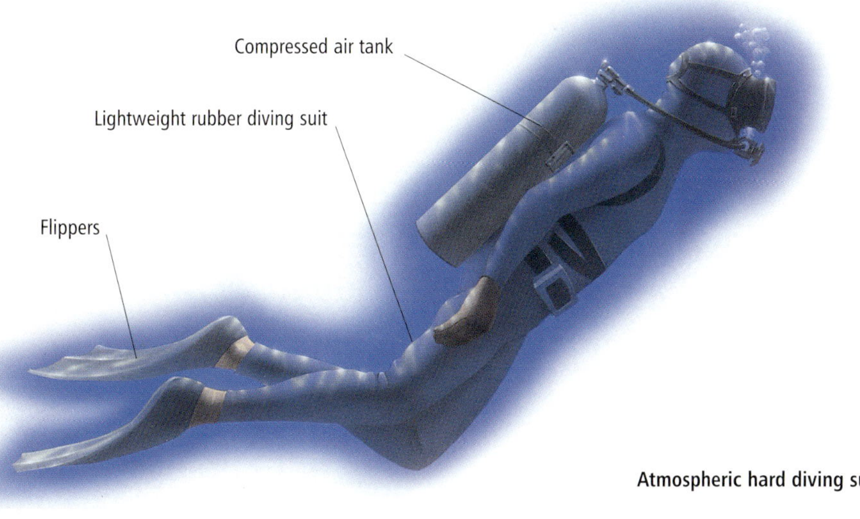

Atmospheric hard diving suit

Underwater Vessels

Submarines and submersibles are vessels that operate underwater. Submarines are often used for military purposes, but most submersibles are engaged in marine research, prospecting for oil and gas, or investigating wrecks.

The world's first submarine was launched in 1620 by Dutch inventor Cornelis Drebbel, who held the position of "court inventor" to King James I of England. Drebbel covered over a rowboat with leather, which he greased to make it watertight. Inside the sealed boat were 12 oarsmen. They took the boat on a journey approximately 13 ft (4 m) below the surface of the River Thames in London. Tubes poking above the water provided air for the crew. When they stopped rowing, the boat slowly started to rise to the surface.

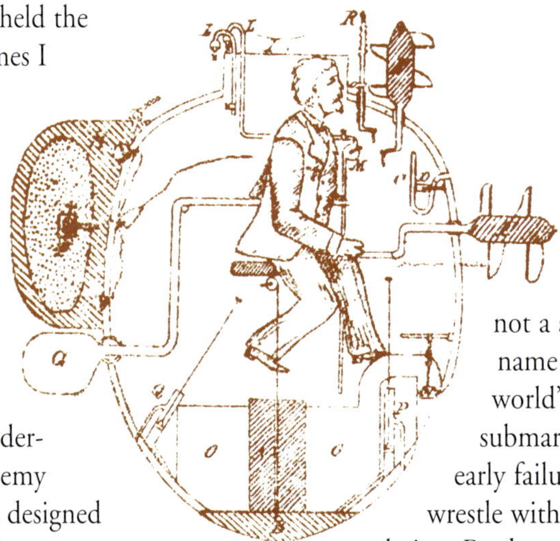

An artist's impression of the *Turtle*, built in 1776. The submariner drove a propeller with his feet to maneuver the vessel against a British warship in New York harbor. He tried unsuccessfully to drill a hole in the ship's side.

Naval engineers dreamed of designing an underwater vessel that could sneak up on and sink enemy ships. In 1776 David Bushnell, a Yale graduate, designed the *Turtle*, the first submarine to make an attack on a warship. In 1800 the American engineer Robert Fulton designed the *Nautilus* submarine for Napoleon I of France to attack British warships. It was not a success, but it later gave its name to the USS *Nautilus*, the world's first nuclear-powered submarine. Undaunted by these early failures, engineers continued to wrestle with the problems of submarine design. By the early 1900s, most navies had submarines armed with torpedoes (explosive missiles). Personnel serving in submarines were called submariners.

Life in a modern submarine

Imagine living under water for months on end. You are in a metal tube that contains everything you need—your food, air supply, laundry facilities, and so on. Everything you and your 100 or so fellow crew needs must be stored in the submarine, or made on board when the submarine is under way. During the day you work in white light. At night the lights turn eerie red to remind you that it is dark on the surface. For much of the time you talk quietly and walk stealthily for fear of making a noise that ships or other submarines could hear. Your voyage is top secret, and it may be weeks before you return home.

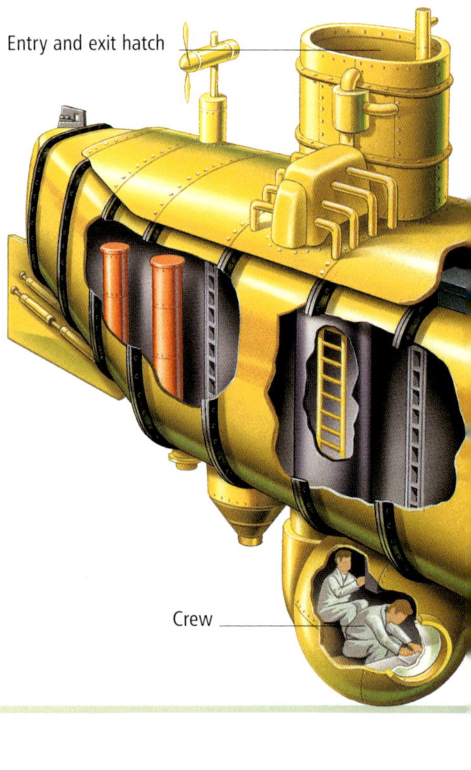

The *Trieste* was a submersible built by the Swiss physicist Auguste Piccard. It was manned by two people who sat in a metal sphere under the vessel. The buoyancy tanks above them were partly filled with gasoline, which is less dense than sea water. The crew let water into the tanks when they wanted the vessel to sink, then released metal weights to make it rise to the surface.

See also **vol 1:** 42–43; **vol 3:** 18–19, 30–31, 34–35

Alvin, a manned submersible owned and operated by Woods Hole Oceanographic Institution in Massachusetts, has completed more than 3,500 dives since its launch in 1964. In 1966 it helped recover a hydrogen bomb lost in the Mediterranean. In the 1970s *Alvin* enabled scientists to examine the amazing animal communities living around hot-water vents in the ocean floor, such as near the Galápagos Islands in the Pacific, reaching depths up to 12,500 ft (3,800 m). In 1986 it explored the wreck of the *Titanic*.

Submersibles

Military submarines are long, cigar-shaped vessels. They are subject to enormous pressure from the weight of the water the deeper they go. The pressure will eventually crush their hulls. Even today, military submarines cannot descend to depths much greater than 1,600 ft (490 m). Submersibles, on the other hand, can reach the ocean floor at depths of 20,000 ft (6,000 m). Many contain cabins that are spherical in shape, so they resist pressure equally on all sides.

In 1934 William Beebe (a zoologist) and Otis Barton (an engineer) made the first successful deep descent in a submersible. In this bathysphere ("deep sphere")—a steel ball with portholes—they descended to 3,028 ft (923 m), allowing them to observe deep-ocean creatures. It was tethered to a mother ship by a steel cable 3,500 ft (1,066 m) long.

In 1948 the Swiss physicist Auguste Piccard designed the bathyscaphe ("deep boat"). Piccard's vessel looked like a shortened submarine, with a pressure-resistant sphere for the crew's living quarters slung underneath. Unlike a bathysphere, the crew could move the vessel about.

In 1960 Piccard's son Jacques and U.S. naval officer Don Walsh descended to the bottom of the Mariana Trench in an updated bathyscaphe called *Trieste*. No one has ever equalled their depth record of 35,802 ft (10,912 m).

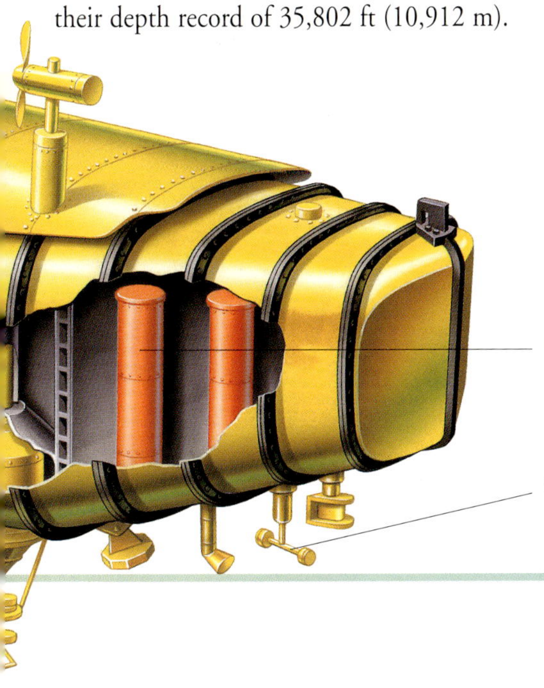

Remotely operated vehicles (ROVs) are unmanned craft that are tethered to either a mother ship or to a piloted submersible such as *Alvin* (see above). ROVs can undertake longer dives than manned vessels and allow oceanographers and others to investigate the secrets of the deep sea world in safety. More than 200 ROVs are in use around the world, particularly in the offshore oil and gas industry.

Remotely piloted underwater craft working without any link to another craft are called **a**utonomous **u**nderwater **v**ehicles (AUVs). They are programed to carry out examinations of the seabed, collect samples, and take readings. The AUVs return to the mother ship from which they launched, and scientists download and analyze their data and samples.

Research at Sea

Modern marine research uses the latest scientific techniques and equipment. Scientists such as biologists, chemists, geologists, and physicists often work together on research projects, each contributing a particular skill.

Two of the most important centers for marine research are the Scripps Institution of Oceanography at La Jolla, California, and the Woods Hole Oceanographic Institution in Massachusetts. Scientists at these and other oceanographic centers around the world work on a wide range of programs, from mapping underwater volcanoes and monitoring phytoplankton blooms to studying the effects of global warming and the preservation of fish stocks.

In order to carry out their research, oceanographers spend long periods at sea on research ships, floating platforms, and underwater laboratories. Each vessel is specially designed for the scientific tasks it is required to carry out. For example, oceanographers at Woods Hole make use of *Atlantis.* This is general-purpose laboratory ship that is 272 ft (83 m) long.

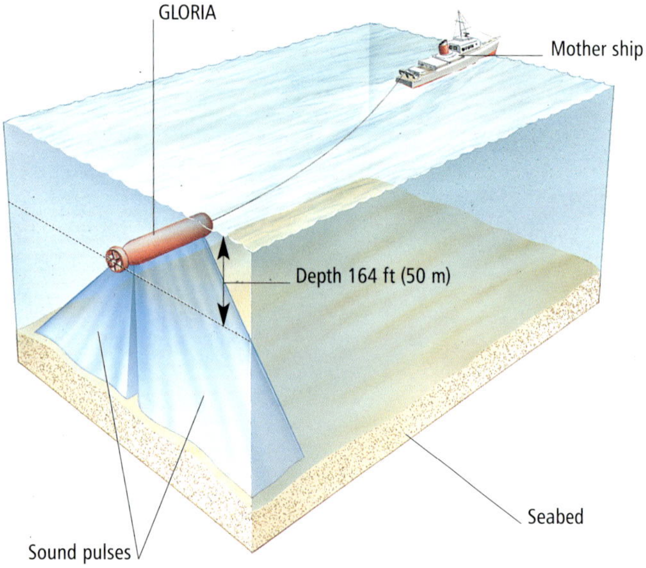

GLORIA (**g**eological **lo**ng **r**ange **i**nclined **a**sdic) is an unmanned, torpedo-shaped sonar device, which has been in use for about 30 years for surveying the ocean floor. GLORIA is 26 ft (8 m) long, and is towed behind its mother ship. Sound pulses are sent out from GLORIA to a distance of 19 mi (30 km) on each side. The echoes that bounce back from the seabed are processed by computers on board the mother ship, enabling the contours of the sea floor to be mapped. GLORIA can help find the best routes for laying undersea cables, and it can also aid in mineral exploration.

Atlantis also acts as a launch platform for the submersible *Alvin* and its remotely operated vehicle (ROV) *Jason.* She has a permanent crew of 23 as well as 13 pilots and technicians who operate and look after the underwater vehicles, and can accommodate up to 23 scientists. *Atlantis* can stay at sea for 60 days, during which time the vessel may travel nearly 20,000 mi (32,200 km).

Like other modern research ships, *Atlantis* is equipped with satellite-linked navigation and communication systems. She also has powerful cranes and winches to lower and raise sampling apparatus (collecting tubes, dredges, nets, and so forth) to different depths of the ocean or to the ocean floor, and to launch her underwater vehicles. Computers that control thrusters make sure that *Atlantis* stays in position, even in high winds and rough seas.

Each research trip *Atlantis* makes has a different scientific purpose. On one mission, *Atlantis* explored Puna Ridge—a chain of extinct and active underwater volcanoes close to the Hawaiian Islands. Researchers used a special kind of sonar to build up a three-dimensional (3-D) map of the volcanoes and used boring machines to extract core samples from the ocean floor. The ROV *Argo II* captured detailed still and video images of the volcanoes in action. These pictures are helping scientists to understand how the Hawaiian Islands were formed, and also how other underwater volcanoes and oceanic islands came into existence.

Standing on end

Scientists at the Scripps Institution of Oceanography do some of their work on FLIP (**fl**oating **i**nstrument **p**latform), designed in the 1960s. FLIP is 350 ft (107 m) long and looks like a giant buoy lying on its side. But when

See also **vol 3:** 28–29, 30–31, 32–33

This collecting tube is attached to a deep-sea ROV. It is used to suck up samples of water or marine life, rather like a vacuum cleaner.

one end is flooded with sea water, it "flips" into a vertical position, leaving only about 50 ft (15 m) of its structure visible above the surface. The remaining 300 ft (90 m) remains submerged. When floating horizontally, FLIP can be towed into position; it has no engine.

FLIP was designed in this way to solve a particular problem. When underwater research is carried out from surface ships, instruments that pick up sound waves below the surface are affected by noises made by the vessel's engines or by the sound of the ocean swell. When FLIP is put in a vertical position, it does not rise and fall much, so noise from movement is reduced and distance measurements are more precise. FLIP forms a stable platform for listening to the sounds of sea creatures, measuring the temperature and saltiness of sea water, and sampling the atmosphere just above the sea.

An underwater laboratory

About three and a half miles (5.5 km) off the Florida coast, *Aquarius*—a yellow, air-filled tube—lies 60 ft (18 m) below sea level. It is the world's only static underwater laboratory. Four scientists can spend up to 10 days at a time on board *Aquarius* studying the marine life on nearby coral reefs. They use scuba diving equipment to swim out to the reefs, where they make their detailed observations and recordings, and return to *Aquarius* regularly when they need replacement oxygen tanks. Because they use *Aquarius* as their base, rather than returning to the surface, the researchers do not have to decompress at the end of each dive to avoid the "bends" (problems associated with decompression), and by working in shifts they can observe the reefs around the clock.

Aquarius's scientists have studied the lives of sea urchins that live off the film of algae growing on the coral reef. They have also investigated the serious problem of coral bleaching—when coral polyps eject their algal partners and die—to find out what might be the cause.

FLIP's crew of 16 live in the section that remains above the surface of the sea. Much of the equipment tilts or swivels, or can be unbolted and repositioned, depending on the way FLIP is floating in the water.

Investigating Shipwrecks

The world's oceans are strewn with thousands of shipwrecks. Many contain valuable items such as cannons, coins, or jewelry. All have much to tell us about shipping, warfare, and trade in past times.

A royal rescue

In 1545 the *Mary Rose*, a warship built for England's King Henry VIII, sank as she was leaving Portsmouth harbor, in southern England, to do battle with the French. More than 400 years later, marine archeologists located the wreck on the seabed. The right-hand (starboard) side of her hull was virtually intact, preserved by the mud in which she was lying. To lift the fragile timbers out of the water, the rescue team built a strong metal cradle and lowered it into the water. Divers gently maneuvered the section of hull onto the cradle and used a crane to raise it slowly from the seabed. Now the hull of the *Mary Rose*, sprayed with preservative to replace the water in the timbers, is on show in Portsmouth, together with the wealth of goods she was carrying.

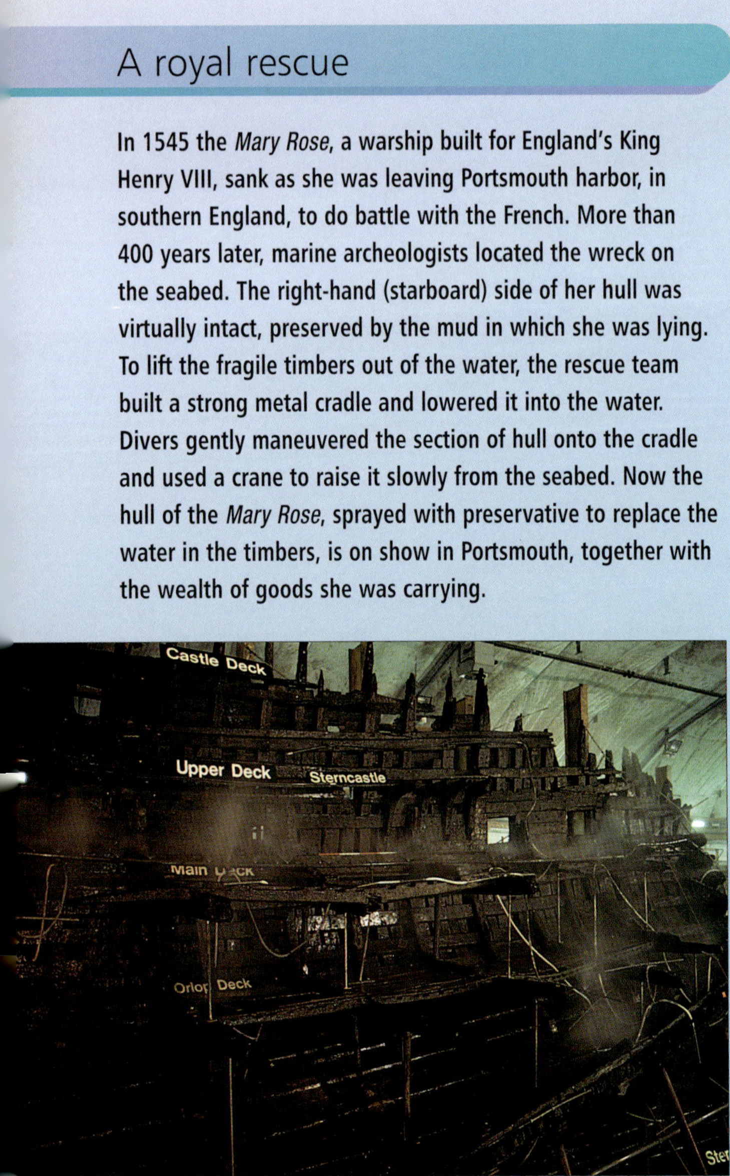

People who investigate shipwrecks and other underwater sites such as ancient harbors are known as marine archeologists. Discovering underwater sites is often a matter of chance. A scuba diver may report seeing a wooden beam or a cannon lying on the seabed. Sometimes people find parts of a wreck washed up on shore, or an exceptionally low tide uncovers the skeleton of a rusting hulk. But marine archeologists also carry out detailed detective work in libraries, studying old charts, ships' logbooks, and company records to find where a wreck might be.

Once the marine archeologist has an approximate idea of a wreck's location, he or she uses sonar equipment to build up a 3-D map of the seabed, looking for unusual features that may point to the presence of a shipwreck. This is how, in 1967, archeologists discovered the wreck of the *Mary Rose*—an English warship that sank in 1545.

Marine archeologists can lower magnetic detectors to reveal the presence of cannons, ships' nails, or other metal objects buried in the sands of the seabed. They also make use of special sonar equipment to locate dense buried objects that may not be made of metal, such as parts of a wooden hull.

Excavating a site

The next stage is to investigate the wreck more closely. In shallow waters, archeologists use scuba equipment to dive down to the site. Working under water to investigate an archeological site is made more difficult by winds, currents, and tides.

First, the archeologists mark off the site with a grid of tape or aluminum. They locate every item in the grid, and

A diver examines barnacle-encrusted earthenware pots known as amphorae—used for storing olive oil or wine—at an underwater archeological site at Bodrum, Turkey.

See also **vol 3:** 18–19, 30–31, 32–33

record its position with notes, drawings, or photographs. They then tag and number all the objects they find in and around the wreck and feed this information into a computer to create a 3-D map of the site. Working on the seabed, the archeologists use tools such as an air dredge (which operates like a giant vacuum cleaner) to remove sediment from the seabed to reveal objects. Then the divers use a small fan or a jet of water to loosen the sediment from individual objects, or they brush it away by hand. Many items are extremely delicate after being under water for so long.

To raise the objects to the surface, the divers place them in air lifts—nets or baskets slung below air-filled bags—and guide them up through the water. They use winches to raise large objects such as a cannon or part of a hull. Objects made of wood, leather, bone, or metal often survive very well in thick mud. But once exposed to the air, they quickly fall apart or corrode. Working in special laboratories, experts conserve small wooden and leather items by freeze-drying

On its maiden voyage from England to New York City in 1912, the British "unsinkable" passenger liner, *Titanic*, struck an iceberg and sank off the Grand Banks of Newfoundland. She lay in very deep water and was remarkably well preserved. This is part of the front of the ship.

them. This immediately turns the water they contain into gas. Larger wooden objects are sprayed with water for months or years to prevent them drying out. Conservators roast iron objects in a furnace to prevent further corrosion. Then the preserved objects go on public display in specially designed cabinets that control the temperature and humidity to prevent any further deterioration.

Exploring the *Titanic*

Marine archeologist Robert D. Ballard has made a specialty of investigating wrecks lying in deep water that cannot be reached by scuba divers. He and his team use sonar to locate the wrecks, and photograph and video them from submersibles and robot vehicles.

One of Ballard's most spectacular feats was to use Woods Hole's submersible *Alvin* to explore the wreck of the *Titanic,* lying in an ocean canyon 2.5 mi (4 km) deep off the coast of Newfoundland, in 1986. Photographs showed that the ship and its contents were very well preserved. Since then, a salvage company has brought up hundreds of artifacts from the wreck. The future of the site is a subject of great debate. Many believe the *Titanic* should be preserved intact as a memorial to those who died when it sank.

Hunting Sea Mammals

For thousands of years, people hunted sea mammals such as whales and seals for their meat, blubber, and bone. Although we have changed our attitudes toward the killing of these large creatures, their numbers are still threatened.

In the early 1700s, sailors from Europe entered the Bering Sea for the first time. There they encountered a marine mammal known as Stellar's sea cow. These slow, ungainly 11-ton (10-tonne) beasts were easy targets for mariners who wanted fresh meat to supplement their meager rations. By the late 1760s, the animal had been hunted to extinction. In spite of this warning, people continued to believe that the seas would provide a never-ending supply of sea mammals.

Hunting whales

Whales in particular were highly prized because of their great size, and because almost all their body parts had value. They yielded tons of meat, which had value as a high-protein food. Indeed, people in parts of the world today—for example, the Japanese and many Scandinavians—still regard whale meat as a delicacy. Whalers boiled down blubber—the layer of insulating fat carried by whales and other marine mammals—for oil, which was burned in lamps before the advent of kerosene and other petroleum-based fuels in the 1800s. The Victorians used whale oil to lubricate industrial machinery, and it was also a major ingredient of soap.

Long before the days of materials such as elastic and lycra, dressmakers used baleen—the thin, flexible plates that hang from the jaws of toothless whales to filter krill from the water—to strengthen the bodices of women's dresses and

A dead baleen whale is hauled out of the water for cutting up into blubber, meat, and other products at Hvalfjorden, on Iceland's coast.

International Whaling Commission

The International Whaling Commission (IWC) was set up in 1946 to control the hunting of whales by fixing quotas for the number of whales each country could kill a year. Some nations such as Japan, Norway, Iceland, and Russia place high value on whale meat, and whaling remains an important part of the economy of some coastal communities. They consistently ignored the quotas, and in the 1970s the international community began to realize that some species of whale were in danger of extinction. This led to stricter rules. Some whaling nations claimed they were hunting whales for "scientific purposes" to study their behavior and preserve stocks. Certain countries or communities continue to hunt whales illegally, ignoring the IWC's rules and quotas.

See also **vol 2:** 36–37, 40–41; **vol 3:** 54–55, 56–57

undergarments. Spermaceti—a waxy substance from the head of sperm whales—was used in the production of ointments, cosmetics, lamp oil, and candles.

Until the mid-1800s, whaling was a very dangerous job. Whaling fleets from the United States, Europe, and Japan followed whale migrations. Once close to a school of whales, the hunters got into small, open boats from which they harpooned (speared) the animals by hand. Sometimes the whales would fight back or pull the whalers' boats under water before dying from exhaustion and loss of blood.

A favorite target of the whaling fleets was the right whale, so-called because it was the "right" whale to catch, being a slow swimmer. It lives in coastal waters rather than the open ocean and floats when dead. Another favorite was the bowhead whale. Atlantic stocks of both species declined drastically between 1600 and 1800.

The heyday of the whaling industry began in the second half of the 19th century, when the whalers of Europe and North America began to use steel-hulled, steam-driven ships that could travel quickly over long distances in pursuit of the whales. By the late 1800s, whaling ships were armed with deadly weapons: a cannon on board ship fired a harpoon that exploded inside the whale, killing it more quickly.

Hunted to extinction

By the 1910s, few whales remained in the North Atlantic, and whalers' attention turned to the antarctic waters of the Southern Ocean. Norwegian whalers set up whaling stations (whale-processing factories) on remote islands such as South Georgia in the South Atlantic. Then, in the 1930s, came the introduction of factory ships, which processed the whales on board. Whaling became so efficient that by the 1970s the large whales in the Southern Ocean had been hunted to commercial extinction, and whalers had to look instead to smaller whales such as the sei and the minke.

In 1986 the International Whaling Commission (IWC) agreed there should be a temporary ban on commercial whaling. This has been largely successful, although several countries still ignore it.

The IWC allows some indigenous peoples such as the Inuit of the Arctic, whose traditional way of life depends on whaling, to catch a specified number of gray whales a year, using traditional hunting methods. The Inuit also continue to hunt seals for food and clothing.

In Greenland, an Inuit hunter loads a dead seal onto his sled. Seals are hunted for both food and clothing. Skins are also used to make kayaks. People living in such inhospitable areas could not survive without being able to hunt. Consequently, they are not subject to the restrictions that apply to commercial operators, who would take many more animals than the Inuit.

Fish from the Sea

Throughout history, people have harvested the seas for fish and other kinds of seafood. Although seafood makes up less than 10 percent of the world's total diet, it is the main source of protein in many seafaring communities.

Fish is a valuable food. Fish are rich in protein, essential for building tissue, and in vitamin D and certain B vitamins. Fish such as tuna, mackerel, and herring contain oils that help lower the body's cholesterol—a fatty substance that can cause heart disease.

Millions of people fish to feed themselves and their families. Here, a fisherman casts his net on the Niger River in West Africa. Such small-scale subsistence fishing rarely poses a threat to fish stocks.

The food chain

Fishers take more than half of the world annual catch of marine fish, by weight, in the North Pacific, North Atlantic, and off the west coast of South America. In these places, rising currents bring cold, nutrient-rich water close to the surface. In warmer months, this encourages phytoplankton to grow rapidly. These microscopic plants form the base of the food chain upon which higher forms of marine life, such as fish, rely. For example, phytoplankton are eaten by small fish and, higher up the chain, small fish are eaten by larger fish. It is the larger fish that are often caught by the commercial fishers.

Fish is the main source of food for more than one billion of the world's people. In many parts of the world, fishing is a small-scale affair. Fishermen use traditional methods such as rod-and-line, nets, or fish traps. They catch only a small number of fish from the oceans each year, which they use to feed themselves and their families, selling the surplus in the local markets.

The picture is very different among the world's richer nations. Commercial fishing boats scoop up vast quantities of fish from the world's oceans. Large fish catches have value not only as human food—small fish such as sardines and anchovies are ground up to produce fish meal, which is used for animal feed, fertilizer, or to make paint, glue, and soap.

Finding fish

The presence of dolphins, seals, tuna, and diving birds such as gannets in a part of the ocean are signs to experienced fishers that shoals of fish are swimming close to the surface. Commercial fishing ships make use of modern technology to locate deeper shoals. Sensitive sonar equipment scans the

Otter trawls (below) are giant, open-mouthed mesh bags. Otter boards force the mouth of the funnel-shaped net to stay open. Drift nets (right) "hang" in the water, trapping fish that swim into them. Purse seine nets (far right) encircle, then trap, entire shoals.

See also **vol 2:** 30–31; **vol 3:** 42–43, 52–53

Commercial fishers take more than half the world's marine catch from shallow waters using large nets. On board the bigger ships, fishers gut the fish, and freeze or chill them, so the catch is still fresh when they return to port. The net of this fishing boat is full of herring. Note the difference in the scale of this operation to that in West Africa (opposite).

contours of the seabed, and a screen in the ship's control room monitors the presence of shoals of fish. The ship's global positioning system (GPS) is able to pinpoint a shoal's position so precisely that the captain can then return to the exact place on a later trip to trawl (catch fish with a trawl net) again in the same place.

Near-surface fishing

When catching the fish near the surface, a fishing boat may lay out a purse seine net to encircle entire shoals of sardines, anchovies, herring, or mackerel. As the net is pulled in, the bottom closes, trapping the fish in a giant mesh bag. Sometimes fishers set gill nets. These hang vertically in the water, and fish swim into them. As their heads push into the mesh, their gills are ensnared.

A giant version of the gill net is called a drift net. Drift nets can cause great harm to many kinds of marine animals. Fishing boats leave nets "drifting" for hours or days. Large creatures such as dolphins, turtles, whales, and seals, can become entangled in the nets, and drown. Drift nets are outlawed in many parts of the world.

Fishers catch large fish such as tuna on baited hooks. They string the hooks by their hundreds onto long fishing lines that stream out for many miles behind the boat.

Bottom fishing

Bottom-living fish include cod and haddock, and flatfish such as flounder and turbot. Fishers use trawl nets—long, bag-shaped nets. A fishing trawler hauls these nets along the sea floor, or just above it. Fish are funneled into the open end of the net and trapped at the bottom of it.

Some trawlers are small and catch only about 1 ton (0.9 tonne) of fish at a time. But supertrawlers have nets large enough to trap 100 tons (90 tonnes). These trawlers process, freeze, and package the fish on board. They stay at sea for weeks or months and often unload their catch thousands of miles from their home port.

Fishers sometimes use trawl nets to catch shellfish such as crabs and clams. Unfortunately, the nets also scoop up small, unwanted fish. Although usually dead, they are thrown overboard because it is illegal to bring in undersized fish.

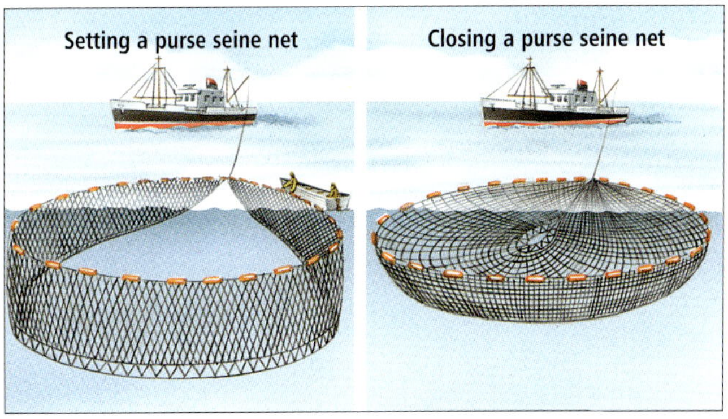

DWINDLING FISH STOCKS

The oceans are being overfished. This means that people are taking more fish out of the sea each year than are being replaced by natural breeding. All around the world people are catching too many fish from dwindling numbers.

In the 1960s, many marine biologists and food experts thought that catching more fish was a good way to help feed the world's rising human population. At that time, the global catch of seafood was about 44 million tons (40 million tonnes) a year. Although fishers in some parts of the world were already recording lower catches of previously abundant species of fish—for example, Californian sardines, Peruvian anchovies, and North Atlantic herring were all in short supply—experts thought fish populations would recover. The annual global catch, they reckoned, could rise to 110 million tons (99 million tonnes).

As a result, governments and companies all over the world bought new fishing fleets, equipped with modern technology to enable them to catch even more fish. But the original estimates were wrong. The world's annual fish catch peaked at 99 million tons (89 million tonnes) in the early 1990s, and has not risen since.

Maintaining stocks

In order to maintain fish stocks so that fishing can continue year after year without decline, the fish being caught must be replaced by young fish that then grow to adult size and breed. But if fishers catch fish more quickly than they can be replaced by younger ones, then numbers gradually decline.

Fisheries scientists study fish populations to calculate how many fish can be caught without harming overall numbers. They measure the size of fish caught on commercial fishing boats as well as those they catch themselves. By doing this they work out the average size of the fish. The scientists also examine the scales and earbones for growth rings; the more rings, the older the fish.

If catches are going down because fish do not live long enough to grow to adult size, it can be a sign that the fish population of a particular species is being overfished. But other factors such as the weather can also affect the success, or failure, of fish breeding. Calculating how many and at what size fish can be caught without seriously affecting stocks is a difficult task.

Pilchard

Peruvian anchovy

Atlantic herring

Atlantic cod

Four of the world's most important commercially caught fish species. By the 1960s, some stocks of these fish were already seriously depleted by overfishing.

See also **vol 2:** 28–29, 30–31; **vol 3:** 54–55, 56–57, 58–59

A southern bluefin tuna, trapped in a net off the coast of South Australia. Sometimes large fish are caught in drift nets intended to catch smaller species, and sometimes they get caught in nets designed to keep sharks away from popular beaches.

Unwanted fish, known as "trash fish," brought up in a shrimp trawler net and left to die on deck. One problem of trawling is that nets scoop up not merely the target fish, but almost everything else in their path.

Dwindling cod numbers

Years of overfishing caused the collapse of cod and haddock populations in the Northwest Atlantic in the early 1990s. More than 40,000 fishers and other workers lost their jobs when Canadian and U.S. fishing grounds had to be closed. Some limited fishing for cod started again in 1999, but the cod population has not recovered as expected. Warming seas may be affecting their breeding success. But another reason may be that dogfish (a type of small shark) have moved in and taken the place of cod in the food chain and are eating the marine life previously eaten by cod.

Overfishing

Scientists may calculate how many fish can safely be caught each year without seriously harming overall numbers and set target quotas for fishers, but this does not always mean that the people in governments who regulate fishing will listen. And even if they do, it is very hard to ensure that fishers follow the regulations. Unless fishing vessels and their catches are checked regularly, fishers may simply try to catch as many fish as possible.

In 1996 the World Conservation Union listed more than 100 species of commercially caught fish and shellfish as being threatened or endangered. In the late 1990s, the United Nations' Food and Agriculture Organization (FAO) estimated that 13 of the world's 17 major fishing grounds were being fished to the limit or were already overfished. Fish populations may never recover once fish stocks have dwindled to the point where adults that can breed are scarce and scattered. When stocks collapse, fishers either go out of business or go in search of new fishing grounds. These, too, then become overfished, and the problem grows.

Ocean Farming

About 10 percent by weight of all the seafood eaten in the world each year is produced in artificial conditions, on marine farms. This percentage is likely to grow while wild fish populations continue to dwindle in size.

The farming of marine life is known as mariculture. The Chinese have been farming fish, shellfish, and seaweeds for food for at least 3,000 years. Oyster farming also has a long history in many parts of the world—in this case, people were more interested in harvesting the valuable pearls that form naturally within the shells of many oysters than in rearing them for food.

In traditional mariculture, farmers rear fish in shallow ponds dug in coastal areas. The ponds are continually replenished with salt water from the incoming tide. Another method is to rear the fish in cages suspended in shallow water. The fish feed on the natural life in the water, but they are also provided with additional food—very often, waste products from agriculture such as rice or wheat husks.

Marine farmers grow shellfish, such as oysters and mussels, in baskets or on ropes or wooden frames in shallow water. Such farming methods rely on natural sources of stock animals. Young fish are placed in the ponds or cages until they grow bigger and are harvested. Young shellfish, taken from the wild, grow larger in the baskets or on the frames.

This shellfish farm on the Atlantic coast of France, near La Rochelle, grows oysters. Oysters and other shellfish—such as clams and mussels—are farmed throughout the world.

Farming for profit

Growing demand for many kinds of seafood has brought about a marked rise in commercial fish farming. This method of rearing fish is carried out under carefully controlled conditions—costly instruments regularly monitor the cleanliness, saltiness, and temperature of the water in which the fish are reared.

To make a profit, marine farmers must stock their tanks and ponds with very large numbers of young fish and shellfish such as shrimp, and bring them to maturity as quickly as possible (called intensive farming). They do this by feeding their stock a nutrient-rich feed to make them grow quickly.

Only a few species of fish can be reared intensively—most do not survive when kept in ponds or cages. One of the most commonly farmed fish are salmon, which are becoming scarce in the wild. Young salmon are reared in fresh water. Later, the farmers move the salmon to floating pens, or "cages," in the sea, usually in fjords and estuaries. In Asia and America, farmers grow intensively farmed shrimp and lobsters in tanks and saltwater ponds.

See also **vol 2:** 22–23, 30–31, 42–43; **vol 3:** 40–41, 42–43

Disease and pollution

When intensively farmed fish are kept in high numbers in comparatively small ponds or tanks diseases spread rapidly among them. To prevent this happening, fish farmers often supplement the fishes' food with antibiotics. Farmed salmon are prone to fish lice (below), parasites that live on the scales of fish and feed on their tissues. There is concern that lice from farmed salmon are now infecting stocks of wild fish. Another danger is pollution. Scientists and environmentalists in Scotland believe that water released from salmon cages contains high levels of ammonia and other wastes, which are polluting the lochs and bays where the fish are farmed.

Fish cages at a salmon farm in a Norwegian fjord. Large pens may contain several thousand individual fish. Normally, the salmon are harvested when they are about 7 to 12 lb (about 3–5 kg) in weight.

Ocean ranching

Ocean ranching is a way of rearing young fish in places such as in ponds, tanks, or in pens in the sea, where they have a much better chance of survival than in the wild. When large enough, the fish go back into the ocean in order to help build up natural stocks. This system is only economically practical if the released fish stay in the area, or return to it later, so that the people who have invested money in rearing them will be able to benefit from catching them.

Salmon return to their source river of their own accord, which makes them highly suitable for ranching. Ocean ranching with other types of fish has usually failed, because the fish migrate, only to be caught by fishers hundreds or even thousands of miles away from the original "ranch."

New developments

One of the biggest advances in mariculture involves altering the way a fish grows or behaves by changing its genetic makeup using modern scientific methods, known as genetic engineering. Marine farmers have altered the genes of some types of fish so that they put all their energy into gaining weight, rather than breeding. Genetic engineering is also being used to produce strains with better flavor, or strains which are more resistant to disease. Eventually a way may be found of growing fish that stay fresh longer after harvesting.

Some people object to farmers altering the natural characteristics of animals in this way. They claim that the technology will only profit people in the richest countries, who will use it to produce food more cheaply for themselves; it may also bring unforeseen consequences. But others maintain that the technology will eventually be of benefit to those in developing countries as well, by helping to increase the food supply.

UNDERSEA MINING

For thousands of years people have harvested the sea for salt, and taken sand and gravel from beaches. We are now mining beneath the ocean floor for petroleum oil and natural gas.

Petroleum oil and natural gas are the most valuable chemical resources we get from the sea. They are used to power ships, automobiles, and aircraft, to heat our homes and offices, and to generate electricity. Petroleum oil also lubricates industrial machinery. It is used as a raw material to produce many plastics and fabrics.

The flare on the left of this oil production rig, is burning off excess oil for safety reasons. A single rig may have a workforce of 400, who are taken to and from the rig by helicopter.

Oil and gas deposits lie trapped between layers of rock in the Earth's crust in many parts of the world, both on land and in offshore deposits beneath the ocean. People first began to extract oil by drilling through the rock in the mid-1800s. They exploited land-based oil deposits first.

Drilling for oil in the sea is much more expensive and difficult than on land. The first commercial offshore drilling operation began in the 1930s, in the Gulf of Mexico. As land-based oil reserves were used up, mining oil and gas beneath the seabed became increasingly common. In 1969 offshore oil accounted for about 13 percent of the world's oil

In shallow waters, oil and gas rigs stand on the seabed, on huge legs. In deeper waters they rest on giant floats called pontoons, but they are anchored to ensure they remain in place. Today, more than 4,000 oil or gas platforms around the world are operating in waters up to 4,000 ft (1,220 m) deep.

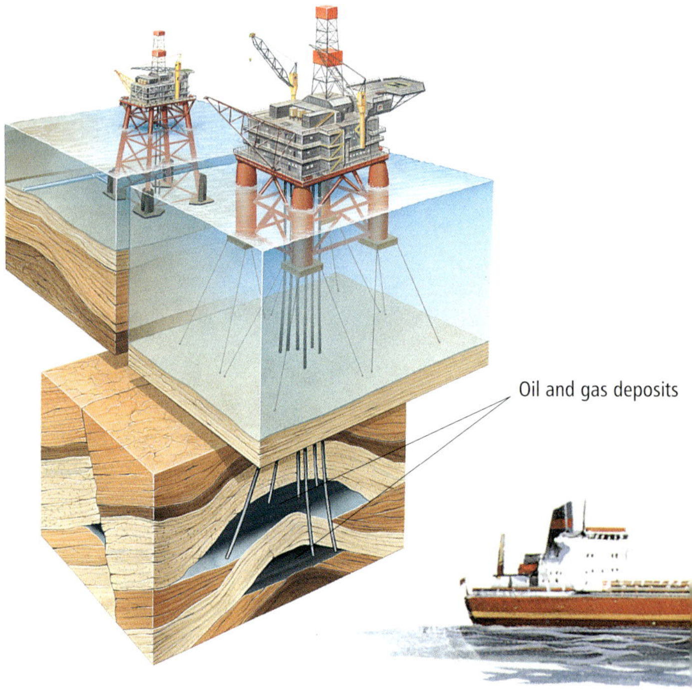

See also **vol 1:** 38–39, 40–41; **vol 3:** 50–51, 52–53

production. By the late 1990s, the figure was approaching 40 percent.

The oil companies send shock waves through the seabed to search for new offshore oil deposits. They create these by pounding the seabed with air-blasting equipment, or by making an explosion. These events set up a series of seismic shock waves, similar to those created by an earthquake. The signals sent back by the shock waves tell geologists whether the underlying rock is likely to contain oil or natural gas. Once a deposit is located, the oil company will use an exploratory drilling rig to see if it contains sufficient quantities of high-quality oil or gas to make setting up a production rig on the ocean floor worthwhile.

Oil and gas deposits frequently lie together under the seabed. Sometimes, natural gas pressure forces the oil or gas up to the rig through pipes bored into the rock. If there is not enough pressure, it may be necessary to pump the oil or gas to the surface. Underwater pipelines carry the oil or gas from the production rig to a terminal on land or to storage tanks on or near the seabed. From there, it travels through pipes to an oil refinery on land, where it is processed for use. Or the oil is piped into storage buoys and from there into supertankers to be shipped farther afield.

Other riches from the sea

We obtain other minerals and valuable metals from the sea. When waves, currents, and tides sift and move sediments of sand and gravel, they sometimes leave deposits of heavy metals and gemstones in hollows on the seabed. These are often called "placer" deposits because they are "placed" there by the action of water.

Miners excavate placer deposits both onshore and in waters up to about 150 ft (45 m) deep. Beaches in Alaska and Oregon yield gold, while Oregon's also produce chromite (for making chrome). The beaches and offshore waters of Namibia in southwest Africa are rich in diamonds.

Valuable minerals, such as the gemstone zircon, are mined at this coastal site in KwaZulu, South Africa. Mining is carried out by cranes called dredgers. Buckets attached to the dredgers scoop up sand containing the minerals. Once the minerals have been separated from the sand, the sand is replaced.

The future for minerals

More than 10 percent of the world's tin, in the form of the mineral cassiterite, is found in marine placer deposits. As mineral reserves on land become used up, more mining will take place beneath the sea—already the case with oil and gas.

At the moment phosphorite (used in agricultural fertilizers) is mined only on land. But massive deposits of phosphorite are known to lie in ocean sediment at depths of at least 3,000 ft (915 m). It may be worth excavating these deposits from the seabed at some time in the future.

Nodules containing the element manganese lie on the seabed at depths of more than 2.5 mi (4 km). Scientists believe the nodules were formed over thousands of years from the activities of deep-sea bacteria. One day it may be possible to suck the nodules up to the surface. However, environmentalists are concerned that we do not understand what role these nodules play in the living communities of the ocean. They argue that they must not be removed until we know more.

Deep-sea mining inevitably churns up the sediment on the sea floor. Some people believe that this disturbance will smother microscopic and larger forms of marine life that live on the seabed, which in turn will impact on the fish and other sea animals that feed on them.

This tanker is moored at a loading buoy. It takes on board oil piped to the loading buoy from the production rig.

Renewable Energy

Fossil fuels such as oil and natural gas may begin to run out by the middle of the 21st century. Other forms of energy will be needed to replace these vital resources. The ocean may be able to provide some of them.

Fossil fuels are "nonrenewable" forms of energy. Once extracted from the rock, they cannot be replaced. They have other disadvantages. When burned, oil and natural gas release gases into the atmosphere: carbon dioxide—a greenhouse gas that may contribute to global warming—and sulfur dioxide and nitrogen oxides, which cause acid rain.

Oil and gas are therefore "dirty" forms of energy. Nuclear power can be used to generate electricity more cleanly, but the Chernobyl nuclear disaster in 1986, when an explosion released high levels of radiation into the atmosphere, emphasized the potential dangers of nuclear power. There is also the problem of disposing safely of radioactive nuclear waste.

We need therefore to look for alternative sources of energy. Many people believe that the oceans could be an abundant source of "renewable" energy.

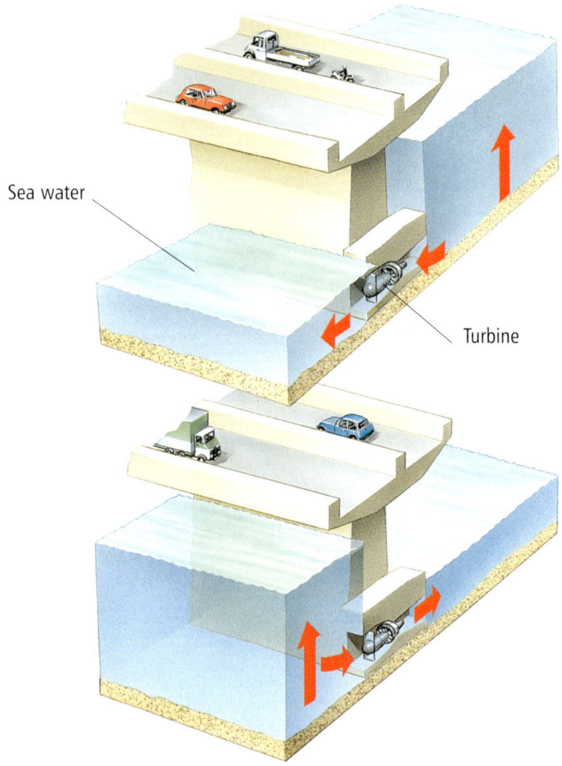

Tidal barrages keep the water level of the estuary low on one side and high on the other. Once the sea level has risen, gates are opened, and the water rushes through from one side to the other, powering turbines.

Sea rushing through the La Rance Tidal Barrage, in northwest France. The world's first successful tidal power station, built in 1966, it generates electricity for tens of thousands of homes.

Tidal power

Renewable energy never runs out. An example is the power produced by high tides, which usually occur twice every 24 hours. Tidal power stations generate electricity by making sea water flow through a dam (barrier) with gates, or sluices, in it. This is known as a barrage. As the water rushes through the sluices, it turns a turbine (a wheel with paddles or vanes). The rotating turbine generates electricity.

Tidal power stations are only practical where the difference between high and low tides (tidal range) is more than 25 ft (7.5 m). The first tidal power station was built in 1966

See also **vol 1:** 42–43, 54–55, 56–57; **vol 3:** 46–47, 60–61

on the La Rance estuary, in France, where the tidal range reaches 44 ft (13.5 m). Other stations operate in Nova Scotia, Canada (the only tidal power station in North America), in China, and in the former Soviet Union. Further stations are being considered for major sites in the United States and Britain.

Tidal power stations are a reliable, cheap, and clean source of electricity. They do have some disadvantages, however, including the high initial costs of construction and the disturbance caused to the marine environment.

Waves and currents

Wind-driven sea waves and ocean currents carry large amounts of kinetic energy (energy resulting from movement). In theory, the power from both could be used to generate electricity, but most proposals for doing so remain on the drawing board. One problem is that waves vary in height. They also change their size and shape with weather conditions and with the state of the tide. Any device sensitive enough to trap energy from medium-size waves, for example, could be swamped and damaged by storm waves.

One suggestion for harnessing wave power is to create a floating boom with sections that turn as the wave washes over them. The rotating sections generate electricity directly. Another proposal is to use waves to force air out of a chamber, or suck air into it. The moving air turns a turbine to generate electricity.

Ocean currents are also a potential source of power. According to one proposal, turbines beneath the sea off the Florida coast could capture some of the Gulf Stream's kinetic energy. About 240 turbines would generate 10,000 megawatts of energy a year—enough, in theory, to provide 10 percent of Florida's electricity needs. The Kuroshio Current, in the northwestern Pacific, is another possible site for ocean-current power schemes. However, there are still a great many technical difficulties to be overcome, and environmental problems to be assessed, before many of these ideas can be put into large-scale operation.

Warm water energy

Conventional power stations fueled by coal or gas heat water to produce steam that drives a turbine to generate electricity. Ocean thermal energy conversion (OTEC) schemes use tropical surface water as the source of energy, and they operate at much lower temperatures than conventional power stations. OTEC schemes take warm, surface water and use it to turn a liquid (usually ammonia) into a gas, which turns turbines to produce electricity. To recycle the gas, cold water from 3,300 ft (1,000 m) down in the ocean is pumped up through pipes alongside the ammonia gas. The water cools the gas and turns it back to liquid. OTEC designers must still overcome technical problems to make the process cheap enough to be economical, however.

Wind farms

Wind-powered turbines are a valuable source of clean, renewable energy that are already in use in many parts of the world. The turbines are sited close together to form wind farms. However, wind farms can be a nuisance. They take up a lot of room, make a lot of noise, and to some they are visually unpleasing. When possible, wind farms are sited in shallow coastal waters, well away from from shipping routes, where they are less of an inconvenience.

Wind farms are often sited in coastal areas, where the winds are likely to be stronger than on land, and where they do not block the view of local residents. These wind turbines are sited next to the sea at Zeebrugge, in Belgium.

Marine Pollution

Even the remotest parts of the ocean are polluted with the debris of human existence. Oil, sewage, and plastic litter are visible forms of pollution. More damaging are unseen chemicals that enter the sea.

On March 24, 1989 the supertanker *Exxon Valdez* ran aground in Prince William Sound, Alaska, spilling thousands of tons of oil into the sea. The accident resulted in the deaths of between 100,000 and 300,000 birds, 3,500 to 5,500 sea otters, and about 200 harbor seals.

Every year, fishing boats discard more than 100,000 tons (90,000 tonnes) of plastic fishing nets and lines in the sea. These drift about on or below the surface of the sea, and inevitably fish, birds, and marine mammals become caught in them, and die. Yachts, cruise ships, and ferries also dump trash such as plastic boxes, bottles, and food waste overboard, adding year by year to the tons of rubbish floating around on the waves. Many plastics can last for centuries before being broken down by bacteria and other living organisms.

Oil pollution

One of the most common, and deadliest, forms of pollution is oil from tankers that run into rocks and spill their cargo of oil into the ocean, or clean out their oil storage tanks at sea. Seepage from oil pipelines is another cause of pollution.

Thick crude oil floating on the surface of the sea causes direct harm to marine life. It clogs the feathers of birds, the gills of fish, and gets into the lungs and hair of seals and other mammals. Less thick or diluted oil kills phytoplankton, zooplankton, and fish eggs. Although bacteria and fungi eventually break down oil, enabling wildlife communities to re-establish themselves, the process may take months or even years.

Fortunately supertankers do not run into rocks very often. But because these supertankers carry such vast amounts of oil, when an accident of this kind does occur, it causes immense local damage to the environment and also kills wildlife.

Attempts are made to trace polluters and make them pay. Remote sensing by satellite can spot ships that are discharging oil. In addition, each cargo of oil contains a "fingerprint" in the form of an identifiable mixture of chemicals. So when an oil spill is found, it can be traced back to a particular ship or coastal refinery. But the authorities prosecute only the worst offenders.

This outlet pipe is discharging industrial waste (effluent) from a nearby factory directly onto a beach.

See also **vol 3:** 20–21, 46–47, 58–59, 60–61

An aircraft sprays an oil slick with special chemicals that break the oil up. Although the chemicals disperse the oil in the short term, they also kill the bacteria that would break down the oil naturally.

In 1979 an oil rig in the Gulf of Mexico accidentally released about 1.1 million tons (990,000 tonnes) of crude oil. This was the single biggest oil pollution incident in history. Because of the potential harm to people and wildlife from spills, oil and gas extraction is now carried out with much greater care.

Poisoning the sea

In many parts of the world, human sewage is emptied directly into the sea, polluting beaches used by tourists. Such waste may contain bacteria that cause human diseases such as cholera. Sewage and water that run into the sea off agricultural land are high in nitrates, which cause seaweed and phytoplankton to grow in an uncontrolled way.

In the late 1970s, a blanket of algae was strangling coral reefs in Kaneohe Bay, Hawaii. Sewage-cleaning projects transformed the bay, and within a few years the coral began to recover. But the problem returned in the 1990s because the levels of sewage flowing into the bay were allowed to rise once more.

Tiny amounts of what are called "heavy metals"—substances such as mercury, lead, and cadmium—enter the sea from rivers, sewage outlets, and in industrial waste discharged from factories and industrial sites. These metals are highly poisonous to most forms of life. Governments in most countries recognize the threat to human and environmental health from the release of poisonous metals. They enforce laws to stop companies discharging waste water contaminated with harmful metals.

Between 1953 and 1956, the Japanese town of Minamata was the site of an ecological catastrophe. A local plastics factory was releasing waste containing amounts of a poisonous substance called dimethyl mercury into the sea. The dimethyl mercury became concentrated in the bodies of minute sea organisms and, in a process called biomagnification, became more and more lethal as it passed up the food chain. The fish caught and eaten by local people contained very high doses of dimethyl mercury. Dozens of people died, and many hundreds developed permanent paralysis, tremors, and severe brain damage.

Most countries banned the synthetic pesticide DDT over 20 years ago. Yet a recent study of antarctic penguins showed that they contain detectable amounts of DDT, which must have traveled thousands of miles to arrive there.

HABITAT DESTRUCTION

A habitat is the place where an animal or plant lives. When humans alter the landscape or seascape they destroy the habitats of many creatures.

Much new development in Singapore, an island state in Southeast Asia, is on land reclaimed from the sea by extensive drainage and infill schemes. This process has drastically altered the natural environment.

About two-thirds of the world's people live within 100 mi (160 km) of the sea. Some of our largest cities are situated on the coast, or on coastal bays or estuaries. When builders create or enlarge a coastal city, they first make seawalls and other sea defenses to prevent erosion and flooding, and then they drain the wetland areas to make the ground firm and dry in order to build strong foundations. In so doing, they destroy marine habitats such as sandy beaches, rocky shores, mudflats, and saltmarshes, which are home to countless animals and plants.

The building of large industrial or tourist complexes along coasts affects the runoff of water from the land into the sea. More water, often containing high levels of industrial and human waste, enters the sea through drainage outlets. These changes affect the saltiness, cloudiness, and temperature of sea water, with damaging results for wildlife.

Mangroves

Mangrove forests once covered about three-quarters of the coastlines of tropical and subtropical countries. Mangrove trees have branching roots that grow out of the water and shelter tiny marine creatures. Today, only half of the original area of mangrove remains. People have felled the forests for timber, or to make room for housing, industry, and tourism.

In Florida, more than one-third of the mangroves in Tampa Bay, and two-thirds of those around Lake Worth, have been lost. Many of Florida's wading birds nest in mangrove trees. The disappearance of their habitat has caused the numbers of wading birds to fall by 75 percent.

Coral reefs and seamounts

People love to snorkel and scuba dive around coral reefs. A careless move or stumble by a diver can destroy a large area of coral. Unscrupulous traders sell pieces of coral as tourist souvenirs. Experts estimate that two-thirds of the damage to coral reefs in the Red Sea is caused by tourism.

At sea, fishers disrupt the seabed by using dredges and trawl nets to catch bottom-living fish and shellfish. Their

See also **vol 2:** 52–53, 54–55; **vol 3:** 54–55

nets scoop up everything in their path. The sediment stirred up by the nets smothers many tiny creatures living on the sea floor.

Seamounts are steep, underwater mountains formed by volcanic action along mid-ocean ridges or at hotspots beneath the seabed. They rise at least 3,300 ft (1,000 m) above the surrounding area of the sea floor. There are between 30,000 and 50,000 seamounts around the world.

The tops of seamounts are often encrusted with soft corals. They are a rich habitat for a wide variety of bottom-living creatures, including sponges, sea anemones, crabs, and sea cucumbers (spiny, cucumber-shaped animals). Harder corals flourish where the sea is shallower and warmer.

Fish are attracted to seamounts. And wherever there are fish, fishers are likely to follow. As they drag their nets across the underwater "mountain tops," the fishers damage the coral and the sealife that lives within the coral. Studies show that unfished seamounts contain up to 50 percent more species than those that are fished.

Just as islands often contain unique species of land plants and animals, so seamounts contain unusually high numbers of marine species that are unique to a particular area. Some seamount species may already have been lost for ever.

Divers swim over an Australian reef. The delicate organisms that make up a reef may be damaged by careless divers and snorkelers.

Alien invaders

Modern ships use water taken from the sea as ballast to adjust their buoyancy and maintain stability while at sea. Before a ship leaves port, it takes on this ballast water. When a ship reaches its destination to unload its cargo, it pumps it out again. This means that sea water teeming with minute organisms is transported thousands of miles across the world before being dumped close to shore in a "foreign" land. Problems arise when "alien invaders" within the ballast water upset the local environment.

In the 1970s a jellyfishlike creature, the American comb jelly, entered the Black Sea, in eastern Europe, in ballast water. In little more than a decade this predator had multiplied to such an extent that it had consumed much of the Black Sea's zooplankton. Shoals of anchovies in the Black Sea once supported a thriving fishing industry. But without zooplankton, the anchovies were starved of food. By 1991, Russia's anchovy fishing fleet in the north of the Black Sea had almost nothing left to catch.

Once changes like this occur, they are difficult, if not impossible, to put right. The International Maritime Organization (IMO) is planning to introduce international regulations on the discharge and treatment of ballast water. These rules could include treating ballast water with eco-friendly chemicals, or bombarding it with ultraviolet light, to kill the organisms it contains.

European zebra mussels invaded North America's Great Lakes in the 1980s. They probably first arrived as tiny, young mussels in the ballast water of European ships. The zebra mussels are competing with local species of shellfish and causing their numbers to be reduced.

Marine Tourism

The sea and the sea shore attract growing numbers of visitors each year. They come to swim and surf in the sea, fish from piers or boats, sail craft, scuba dive, or observe marine animals such as whales in their natural habitat.

Within the last 30 years, marine tourism has become a highly organized industry. People in coastal communities rely on the income from tourists for their livelihoods. Tourism contributes vitally to the economy of many poorer countries from the Caribbean islands to the Indian Ocean and the South Pacific.

We all dream of finding our island in the sun. But while marine tourism gives great pleasure to millions each year, it also carries dangers. Unless tourism is properly managed, the things that attract people in the first place, such as clean sandy beaches or vibrant coral reefs, will be destroyed.

A delicate balancing act

Wherever there is unspoiled coastline, there is potential for tourism. Coastal resorts affect the marine environment in many ways—new airports, hotels, and roads bring problems of litter, sewage disposal, and pollution of sea water. The challenge for developers is to strike the right balance between attracting large numbers of visitors and preserving the coastline, and its wildlife, for future generations to enjoy.

An example of the explosion in tourist numbers can be seen at the seaside resort of Hurghada, Egypt. Twenty years ago, there were fewer than six hotels in what was then a fishing village on the shore of the Red Sea. Hardly anyone visited the nearby coral reefs. Now there are more than a hundred hotels, and every day fleets of boats take tourists to dive and snorkel around the reefs. This experience is being repeated in tourist resorts around the world.

If people are to go on enjoying coral reefs, there must be regulated access to them. One way of doing this is to make the reef a marine protected area (MPA), as is the case with the Great Barrier Reef, off the coast of Queensland, Australia. The Great Barrier Reef, created by living organisms, is the biggest recognizable structure in the ocean. It is about 1,250 mi (2,010 km) long and is home to many thousands of species. Access to some parts of the reef is completely prohibited except for scientific research. Tourists are able to visit less vulnerable areas—"general use" and "look don't touch" areas—where they can observe the brilliant colors of the coral and its marine life by snorkeling, scuba diving, or from glass-bottomed boats.

Tourists watch a gray whale from a small boat. When people see creatures in this way they learn to respect them and their environment.

Under the sea

The world beneath the sea was once a place where only a few scientists ventured. Now it is becoming accessible to thousands of people. More than a dozen tourist submarines operate across the world, taking visitors to depths of more than 150 ft (45 m) and giving them the opportunity to encounter underwater life in the wild.

In Western Australia, diving operators allow wildlife tourists to snorkel in the ocean with a type of shark, the harmless, plankton-eating whale shark—the largest fish in

See also **vol 3:** 56–57, 60–61

Cruise ships

Cruise ships are an increasingly popular way for people to experience the marine environment. Many cruise ships are like floating hotels, complete with luxury cabins, restaurants and bars, swimming pools, shops, and theaters. The biggest carry thousands of passengers and have crews of many hundreds.

Most cruises take vacationers in search of the sun, island-hopping in the Caribbean, for example. But some go to more demanding destinations, such as Alaska or the Antarctic. They offer a chance for people to spend many days at sea learning about whales and other wildlife. The opportunity to sail among a school of whales as they come up to the surface to blow is a never-to-be-forgotten experience.

the world at 40 ft (12 m) in length. The chance to do so attracts people from all over the world, and the local tourist industry benefits.

More importantly, the whale sharks benefit as well. When an international ban was passed against large whale hunting a few years ago, some whalers turned to hunting whale sharks instead. The fish's numbers began to fall. But now tourist centers in parts of Southeast Asia have caught on to the idea of making whale sharks a wildlife tourist attraction.

Hunting whale sharks for meat makes money for a few people, but only as long as stocks last. Live whale sharks can provide an unforgettable wildlife adventure year after year for hundreds of tourists, all of whom will pay for the experience. This way, a live whale shark becomes worth much more to the local economy than a dead one.

Environmental education

Often, it is ignorance and carelessness that harms marine life. For example, every year in Florida dozens of manatees—large, slow-moving marine mammals—are struck by power boats and jetskis that speed across the waters where these creatures live. The manatees suffer great injuries, and many die. Manatees are already threatened by loss of habitat as dredging and pollution destroy the seagrass meadows in which they graze.

Marine protected areas (MPAs) play an important role in protecting wildlife and educating the public in ways of conserving the sea and its resources. Tourists can be made to pay some of the costs of protection. For example, all visitors to the small Central American country of Belize are charged a small tax that pays for wildlife conservation. Many other countries are adopting similar schemes.

Large, panoramic aquariums, like the one shown here, give tourists the chance to see many marine animals in an environment similar to the one they inhabit in the wild. Visitors experience the world of the ocean almost at first hand and learn about its different living communities.

PROTECTING THE OCEANS

Conservation of the marine environment is concerned with maintaining wildlife habitats and with ensuring that plant and animal life is not over-exploited or needlessly destroyed.

Scientists use the term biodiversity, or biological diversity, to describe the rich variety of plant and animal life on Earth. The greater the biodiversity, the more species there are, and the more habitats in which communities of creatures can live.

Importance of biodiversity

Marine ecosystems—communities of plant and animal life in their natural habitats—provide benefits we take for granted. For example, a saltmarsh naturally cleans the water that passes through it. The mud particles trap harmful substances such as lead. Marsh grasses also absorb nitrate fertilizers that wash into the sea from rivers and cause algal blooms (the uncontrolled growth of phytoplankton). Without the cleansing effect of the saltmarshes, water that enters the sea is less pure, and this can damage the local marine ecosystems.

Phytoplankton in the ocean help to clean the atmosphere. These minute plants remove carbon dioxide from the air and replace it with the oxygen that we, and other animals, breathe.

If left undisturbed, ecosystems achieve their own natural balance. Interference from people can upset this balance, often with unforeseen consequences. If we reduce biodiversity by destroying habitats or causing animal and plant species to

Endangered marine animals

The World Conservation Union (WCU) lists more than 200 species of marine vertebrates as Vulnerable or Endangered. For example:

- **31 species of seahorse are vulnerable to pollution, habitat loss, and collectors—dealers sell dried seahorse bodies as souvenirs or for grinding up in Chinese medicines.**
- **Right whales have been hunted in the North Atlantic for hundreds of years. Since becoming protected in 1935, their numbers have failed to recover. There are probably only 300 to 400 remaining. Pollution, inbreeding, collisions with ships, and fishing nets continue to endanger these slow-swimming whales.**
- **The vaquita is a small porpoise that lives in the northern part of the Gulf of California, Mexico. Pollution, entanglement in fishing nets, and illegal hunting all threaten its survival. There are probably less than 500 left.**

See also **vol 3:** 58–59, 60–61

A Marine Park sign, Western Australia, asks visitors to make brief landings only. This ensures that habitats are not unduly disturbed.

die out, then we leave a planet with fewer natural wonders for future generations to enjoy. Even more crucially, the oceans are a storehouse of medicinal riches. In the last 20 years, dozens of chemicals have been found in marine organisms, ranging from anticancer drugs to natural insecticides. But if our actions make species extinct, then such important chemicals may be lost forever.

Threats to conservation

There are five main threats to marine biodiversity. The first is large-scale commercial fishing, by which fish stocks are depleted. The second is chemical pollution—from land, fresh water, and air—that finds its way into the ocean. The third concerns changes to marine habitats, caused by factors such as fishing trawls being dragged across the seabed, or coastal developments. The fourth is invasion by "alien" species. The fifth is global climate change, where "greenhouse gases" are believed to be causing environmental temperatures to rise and weather patterns to alter.

Marine protected areas

One way to help preserve marine biodiversity is to make an area where a particular species or habitat is under threat a marine protected area (MPA). There are at present more than 1,300 MPAs throughout the world. The largest is the Great Barrier Reef Marine Park, off the coast of Queensland, Australia. It is also one of the oldest and best managed. The public are kept out of many areas of the reef, and other areas close in the breeding season to protect turtles and seabirds. Even so, many parts of the reef have been damaged by pollution, and by fishing and boating activities. There is also a threat that oil companies may begin offshore oil drilling near the reef area.

Many MPAs are very small, and funds are not always available to manage them efficiently. Some countries cannot even protect their MPAs against fishing. Other MPAs may provide protection against some activities—such as mining or oil exploration—but not others, such as netting. There is another drawback. An MPA may not contain all the feeding, breeding, and nursery areas of the species it has been set up to protect. So an animal may spend the early part of its life cycle within an MPA, but then move outside to breed, when it falls victim to fishers.

Much more needs to be done to extend marine protection. Now, in the early 21st century, there are still only 13 marine sanctuaries in the United States, covering just 1 percent of U.S. coastal waters. Environmentalists are increasing pressure to create new MPAs and improve conservation measures. A recent move was made to combine and extend the MPAs in the Florida Keys area to create the Florida Keys National Marine Sanctuary.

The environmental group Greenpeace was one of the first organizations to call attention to the plight of the world's oceans. Its vessel, also named *Greenpeace*, keeps a check on the movement of nuclear waste, illegal whaling, dumping at sea, and other harmful activities.

MANAGING THE OCEANS

For centuries, nations have disputed who has the right to fish in their territorial waters. The Law of the Sea helps to regulate which countries have use of which parts of the ocean and what their responsibilities are.

Fishing disputes

Some of the most recent fishing disputes are listed below.

- **In the early 1970s, Iceland extended her exclusive fishing zone to 50 mi (80 km), disregarding a ruling from the World Court. British trawlers went ahead and fished within this area, sparking off the so-called "cod war." British trawlers were sunk, and in 1973 an Icelandic gunboat engineer died after a collision with a British vessel.**
- **In March 1995 Canadian fisheries officers seized the Spanish trawler *Estai* while it was fishing on the Grand Banks off the coast of Newfoundland. They claimed the fishers were catching more turbot than the law allowed.**
- **In 1998 Yemeni forces captured five Egyptian fishing boats that entered Yemeni waters without permission.**

In the early 1600s, Dutch politician Hugo Grotius suggested that the seas far from land—the high seas—were open to anyone who could reach them. This idea of "freedom of the high seas" suited the European nations. They wanted to expand their trade routes, especially to the Far East. At about the same time, these nations began claiming local coastal waters as their own. They called the coastal waters "territorial waters." Territorial waters extended out from the coastline for 3 nautical mi (5.5 km). This was the distance that could be protected by cannon fire. Nations fished in their own territorial waters (and on the high seas), but were not allowed to fish in another country's territorial waters unless they had been given permission.

These two traditions (freedom of the high seas for all and territorial waters for one nation) remained mostly unchanged until the 1950s. Then countries that wanted larger fishing grounds began to claim rights to increased territorial waters. Different countries set different limits. This is one of the reasons why representatives from many countries met to agree on a Law of the Sea.

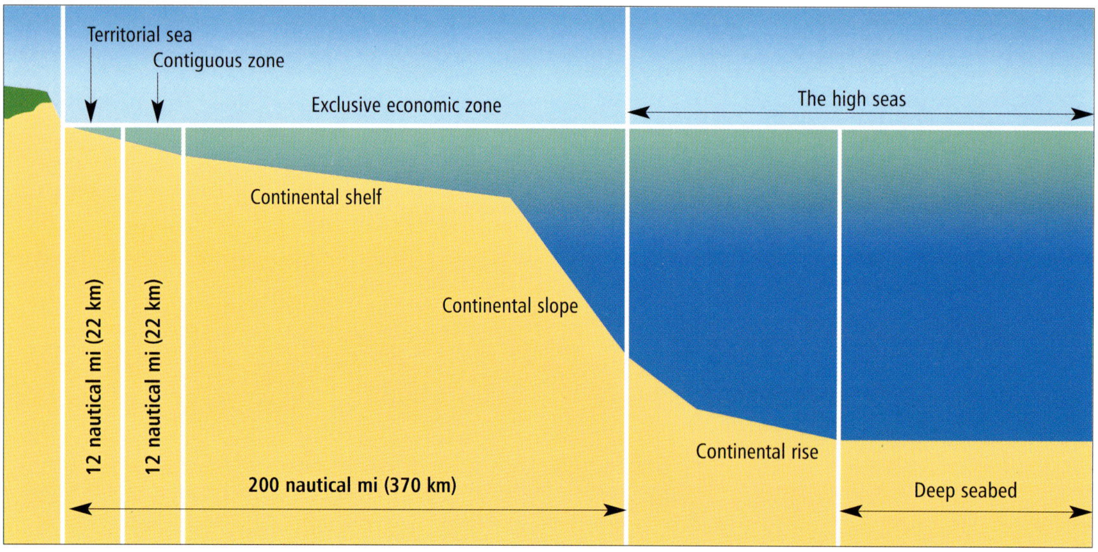

Coastal nations have a "territorial sea" that extends 12 nautical mi (22 km) from shore. In addition, they can fish, or drill for minerals in an "exclusive economic zone" that extends up to 200 nautical mi (370 km) from shore. Beyond that are "the high seas."

See also **vol 3:** 50–51, 52–53, 60–61

Modern fishing vessels, like these Japanese ships, are overfishing the oceans—hence the need for internationally agreed rules for the management of fisheries.

The Law of the Sea

After many years of discussion, the Law of the Sea became international law in 1994. It aims to be fair to nations of different sizes, to ensure that fishing fleets, navies, and those mining for offshore oil and gas deposits, can go about their business lawfully. The Law of the Sea also ended years of disputes between countries claiming different territorial limits.

Two of the Law of the Sea's many rules are concerned with territorial waters and with the deep seabed. The first of these states that maritime countries have an "exclusive economic zone" (EEZ). Within this zone, nations control commercial activities—fishing, drilling for oil, and so on. They are also expected to look after and protect these resources.

Further out to sea, the deep seabed, lying beyond the shallower waters of the continental shelf, is called "The Area," and is "the common heritage of humankind." If anyone wants to mine its mineral riches, that person must get permission from the International Seabed Authority.

Managing fisheries

Some countries that police fishing in their EEZs (others do not) use remote sensing by satellite for this task. Countries of the European Union (EU) use this method as a means of looking for illegal activity, such as fishing in the wrong place at the wrong time.

Managing a fishing ground so that it is not overfished can involve establishing quotas (in other words, laying down rules about what can be brought ashore). The rules are concerned with the type of species caught, minimum sizes, maximum numbers, and the seasons of the year in which different fish species can be caught.

Rules regarding sizes, unfortunately, do not work well in practice. Too-small fish must not be landed, but there is no way a trawl net can prevent them being caught in the first place. The regulations state that small fish must be thrown back in the sea, but most of them are already dead.

An alternative to quotas is a ban on all fishing within an area called a "no take" zone. In this area, fish can grow larger and produce plenty of larvae. Some of the larvae drift outside the zone and boost fish stocks elsewhere. This approach can work. When tried in Kenya fishers had a reduced area in which to fish but the overall weight of their catch increased.

Eliminating pollution

The United Nations Environment Program (UNEP) is seeking to find ways to monitor the oceans. For example, UNEP finds areas affected by pollution or overfishing and then arranges for scientists, politicians, and planners from different countries to work together to solve the problems. Progress is slow, and pollution remains in the Black Sea, Mediterranean, Caribbean, and elsewhere. Countries often take action only when pollution is visible (like oil) or poses an obvious threat to health (like sewage). It is harder to reach agreement on "invisible" pollution, like acid rain.

Prospects for the Future

What will the vast oceans of our planet be like in 20 years? The answer depends on us. We can all make a difference to the health and well-being of marine life and the habitats in which they live.

Use of the sea will increase over the next 20 years. More and more of the shallow continental shelf next to crowded coastal cities will be reclaimed for building projects. Already Hong Kong's new airport is built on reclaimed land, and in Tokyo developers have created several artificial islands on which to build offices and housing.

A hotel in Florida boasts unique "ocean views." The hotel is under water, and the views from the bedrooms are of marine life. The world's most expensive properties could soon be built on floating islands. They could even be towed from one place to another, to make the most of the seasons.

An even more advanced possibility is the "growing" of buildings under water. Some engineers believe that by using electricity generated from solar energy, concretelike walls and ceilings could be formed from chemicals that exist naturally in the sea. This would be similar to the way in which hard corals lay down a chalky skeleton as "concrete," but on a much bigger scale.

It is possible for us to protect the marine environment, and to put right some of the damage we have already caused. But if we fail to act, then future building developments,

What you can do

- Clear up beach litter and do not leave any of your own.
- Join environmental groups that campaign to protect marine wildlife and organize action to make the seas cleaner.
- Ask political representatives what they are doing to improve the quality of our seas. What are their views on global warming? Or on the burning of fossil fuels?
- Find out where your seafood comes from. Is it caught or farmed in a way that does not harm the environment?
- Visit aquariums or sealife centers and find out what they are doing to conserve marine creatures. If they keep marine mammals, do they have plans to return them to the wild?
- Find out more by reading *Exploring the Ocean*, and by researching the books and websites listed in Volume 4.

Nature is quick to improvize on opportunities when it can. The wreck of the *Duane*, a cutter (type of sailing boat) scuttled in Key Largo, Florida, in 1987, makes an acceptable reef habitat for marine animals.

See also **vol 1:** 60–61

whether they are on, under, or beside the sea, are likely to increase pollution and destroy even more habitats.

Global warming

Many people think that global warming is the greatest short-term threat to the planet—to life on land and in the sea. It might be caused by the uncontrolled burning of fossil fuels adding too much carbon dioxide to the atmosphere, trapping infrared rays (heat energy emitted from the Earth), and causing the atmosphere to heat up. Cutting back on the burning of fossil fuels, such as coal, oil, and natural gas, may be the only way to solve the problem, but this would require people to use environmentally friendly forms of energy such as solar power to heat their homes, and to use their automobiles less. Today, the United States produces about 25 percent of the world's industrial greenhouse gases—the gases that scientists think contribute to global warming.

Others claim that global warming is not happening or if it is, a period of warming will soon be followed, naturally, by a period of cooling. But if global warming is happening, it will change ocean currents, and weather and climate patterns, in ways we cannot predict. If the seas warm, and sea levels rise by only 20 in (50 cm) this century—as many scientists suspect they will—then low-lying regions of the world, like parts of Florida, the Maldives (in the Indian Ocean), and Bangladesh, would experience severe flooding.

Taking action

There is much we still have to learn about the oceans. We do not fully understand how different aspects of the ocean—the climate, ocean currents, and marine life—affect each another. We must learn how to manage the stocks of important species of fish, such as the ones we rely on for food. For example, North Atlantic cod were once so abundant that supplies were thought to be inexhaustible. This species is now endangered.

Scientists are working together across the globe to raise our understanding. But everyone can play a part. At home, in school or college, and at the seaside, choices can be made that reduce waste, lower energy consumption, and cut down pollution. Everyone can set a good example to others and try to ensure their actions are leaving a better world for the next generation, rather than a worse one.

Cleaner oceans, and less pollution, will lead to cleaner beaches and a more healthy and enjoyable environment.

INDEX

Page numbers in *italics* refer to illustrations

Aborigines 6
acid rain 48, 59
Africa, sailing around 9
Africans, slaves 7
Agassiz, Alexander *26*, 27
Agassiz, Louis 27
aircraft carriers 4, 19, *19*
Alaska 6, 7, 12, 13, *21*, 55
Albert, Prince, of Monaco 27
algae 51, 56
algal bloom 56
Alvin (submersible) *33*, 34, 37
amber 9
Amundsen, Roald 12
anchovies 41, 42, *42*, 53
Antarctic Circle 13
Antarctica 13, 55
aqualung 31, *31*
aquariums *55*
Aquarius (underwater laboratory) 35
Arabian Gulf 5
Arabs 10
archeologists, marine 36
Arctic 25
Arctic Circle 13
Argo II (ROV) 34
Aristotle 24, 30
astrolabe 11, *11*
Atlantic, seabed mapping 28
Atlantis 16
Atlantis (research ship) 34
atomic weapons 19
Australia 6, 13, *14*
autonomous underwater vehicles (AUVs) 33
Aztecs 7

Baffin, William 12
Baffin Bay 22
Baffin Island 12
Bahamas 11, 20
baleen 38
Ballard, Robert D. 37
ballast water, organisms in 53
Bangladesh 61
barnacles 25
barrages, tidal 48–9, *48*
Barton, Otis 33
bathyscaphe 33
bathysphere 33
battleships 19
beaches
  pollution 51
  for tourists 54
  vacationers on *4*
Beagle, HMS 25, *25*
Beaufort Sea 12
Beebe, William 33
"bends" 30, 35
Bering Sea 7, 38
Bering Strait 12
Bering, Vitus 12
Bermuda Triangle *16–17*, 17
biodiversity 56–7
biology, marine 24, 25
birds
  harm caused to by oil 50
  seabirds 57
    threat to 5, 52
Black Sea 10, 53, 59
blubber 38
bottom fishing 41
boums 10
Boyle, Robert 24
Brunel, Isambard Kingdom 15
buildings, underwater 60
buoys, marker 23
Bushnell, David 32

Cabot, John 12
Canada
  fishing rights 58
  inhabitants 7
canoes
  double-hulled 6
  hollowed-out trunks *7*
caravels 14
carbon dioxide 48, 61
cargo 20–1
cargo ships 4, *8–9*, 9, 19
Caribbean 11, 59
Carpenter, William 26
cassiterite 47
Challenger, HMS 26–7
Champlain, Samuel de 7
charts, marine 24
chemicals
  medical 57
  pollution from 50, 57
Chernobyl 48
Chinese
  explorers 10
  farming fish 44
  inventions 11
  migration to U.S. 7
cholera 51
chronometers 12, *12*, 22, 24
cities, coastal, effect on habitats 52
clams
  catching 41
  threats to 5
climate, changes 57
clippers 14, *14*, 15
coal 19, 61
coastal resorts 52, 54
cod 27, *42*
  dwindling numbers 43, 61
Columbus, Christopher 11, 14
comb jellies 53
compass 11, 22, *22*
  gyroscopic 22, *22*
computers
  controlling thrusters 34
  use in navigation 23
conservation 56–7
container ships 4, 20–1, *20*
continental drift 28
convoys 19
Cook, Captain James 13, 22, 24
copper 9
coral atolls 25
coral reefs *52–3*
  research into 35
  strangled 51
  threats to 52
  tourism and 54
corals
  bleaching 35
  threats to 5, 52, 53
Cousteau, Jacques 31
crabs 53
  catching 41
cranes *see* derricks
Crete 16
cruise ships *21*, 55
cruisers 19
CTD sampling device 28–9
currents, identifying source of 9
  ocean, generating power by 49

Darwin, Charles 25
DDT 51
decompression sickness 30, 35
depth charges 19
derricks 21
destroyers 19, *23*
diamonds 47
diesel engines 15
dimethyl mercury 51
diving 30–1, *30–1*
  tourism and *52–3*, 54–5
diving bell 30
diving suits 30–1, *30–1*
dockers 21
dogfish 43
dolphins, in fishing nets 41
Dreadnought, HMS 19
dreadnoughts 18–19
Drebbel, Cornelis 32
dredging 25, 26, 52
drift nets 41, *41*
drilling ships 28
drugs, in marine organisms 57
Dutch East India Trading Company 14

Easter Island 6
echo sounding *see* sonar
ecosystems, marine 56–7
Egypt 54
Egyptians, ocean-going ships 8, *8*
El Niño 29
electricity
  by ocean currents 49
  by tidal power 48–9, *48*
  by wave power 49
  by wind power 49, *49*
  ocean thermal energy conversion (OTEC) 49
electronic charts 23
Endeavour, HMS 13, *13*
energy, renewable 48–9, 61
English settlers, in North America 7
Erik the Red 10
Eriksson, Lief 10
evolution, theory of 25
exclusive economic zone (EEZ) *58*, 59
explorers, voyages 8–13, *10*, *11*
Exxon Valdez 50

factory ships 39
"famine ships" 7
ferries 20
Fiji 6
fish
  as food 40
  genetic engineering 45
  harm caused to, by oil 50
  ocean farming 44–5, *44*, *45*
fishing
  catches 5
  commercial 40–1, 57
  disputes over 58
  dwindling stocks 42–3
  excessive 27
  habitat damage caused by 53
  locating shoals 40–1
  methods 40–1
  for pearls *31*
  policing 59
  preservation of stocks 34, 42
  quotas 59
fishing nets *40*, 41
fishing vessels 4–5, *59*
Fitzroy, Robert 25
FLIP (floating instrument platform) 34–5, *35*
flooding 61
Food and Agriculture Organization (FAO) 43
food chain 40, 51
fossil fuels 5, 48, 61
Fram (steam sailing ship) *13*
Franklin, Sir John 12
Frobisher, Martin 12
Fulton, Robert 15, 32

Gagnan, Emile 31
galleys 18
Gama, Vasco da 11, 14
gas
  drilling for 46–7, *46*
  greenhouse 48
  land-based deposits 46
  natural 61
gas platforms *46*
genetic engineering 45
ghost ships 17
gill nets 41
Gjöa (steam sailing ship) 12
global positioning system (GPS) *23*, 41
global warming 5, 29, 34, 48, 61

Globtik Tokyo 21
Glomar Challenger 28
GLORIA (geological long range inclined asdic) *34*
gold 47
  prospecting for 14
  trade in 8, 9
Grand Banks, Newfoundland 37, 58
Great Barrier Reef 54
Great Barrier Reef Marine Park 57
Great Britain 14–15, 15
Greece 16, 20
Greeks 4, 9, 18
  migration to U.S. 7
greenhouse gases 48, 57
Greenland 10, *39*
Greenpeace *56–7*
Greenwich Mean Time (GMT) 12
Greenwich Meridian 22
Gulf Stream 49
guns, used at sea 18–19
gyrocompass 22, *22*

habitats, destruction 52–3, 54–5
haddock, dwindling numbers 43
Halley, Edmond 30
harpoons 39
Hawaiian Islands 6, 13, 34
He, Admiral Zheng 10
heavy metals, pollution from 51
Heezen, Bruce 28
Henry the Navigator 11
Henry VIII of England 36
herring 27, 41, *41*, 42, *42*
Heyerdahl, Thor 8
high seas 58, *58*
highways, oceans as *5*, 6–7, *6*
Hong Kong *11*, 20, 60
Hudson, Henry 12
Hudson Bay 12
humans, migrations 6–7, *6, 7*
hunting, of sea mammals 38–9, *38, 39*
hydrothermal vents 33

ice ages 6, 27
Iceland, fishing rights 58
Incas 7
incense 8
India 10, 11
Indian Ocean 10, 11
Indonesia 6, 10
industrial waste, pollution from *50–1*, 51
insecticides 57
International Council for the Exploration of the Sea (ICES) 27
International Marine Organization (IMO) 53
International Seabed Authority 59
International Whaling Commission (IWC) 38, 39
Inuit 7, *25*, 39, *39*
Irish, migrants 7
iron, cargo of 19
islands, Pacific, people of 6–7
Italians, migration to U.S. 7
ivory 8

Jahre Viking 21
Japan 10
Japanese, migration to U.S. 7
Jason (ROV) 34
jewelry 9, 14
Jews, migration to U.S. 7
junks 10, *11*

Kaneohe Bay, Hawaii 51
kinetic energy 49
kraken 16
krill 16
Kuroshio Current 49

La Jolla, California 27, 34
Labrador 10
land, signs of at sea 6, 9
land bridges 6, 7
land reclamation 60
latitude 11
Law of the Sea 58–9
lead 51, 56
Lebanon 8
Liberia 20
lice, fish 45, *45*
lighthouses 23
lightships 23
lobsters, farming 44
lodestone 11
longitude 11, 12, 22, 24
longships 10
Luxor 8

mackerel 41
Magellan, Ferdinand 11, 12, 14
magnetite 11
Maldives 61
mammals
  harm caused to, by oil 50
  hunting 38–9
  threat to 5
manatees 55
manganese 47
mangroves 52
mapping 27
Mariana Trench 27, 33
mariculture 44–5, *44, 45*
marinas 5
marine life *24*
  studying 24–5
marine protected areas (MPAs) 54, 55, 57
Mary Celeste 17
Mary Rose 36, *36*
Massachusetts 7
Maury, Matthew 24
Mayans 7
medicines 57
Mediterranean 8, 9, 10, 15, 59
megamouth 16
merchant vessels 20
mercury 51
metals 47
  heavy, pollution from 51
Meteor (research vessel) 27
Mexico, Gulf of 5, 46, 51
Mid-Atlantic Ridge 28
migrations, human 6–7, *6, 7*
Minamata, Japan 51
minerals 47
  placer deposits 47
mining, undersea 46–7, *46,*
Minoan civilization 16
Mississippi River 20
monsters, sea 16–17
mother-of-pearl shells 30
mussels,
  farming 44
  zebra 53
myths 16–17

Namibia, diamonds 47
Native Americans 7
natural selection 25
Nautilus (early submarine) 32
Nautilus, USS (nuclear submarine) 19, 32
navies 4, 19
navigation 9, 22–3, *23*
  instruments of 11, 22–3, *23*
New Guinea 6
New York City 20
New Zealand 7, 13
Newfoundland 10
Niger, River *40*
Nile, River 8
Nimitz, USS 19
nitrates 45, 51, 56
nitrogen narcosis 30
North America, peoples of 7
North Star 9
Northwest Passage 12
Norway, salmon farming *45*
Norwegians 16
Nova Scotia, Canada, tidal power 49
nuclear power 48
nuclear weapons 19

ocean, seabed, law concerning 59
ocean thermal energy conversion schemes (OTEC) 49
oceanography 26–7
oceans
  depth, measuring 26
  floor
    formation 28
    mapping 27
    managing 58–9
    protecting 56–7
    surveying *34*
    temperatures, measuring 28–9
octopuses, giant 16
oil 61
  crude, transporting 21, *21*
  deposits, searching for 47
  drilling for 46–7, *46*, 57
  extracting 5
  land-based deposits 46
  pollution from 5, 50–1, *50*
  spraying *51*
oil rigs *46*, 51
olive oil 9
otter trawls *40*
overfishing 27, 42–3, *59*
oysters, farming 44, *44*

Pacific, island inhabitants 6–7
Panama 20
Panama Canal 5
pearl fishermen *31*
pearls 44
penguins 51
Persian Gulf 10
Persians 18
petroleum oil *see* oil
Philippines 11
Phoenicians 9
phytoplankton
  benefits of 56
  bloom 56
  distribution, on satellite image 29, *29*
  effect of nitrates on 51
  harm caused to, by oil 50
  monitoring 34
  uncontrolled growth 51, 56
Piccard, Auguste 33
Piccard, Jacques 33
pilchard *42*
Pilgrims 7
pipelines, underwater 5
pirates 14
placer deposits 47
plastic, pollution from 50
plate tectonics 28
Poles, migration to U.S. 7
pollution 50–1
  chemical 50, 57
  eliminating 59
  from fish farms 45
  industrial waste *50*–1, 51
  metal 51
  oil 5, 50–1, *50,* 51
  plastic 50
  sewage 51
Polynesians 6–7
porpoises,
  threats to 5
  vaquita 56
ports 5, 20
Portuguese sailors 11
Punt 8
purse seine nets *40*, 41, *41*

Queen Mary 15

Ra (reed boat) 8, *8*
radar 22–3, *23, 29*
radar altimetry 29
radioactive waste 5
Radiolaria 26
rafts, log 8
La Rance, France, tidal barrier *48,* 49
ranching, ocean 45
Red Sea 8, 10, 15, 52
reed boats 8, *8*
remote-sensing 29
remotely operated vehicles (ROVs) 33, *35*, 45
research, marine 34–5
resorts, coastal 52, 54
Revolution 13
Ritter, William 27
rivers, as highways 8
La Rochelle, France, mariculture *44*

Romans 4
Ross, James Clark 25, *25*
Ross, John 25, *25*
rowing ships 18
Russians, migration to U.S. 7

sailing vessels 14–15
  reed 8, *8*
salmon, farming 44, 45, *45*
saltiness, measuring 28–9
saltmarsh 56
Samoa 6
sardines 41, 42
satellite altimetry 29, *29*
satellite images, "false-color" *29*
satellite, signals from 23, 29, *29*
Scripps Institution for Biological Research 27, 34
scuba diving 30, 31, *31*
sea anemones 53
sea cows, Stellar's 38
sea cucumbers 53
sea floor *see* oceans
sea otters 5
sea urchins 35
seabed, law concerning 59
seahorses 56
seals
  in fishing nets 41
  harm caused to, by oil 50
  hunting 39, *39*
seas
  exclusive economic zone *58*, 59
  law of 58–9
  territorial *58*, 58
Seasat satellite *29*
seaweeds
  effect of nitrates on 51
  farming 44
seine nets 41, *41*
sewage 5, 51
sharks
  dogfish 43
  megamouth 16
  whale 54–5
shellfish
  catching 41
  farming 44, *44*
ships 4–5, 10, 11, 14–15, 20
  bridge of 22, *22*
  transporting water as ballast 53
shipwrecks 36–7, *36, 37*
shrimp, farming 44
Siberia 6
Siebe, Augustus 30–1
silks 14
silver 9
Singapore *52*
slaves, transported from Africa 7
solar power 61
sonar (echo sounding) 27, 28, *28*, 34, *34*, 36, 37
  use by fishing ships 40–1
soundings 26
South Georgia 39
Southern Cross 9
Southern Ocean 39
Spanish settlers 7
spermaceti 38–9
Sperry, Elmer 22
spices 14
sponges 53
squid, giant 16, *17*
stars, navigation by 9
steam turbines 15
steamships 4, 15, *15*, 18–19
submariners 32
submarines 4, 15, 19, 32
  military 33
  nuclear-powered 19, *19*
  tourist 54
submersibles *27*, 32, *32–3*, 33, 37
Suez Canal 15, 21
sulfur dioxide 48
Sumerians 8
Sun, navigation by 9
supertankers 4, 5, 21, *21*
  oil spills from 50, *50*
supertrawlers 41

Tahiti 6, 13
Tampa Bay, Florida 52
tankers 21, 21, *47*
  oil spills from 50, *50*
Tasmania 13
tea 14
territorial waters 58, *58*
Tharp, Marie 28
tidal barrages 48–9, *48*
tidal power 48–9, *48*
Timor 6
tin 9, 47
Titanic 37, *37*
Tonga
torpedoes 19, 32
tourism 5, 20, 54–5
  damaging habitats 52–3
toxic substances 5
trade 4, 8–9, 10, 11, 14, 18, 20
trawl nets *40*, 41, 52–3, 57
trawlers 41
Trieste (submersible) *32–3*, 33
triremes 4, 18
tuna
  bluefin *43*
  catching 41
  trapped in net *43*
turbines 15
turbot 58
Turtle (submarine) 32, *32*
turtles 57
  in fishing nets 41
  threats to 5

U-boats 19
United Nations,
  Environment Program 59
  Food and Agriculture Organization (FAO) 43
United States 15
U.S. immigration 7, *7*

Van Dieman's Land 13
vertebrates, endangered 56
Vikings 10
volcanoes, underwater 34
voyages 4
  of ancient peoples 6–7
  of discovery 8–13, *10, 11*
  of scientific exploration 24

Walsh, Don 33
warships 4, 10, 15, 18–19, *18, 19*
  submarine attacks on 32,
  used for research 28
wave power 49
weapons, modern 19
wetlands 52
whale sharks 54–5
whales
  baleen 38
  in fishing nets 41
  hunting 38–9
  schools of 55
  species,
    bow-head 39
    gray 39, 54
    minke 39
    right 39, 56
    sei 39
    sperm 17, 39
  threats to 5
  uses 38–9
whaling 39, 57
whaling stations 39
wind farms 49, 49
wine 9
Woods Hole Oceanographic Institution 27, *27*, 34
wool 15
World Conservation Union (WCU) 43, 56

Yemen, fishing rights 58

zebra mussels 53
Zeebrugge, Belgium, wind farm 49
zooplankton,
  and alien organisms 53
  harm caused to by oil 5

Picture Credits

Photographs

Abbreviations

C Corbis; FLPA Frank Lane Picture Agency; NHPA Natural History Photographic Agency; OSF Oxford Scientific Films; SPL Science Photo Library

1 Craig Aurness/C; 3 Judy Griesedieck/C; 4 K. Aitken/Panda/FLPA; 5 Danny Lehman/C; 7t Anthony Bannister/NHPA; 7b Library of Congress; 8 R. Sorensen & J. Olsen/NHPA; 11 Vince Streano/C; 13t Mary Evans Picture Library; 13b G.I. Bernard/NHPA; 14 G.I. Bernard/SPL; 19c S.H.K./Corbis Sygma; 19b Jonathan Eastland/Ajax News & Feature Service; 20 Ernie Janes/NHPA; 21 Matthias Breiter/OSF; 22–23 Nigel Bailey; 23 Lucidio Studio Inc./C; 24 The Natural History Museum, London; 25t G.I. Bernard/NHPA; 25b Mary Evans Picture Library; 26l Bettmann/C; 26r G.I. Bernard/NHPA; 27 Arthur Tress/SPL; 29 Dr. Gene Feldman, NASA GSFC/SPL; 30 Sheila Terry/SPL; 31t Michael Freeman/C; 32 SPL; 33 Michael Pitts/SAL/OSF; 35t Mark Deeble & Victoria Stone/OSF; 35b Jim Amos/SPL; 36 Roger Wilmshurst/FLPA; 37t Ralph White/C; 37b S. Navarrini/Panda/FLPA; 38 Dave G. Houser/C; 39 B. & C. Alexander/NHPA; 40 Martin, Jerrican/SPL; 41 S. Jonasson/FLPA; 43l David Fleetham/OSF; 43r Howard Hall/OSF; 44 Daniel Heuclin/NHPA; 45l G.I. Bernard/NHPA; 45r W. Broadhurst/FLPA; 46 Mark Webster/OSF; 47 Anthony Bannister/NHPA; 48 Ronald Toms/OSF; 49 David T. Grewcock/FLPA; 50–51 Simon Fraser/SPL; 51 Hermann Buttigieg/FLPA; 52 Morten Strange/NHPA; 52–53 A.N.T./NHPA; 53 W. Meinderts/Foto Natura/FLPA; 54 Natalie Fobes/C; 55 Dr. Jurgen Scriba/SPL; 56–57 Doug Allan/SPL; 57 Bob Goodale/OSF; 59 Andy Rouse/NHPA; 60 Trevor McDonald/NHPA; 61 Johan De Meester/Ardea; cover photograph: Powerstock.

Illustrations

Andromeda Oxford Ltd 6, 8, 9, 10, 11, 16–17, 17t, 18, 21, 23, 28–29, 30, 31, 32–33, 42, 46, 47; Alan Hancocks 12, 14, 15, 16–17, 22, 34, 40, 41, 48; Maltings Partnership 28; Tony Truscott Designs 58.

EXPLORING THE OCEAN

VOLUME 2

LIFE IN THE OCEAN

Trevor Day

OXFORD
UNIVERSITY PRESS

Consulting Editor Professor Dorrik Stow
University of Southampton Oceanography Centre, UK

Published in the United States of America by
Oxford University Press, Inc.
198 Madison Avenue
New York, NY 10016
www.oup.com

Oxford is a registered trademark of
Oxford University Press

Planned and produced by
Andromeda Oxford Limited
11-13 The Vineyard
Abingdon
Oxon OX14 3PX
UK
www.andromeda.co.uk

Library of Congress Cataloging-in-Publication Data
Day, Trevor.
Exploring the ocean / Trevor Day.
p . cm.
Summary: Explores various aspects of the world's oceans including currents, tides, weather patterns, the sea floor, marine life, uses by humans, and ecological issues.
Contents: v. 1. The physical ocean -- v. 2. Life in the ocean -- v. 3. Uses of the ocean -- v. 4. Reference and set index.
ISBN 0-19-515738-9 (set : alk. paper)
1. Oceanography--Juvenile literature. 2. Marine organisms--Juvenile literature. 3. Marine biology--Juvenile literature. 4. Ocean [1. Oceanography. 2. Marine biology. 3. Ocean.] I. Title.

GC21.5.D385 2003
551.46--dc21 2002042583

Project Director Susan Kennedy
Project Manager Derek Hall
Design Tony Truscott
Editors Marion Dent, Deborah Fox, Suzanne Juby
Cartographers Tim Williams, Richard Watts
Picture Manager Claire Turner
Picture Research Alison Floyd
Production Clive Sparling

Printed in the U.A.E. by Oriental Press Limited, Dubai

USING THIS SET

This four-volume set describes the fascinating world of the ocean through informative text and colorful illustrations and maps. Each volume explores a different theme. **Volume 1** is concerned with the physical ocean: its origin and structure, and associated features such as tides, weather systems, islands, and coral reefs; **volume 2** describes the varied natural life in the ocean: evolution, marine animals and plants, food webs, ocean zones, communities of creatures, and conservation; **volume 3** looks at the way humans use the ocean: exploration from earliest times to the present day, exploitation of ocean resources, pollution, and ocean management.

In each volume, feature boxes highlight items of special interest. Cross-references at the bottom of each left-hand page point the reader to further information on related topics in other sections and volumes: the relevant volume number is given in bold type, followed by the page numbers. Each volume also has its own subject index.

Volume 4 contains a comprehensive series index, together with further reference resources, including maps, datafiles, full descriptions of the five oceans, biographies of men and women who have contributed to our knowledge of the oceans, timelines, places to visit, websites, further reading, and a glossary.

Throughout the set the following abbreviations are used for measurements:–
Length in (inches), cm (centimeters), ft (feet), m (meters), mi (miles), km (kilometers)
Area sq in (square inches), sq cm (square centimeters), sq ft (square feet), sq m (square meters), sq mi (square miles), sq km (square kilometers)
Volume cu mi (cubic miles), cu km (cubic kilometers)
Temperature F (Fahrenheit), C (Celsius)
Weight Given in U.S. tons, followed by metric tonnes

Previous page: An emperor angelfish, found around coral reefs in the Pacific and Indian Oceans.

Right: Primitive, rocklike, marine organisms called stromatolites growing in shallow water in Western Australia.

Contents

Life in the Ocean	4
Earliest Life	6
Explosion of Life	8
Mass Extinctions	10
Phytoplankton	12
Seaweeds and Seagrasses	14
Food Chains and Webs	16
Depth Zones	18
Zooplankton	20
Spineless Life	22
Jawless Fish	24
Fish with Jaws	26
Sharks, Skates, and Rays	28
Bony Fish	30
Marine Reptiles	32
Seabirds	34
Marine Mammals	36
Toothed Whales	38
Baleen Whales	40
Marine Migrations	42
Life on a Rocky Shore	44
Life on a Sandy Shore	46
Life in an Estuary	48
The Shallow Sea Floor	50
The Crowded Reef	52
Threatened Communities	54
Deep-Sea Life	56
The Dark Zone	58
Living Together	60
Index	62
Picture Credits	64

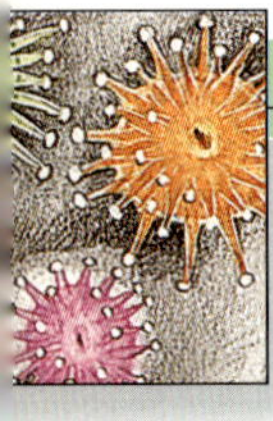

LIFE IN THE OCEAN

Most of the space available for life on Earth is in the ocean. The watery ocean environment supports a huge variety of plant and animal species.

The ocean is probably the place where life on Earth began, more than 3.6 billion years ago. For the next 3 billion years, the most important stages in the evolution of life occurred in the sea; animals and plants have conquered the land only within the last 500 million years or so. Today, the oceans are home to the largest animal that has ever lived—the giant, 200-ton (180-tonne) blue whale. The oceans also contain some of the world's smallest organisms. For example, you could fit hundreds of tiny primitive organisms called cyanobacteria on the point of a needle.

Microscopic plants called phytoplankton float in the surface waters of the oceans. Each year, countless numbers of phytoplankton produce about half the world's oxygen. The total weight of this phytoplankton is the equivalent of one billion elephants. When these plants photosynthesize (make their own food from simple substances using the energy from sunlight), they also take carbon dioxide out of the atmosphere. High levels of this gas in the atmosphere may be helping to overheat our planet, so phytoplankton perform a very useful function.

How many species?

Biologists think there may be between 2 and 20 million species of plants, animals, and microbes (microscopic single-celled organisms such as bacteria) living in the oceans. So far, however, scientists have only been able to identify about 300,000 of them. Until recently, biologists thought there were more creatures living on land than in the sea. Now scientists are not so sure. In the 1990s marine scientists dredged samples from the deep-sea floor of the North Atlantic and found, on average, one new animal species in each sample.

See also **vol 1:** 4–5, 8–9; **vol 3:** 26–27, 28–29, 33

Studying marine life

The undersea world is an alien environment to land-living, air-breathing creatures like ourselves. If we wish to dive beneath the sea, we must bring our own air supplies to breathe. When we dive deeply, we must also bring artificial

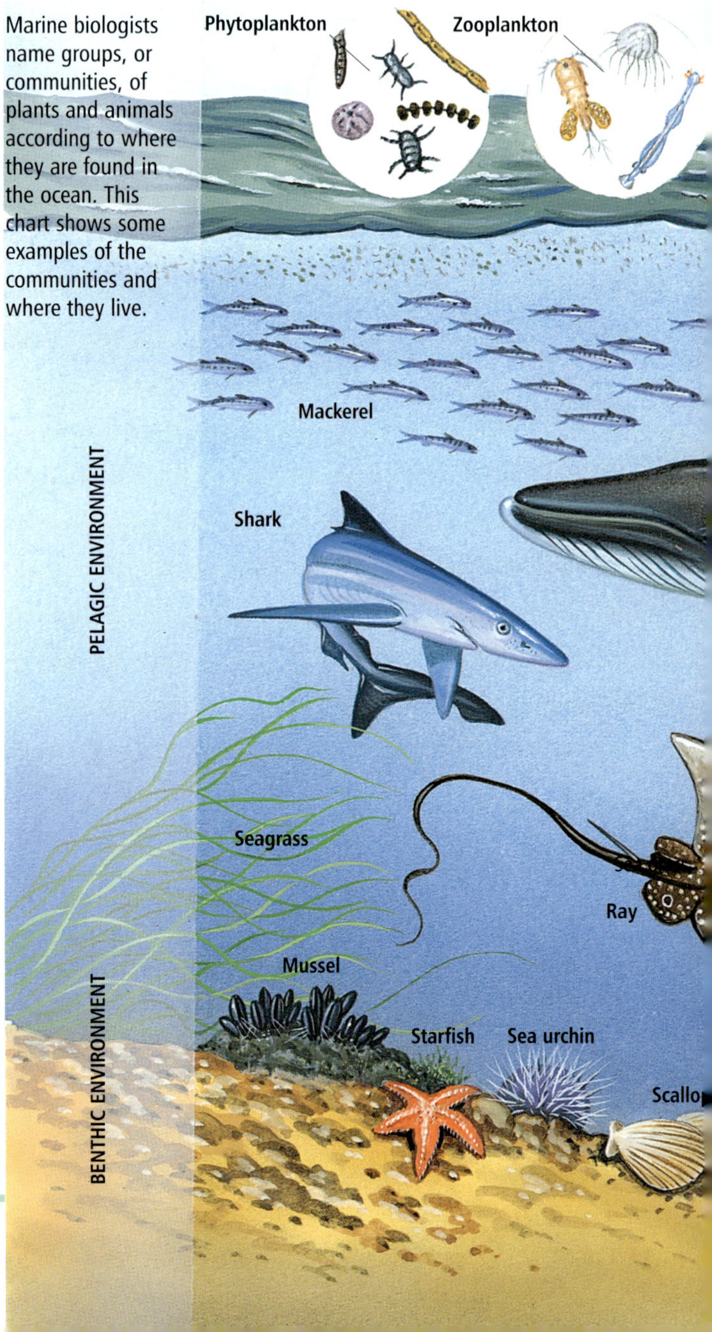

Marine biologists name groups, or communities, of plants and animals according to where they are found in the ocean. This chart shows some examples of the communities and where they live.

light so we can see in the gloom. We must be insulated from the cold, and we need to be protected from the great pressure of water that bears down on us.

Our knowledge of marine life is very incomplete—this is partly because we cannot live easily under water but also because the oceans are vast and varied. Biologists explore marine life using nets to bring creatures to the surface or by diving below the surface. These methods often reveal only a tiny part of the ocean's varied life. Also, the ocean's water is constantly moving, and so are all the organisms that live in it. Because of this, marine scientists find it hard to work out how many organisms there are in the world's oceans and where they all live.

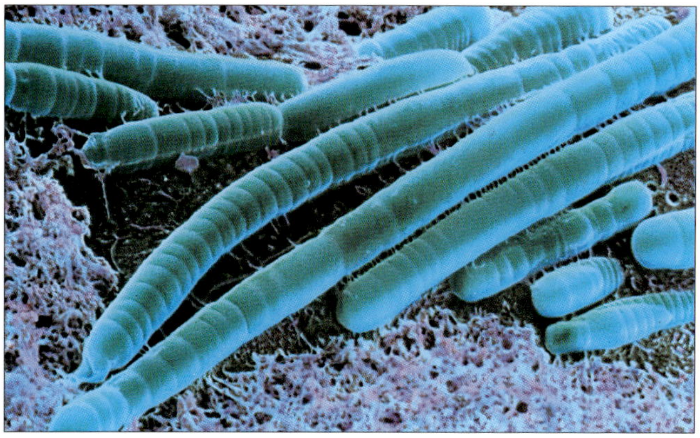

Groups of cyanobacteria, magnified 600 times. Each elongated strand is made up of rows of single-celled organisms. This species of cyanobacteria is common in the sea as well as in mangrove swamps.

Uneven distribution

Marine life is not spread uniformly throughout the oceans. Most of it is concentrated in the top few hundred feet of the sea, or on and within the sediment covering the ocean floor. But even where life is at its most abundant, it is not evenly distributed. For example, across the surface waters there are places called "living hotspots" where life is particularly abundant. These places include regions where cool, nutrient-rich waters rise to the surface (called upwelling), encouraging the growth of phytoplankton and the animals that feed on them and on each other. Many of the world's greatest fisheries, such as those off the west coasts of Africa and North and South America, are centered around areas of upwelling.

Floating and swimming

Scientists divide the organisms that live in the ocean into two groups—those that swim or float in the water (pelagic), and those that live on the ocean floor (benthic). "Pelagic" is from the Greek word for "ocean," and "benthic" from the Greek word for "depth."

Some pelagic organisms drift with the current. They are called plankton (from the Greek word for drifting). Plankton, which consists of both plants (phytoplankton) and animals (zooplankton), range in size from microscopic bacteria to jellyfish and floating seaweeds, some of which grow to many feet in length. Creatures that swim actively and do not need the currents to carry them along are called nekton (from the Greek word for swimming). Nekton includes fish, squid, and marine mammals.

Earliest Life

Life has existed on Earth for more than 3.6 billion years. For more than 2 billion years, however, ocean-living bacteria were the only form of life.

There are good reasons to think that the first organisms evolved in the sea, rather than on land or in fresh water. All living things are made up of at least 65 percent water. Water is relatively scarce on land, whereas it is superabundant in the ocean. The mix of chemicals found in the body fluids of animals and plants is much more like that found in sea water than in fresh water.

For most of the Earth's history, conditions were unsuitable for life on land. Initially, the atmosphere—the envelope of gases surrounding the Earth—did not shield the Earth's surface from most of the Sun's harmful ultraviolet (U.V.) radiation. High levels of U.V. radiation cause mutations (genetic changes in the cells of living organisms). Just a few feet of water is enough to filter out most U.V. radiation, so U.V. light poses little danger to organisms living below this depth. Within the last 500 million years or so levels of U.V. radiation have fallen as levels of oxygen in the atmosphere have risen, enabling life to survive on land.

Cushion-shaped stromatolites are rocklike structures built up of matted layers of cyanobacteria and chalky sediment that formed on shallow sea floors. Cyanobacteria were one of the earliest forms of life to appear on Earth, and they still exist today.

Early fossils—the preserved remains of past life forms—are found in rocks 3.6 billion years old. These fossils, called stromatolites, are made from layer upon layer of tiny, bacterialike organisms called cyanobacteria that grow in shallow sea waters.

The first organisms

The first organisms were probably types of bacteria similar to those found in deep-sea hot-water vents or in sulfur springs on land today. Bacteria are primitive cells without a proper nucleus. The earliest types of bacteria got their energy by

What is life?

The bodies of all living creatures on Earth are formed mainly from the element carbon and are composed of at least one cell. The cell is a structure enclosed within a covering called a membrane. All the organism's life processes take place in cells. The simplest organisms such as bacteria consist of a single cell. The most complex organisms such as trees or human beings contain billions of cells.

Living organisms grow and reproduce (make copies of themselves). They do so by making or consuming complex chemicals. They then break down the chemicals to release energy. Biologists call these processes nutrition and respiration. Organisms use energy for life processes such as movement or manufacturing new cells.

See also **vol 1:** 4–5, 8–9, 10–11, 15; **vol 2:** 8–9, 10–11, 57

breaking down chemicals such as hydrogen sulfide and methane. Scientists call this method of obtaining energy chemosynthesis. Bacteria used the energy to make the substances that living organisms need for life, such as carbohydrates, proteins, fats, as well as nucleic acids like DNA that carry the instructions for building organisms.

New life forms

The earliest organisms began to alter the environment in which they lived. During chemosynthesis they used up certain chemicals and released others as gases. Over millions of years, these organisms gradually changed the mixture of gases in the Earth's atmosphere. Bacteria also released acids that slowly ate away rocks.

By 2.5 billion years ago, some bacteria had evolved another method for obtaining energy by trapping sunlight. This method of harnessing energy, called photosynthesis, was more efficient than chemosynthesis, and soon bacteria that could photosynthesize multiplied and flourished. Photosynthesizing bacteria release oxygen gas, and the gas slowly began to accumulate in the atmosphere. This allowed more complex life forms, which need oxygen to live, to evolve.

By 1.2 billion years ago, some bacteria had joined together to form more complex life forms called eukaryotes. Unlike bacteria, the cells of eukaryotes contain a distinct, membrane-bound structure called the nucleus, which contains most of the cell's DNA. In time, some eukaryotes came together to form colonies or clusters of cells. By 850 million years ago, the first true multicelled organisms, the ancestors of all the living creatures except bacteria and cyanobacteria, had appeared.

By 600 million years ago, the oxygen concentration in the atmosphere was high enough to allow oxygen-breathing animals to evolve. To breathe under water, some animals developed gills—delicate, feathery tissue that removes oxygen from water. The stage was now set for an evolutionary explosion of life forms. By 550 million years ago, all the major groups of animals including the earliest ancestors of vertebrates, the group to which we belong, had evolved in the ocean.

Life on Earth began about 3.6 billion years ago. Slowly, over many millions of years, conditions on the planet became suitable for other creatures to evolve.

Explosion of Life

Fossils—the remains of creatures that have been preserved in rocks—tell scientists much about the life forms that once existed on Earth. Fossils reflect the wealth of plant and animal life buried in the geological past.

Some 600 million years ago soft-bodied creatures, perhaps resembling jellyfish, floated in the Earth's shallow seas. When the creatures died, their soft bodies settled on the sea floor and were quickly covered with fresh layers of sandy mud before they had completely rotted away. Over millions of years, huge geological pressures compressed the sand into rock. Continents formed, broke up, and drifted across the surface of the Earth, carrying the rocks—and the fossils they contained—with them.

Today, what was once a shallow sandy lagoon 600 million years ago now forms part of the Ediacara Hills of southern Australia. In 1946 scientists found the impressions of soft-bodied creatures, with fossilized tracks and burrows, in the Ediacara Hills. These fossils are the earliest evidence we have for the existence of large, multicelled animals on Earth. Some scientists think that these creatures were similar to jellyfish, others that they were more like sea urchins, worms, or crustaceans. Some experts believe they belonged to major animal groups now extinct.

Fossils of similar soft-bodied animals have been found in rocks on all the continents except Antarctica. The creatures were evidently once abundant and widespread. But their fossils have not been found in rocks later than 570 million years old. They vanished—perhaps because of an environmental disaster, or because they were unable to compete with more successful creatures. An entirely new range of marine animals replaced them.

The strange case of *Hallucigenia*

Fossil experts spend months, sometimes years, gathering tiny fossil fragments to make a whole fossil. Sometimes they get things wrong. Charles D. Walcott thought one of the fragmented fossils he found in the Burgess Shale, in the Canadian Rockies, was a type of worm. In the 1960s British researchers re-examined these fossil fragments and discovered rows of spines along the creature's body. They reconstructed the fossil as a creature that walked on spines and had waving tentacles on its back. They called it *Hallucigenia* because it looked like an hallucination. Specimens discovered in China in the 1990s suggested that the British scientists had got their creature upside down. It seems that *Hallucigenia* looked more like a caterpillar. It probably walked on stiltlike tentacles, with its spines sticking upward in the form of a defensive array.

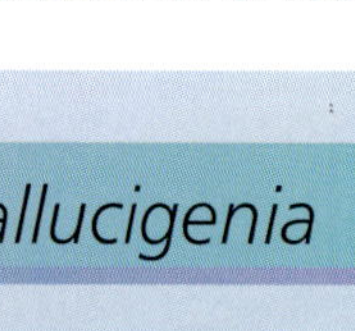

The Burgess Shale

One of the richest areas for fossils in the world is the Burgess Shale, an outcrop of rock in the Canadian Rockies of British Columbia. In 1909 American paleobiologist Charles D. Walcott made some amazing discoveries here. He dug up some fossils of marine creatures that had a hard outer shell or skeleton from the shale. Walcott calculated that they were 550 million years old. Eventually, scientists recovered more than 65,000 fossils from the shale.

The creatures that Walcott found were a rich mixture of forms—some wormlike or clamlike, and others like sea urchins, sponges, corals, and crustaceans. Many of the animals looked nothing like any animals living today. It seems that the Burgess Shale fossils were created by an underwater mud slide that swamped more than 100 species of hard- and soft-bodied animals living on an ancient coral reef. Some of the fossils were trilobites—animals with a plated, segmented body and many legs, that looked rather like modern woodlice. Trilobites were arthropods—a group of animals

See also **vol 1:** 6–7, 10–11; **vol 2:** 6–7, 10–11, 22–23

These strange creatures lived in the ocean about 550 million years ago. *Anomalocaris* had a pair of eyes on the top of its head and a rounded mouth. *Olenoides* was a tiny trilobite.

Anomalocaris

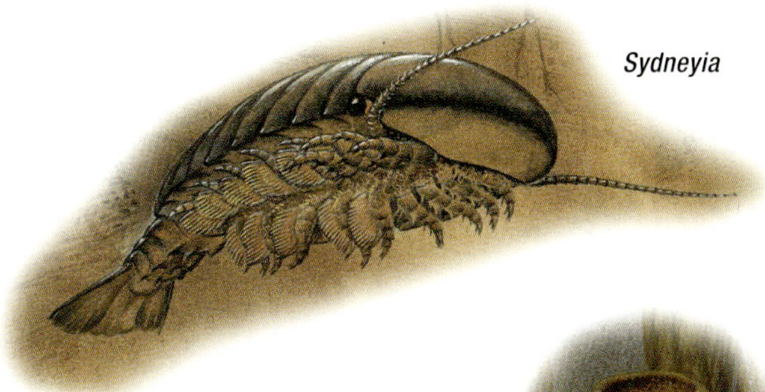

Sydneyia

Odaraia

Aysheaia

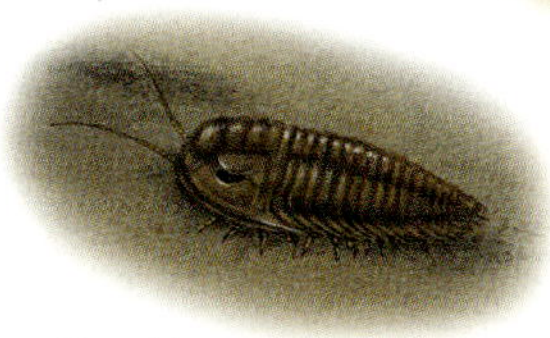

Olenoides

Odontogryphus

that includes crustaceans, spiders, and mites. The largest creature from the Burgess Shale is probably *Anomalocaris*—a 2-ft (0.6-m) long, armor-plated predator.

Around 550 million years ago, hundreds of new species burst into life. This period is known as the Cambrian explosion. By the middle of the Cambrian period, some 520 million years ago, sponges, corals and jellyfish, worms, and arthropods were abundant in the ocean. More advanced animals such as chordates—the direct ancestors of fish—were also present.

Why the explosion of life?

Scientists believe that a single supercontinent, which they call Rodinia, had emerged from the sea by 650 million years ago. By 570 million years ago it tore apart, creating vast areas of shallow water around its broken edges. Water temperatures at this time reached 68–104° F (20–40° C). The atmosphere was becoming oxygen-rich, and the chemical makeup of the ancient sea water was similar to that of modern sea water. All these conditions were ideal for marine life. A wide variety of creatures could flourish.

This fossil trilobite, with its head at the left, shows the body segments that enabled the animal to roll itself into a ball for protection.

Mass Extinctions

We know from study of marine fossils that living creatures on Earth have been wiped out by catastrophic events five times in the last 500 million years.

Scientists who study fossils know that major groups of plants and animals suddenly vanish at the same time from the fossil record—in other words, no more fossils of these plants and animals are found in rocks of a particular age. They call the sudden disappearance of living creatures a mass extinction event. Fossil records show there were five mass extinction events between 440 and 65 million years ago.

During the last mass extinction event, dinosaurs on land became extinct and so did many marine creatures. Marine plants and animals provide better clues than land creatures to the size of an extinction event because many survive better as fossils. Organisms usually become preserved as fossils by becoming smothered in sediment, and most of the world's sediment accumulates in oceans.

The biggest extinction

About 250 million years ago, at the end of what geologists call the Permian period, the biggest mass extinction event ever took place. More than 90 percent of marine species were wiped out. Only a catastrophe on a massive scale could have led to so many creatures dying at the same time. Some scientists believe that a huge increase in volcanic activity caused the late Permian extinction. Evidence for this comes from Siberia, in northern Eurasia, where geologists have found vast sheets of solidified lava—molten rock spewed out by volcanoes 250 million years ago. Such volcanic eruptions would have sent huge clouds of volcanic dust and carbon dioxide and sulfur dioxide gas into the atmosphere. The dust would have blotted out the Sun's rays and cooled the planet. The gases mixing with moisture in air would have formed poisonous acid rain that fell back to Earth, killing plants and animals. After cooling, the gases may have triggered more climate changes by trapping heat radiated from Earth in the atmosphere and raising temperatures on the planet.

The sixth extinction?

Some biologists believe that species are becoming extinct at a faster rate today than at any other time in the Earth's history. They think humans are the cause of this "sixth extinction." Human activities such as the overharvesting of many plant and animal species, the release of pollutants, and the destruction of habitats are causing hundreds or even thousands of species to become extinct each year.

See also **vol 1:** 8, 10–11, 60–61; **vol 2:** 8–9, 26–27; **vol 3:** 57

Coelacanth: a living fossil

Scientists once believed that the coelacanth, a primitive form of fish, had been extinct for 60 million years. Then, in 1938, an amazing discovery was made. A boat fishing in deep water in the Indian Ocean off South Africa caught a very unusual 5-ft (1.5-m) long fish. When a fish expert saw it, he immediately recognized that it was a coelacanth.

Other coelacanths have been caught in the Indian Ocean since then. In 2000 a scuba diving expedition discovered living coelacanths at a depth of 355 ft (108 m) just off the South African coast—so it is possible that there are many more of these strange, ancient fish still alive today. There may be many other creatures, as yet undiscovered, in the ocean depths.

Blotting out life

A large meteorite hitting the Earth with enormous speed would have very similar effects to that of intense volcanic activity. Shock waves from the force of the impact would wipe out anything in the immediate area and launch huge amounts of rock particles into the air that would have darkened the skies for months or years. Deprived of sunlight, plants could not photosynthesize (make food), causing other creatures to starve. Some scientists think a meteorite strike may have caused the last major mass extinction, when the dinosaurs and about 75 percent of all marine animals suddenly died out.

Some mass extinctions coincided with a major drop in sea level, which occurs when the Earth cools rapidly and water turns to ice instead of filling the oceans. The massive collision or break-up of huge landmasses drifting across the Earth's surface may also have triggered catastrophic global climate changes by altering the strength and direction of winds and ocean currents. Many species, both on land and in the sea, would have been unable to cope with the rapid temperature changes and would have simply died out.

This massive crater in Australia's Northern Territory was caused by a meteorite impact 130 million years ago. The crater is 2.5 mi (4 km) wide, and the walls rise to a height of 820 ft (250 m).

PHYTOPLANKTON

On land, plants such as grasses and trees that sustain animal life are mostly large and visible. Most marine plants are microscopic, but they produce as much food and oxygen as the plants on land.

The microscopic plants and animals that float mostly in the surface water of the ocean are called plankton. Plant plankton is called phytoplankton. "Phyto" is from the Greek word for "plant," and "plankton" from the Greek word for "drifting." Phytoplankton are the main food source for most marine creatures. Some creatures feed directly on phytoplankton. Larger animals may then eat them. All phytoplankton absorb energy from sunlight, which they use to make food. This process is called photosynthesis.

Red tides

When dense blooms of phytoplankton occur in the ocean, usually as a result of increased nutrient levels, they make the sea water red or brown. The areas of stained sea water are called red tides. Red tides sometimes release poisons that are harmful to animals. For example, a poison-making dinoflagellate called *Pfiesteria* is often present in red tides and it can kill fish and shellfish that eat it. In the 1990s red tides of *Pfiesteria* off the Atlantic and Pacific American coasts killed fish and caused a halt in fishing. Scientists who studied the red tides complained of skin rashes and mental disturbances such as dizziness and short-term memory loss.

Types of phytoplankton

The smallest of the phytoplankton are the cyanobacteria. They get their name because they appear blue, or "cyan." Cyanobacteria contain pigments that trap visible light. They can also trap nitrogen gas dissolved in the water and then use it to make their own nitrogen-based nutrients. These nutrients are vital for building proteins and DNA in their bodies. Trapping nitrogen gas is something very few organisms can

There are 6,000 or more species of diatoms. They come in an enormous variety of shapes. Some are spherical and others are like ornate pillboxes or glass slippers. These diatoms have been magnified 600 times.

See also **vol 1:** 4, 38–39, 43, 61; **vol 2:** 16–17, 18, 20–21; **vol 3:** 29

do, and it means that cyanobacteria, unlike most phytoplankton, do not need to take in nitrogen-based nutrients from sea water. Until recently, marine scientists were unaware of these important organisms. Cyanobacteria are so small they simply slipped through collecting nets. Scientists now use special filters to capture cyanobacteria. Some patches of the ocean teem with countless billions of cyanobacteria.

Another sort of phytoplankton, the diatoms, are bigger than cyanobacteria. Diatoms have a two-part outer skeleton made of a glasslike substance called silicate. During late summer, diatom numbers swell and they become the most common large phytoplankton in temperate and polar waters.

Dinoflagellates are also phytoplankton, but they have some curious animal-like features. For example, they have two hairlike structures—called flagella—to row themselves through the water. Most dinoflagellates are smaller than diatoms. In tropical and subtropical waters, dinoflagellates are the most abundant of the larger phytoplankton.

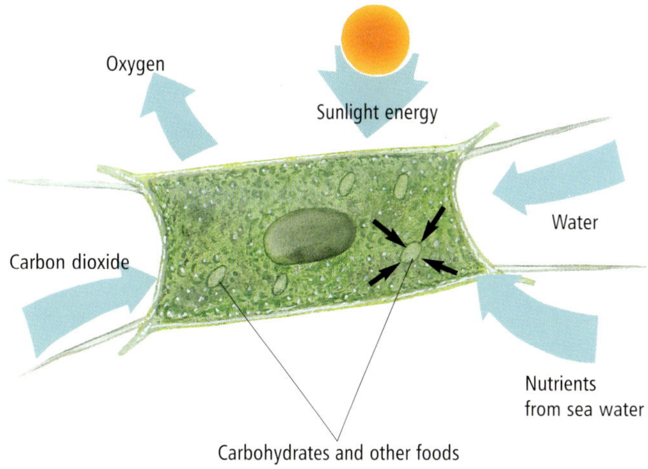

This diagram shows how photosynthesis works. Plants, like the diatom shown here, trap sunlight and use the light energy to combine simple substances, such as water and carbon dioxide, to make more complex, energy-rich substances, such as carbohydrates and fats. Plants then use the carbohydrates and fats as sources of food.

Too many nutrients

Phytoplankton thrive where nutrients and sunlight are abundant. In some coastal areas, sewage and chemical fertilizers are washed into the sea from the land. These waste matters are high in nitrogen. When nitrogen levels become especially high, phytoplankton grow so rapidly that animals cannot eat them quickly enough. The phytoplankton multiply, coloring the sea water red or brown. This is called a plankton bloom. They may cause a thick scum on the surface, which stops sunlight and oxygen from penetrating the water. Soon the phytoplankton begin to die and decompose. Bacteria that feed on the dead phytoplankton use up what little oxygen remains in the water. Most animals need oxygen to live. Some animals can swim away to clearer waters, but creatures that are fixed in one place, are doomed.

During summer in a deep-water region of the Gulf of Mexico, summer plankton blooms regularly create a "dead zone" on the sea floor, where almost nothing can live. The zone sometimes extends to an area of 7,000 square mi (18,000 sq km)—about the size of New Jersey.

Monitoring from the skies

The color of the oceans, when viewed from the sky, gives scientists valuable information about what is happening beneath the surface. Various types of phytoplankton stain sea water different colors. Satellites monitor the hue of sea water, while scientists take plankton samples at sea. By matching the two, scientists can calculate where different types of phytoplankton are growing, and how quickly. This can tell them whether plankton blooms are healthy, and support rich food chains, or whether they are poisonous.

Seaweeds and Seagrasses

Some ocean plants, seaweeds and seagrasses, are larger than phytoplankton. Seaweeds form underwater forests, and seagrasses form underwater meadows.

Seaweeds are a familiar sight on many sheltered rocky shores in temperate or cooler parts of the world. They often decorate the rocks and rockpools in slippery brown, green, or red clumps, although seaweeds also grow under water, provided there is enough light for photosynthesis and a firm, rocky seabed for attachment. Seaweeds are algae, a simple form of plant. The main body of a seaweed consists of fronds. Fronds are flexible, and they sway with waves and currents. In some species, the fronds contain air sacs so they float up to capture the sunlight. The stem of a seaweed is a flexible stalk. Most seaweeds anchor themselves to rocks with a rootlike structure called a holdfast.

Kelps

The largest seaweeds are the brown algae called kelps. The giant kelp of the North Pacific reaches 330 ft (100 m) in length, and can grow by 24 in (60 cm) in a single day. Great masses of kelp form underwater forests that support much other life. Many microscopic plants and animals live on the kelps' fronds. These are eaten by some sea snails, and crustaceans such as shrimp. Many of the animals anchored to the fronds filter sea water to obtain plankton. Worms, brittle stars (animals related to starfish), and crabs also live on and around the holdfasts—filtering the water for food, sifting the sediment, or hunting other creatures. Fish are found throughout the kelp forest. Other animals, including seals, live among the fronds to hide from predators such as sharks.

A few animals graze the kelp itself. These include sea urchins and some snails and sea slugs. Humans harvest seaweeds as a valuable source of food. Agar, a food preservative, is made from seaweed, as are many agricultural fertilizers. The weight of giant kelp harvested yearly in North America equals 10,000 elephants.

Seagrasses

Seagrasses are related to lilies. They grow in soft seabeds in shallow water. Seagrasses stay fully submerged and even flower under water. They have true roots, stems, and leaves,

The main parts of a seaweed are shown on this bladderwrack—a typical brown seaweed.

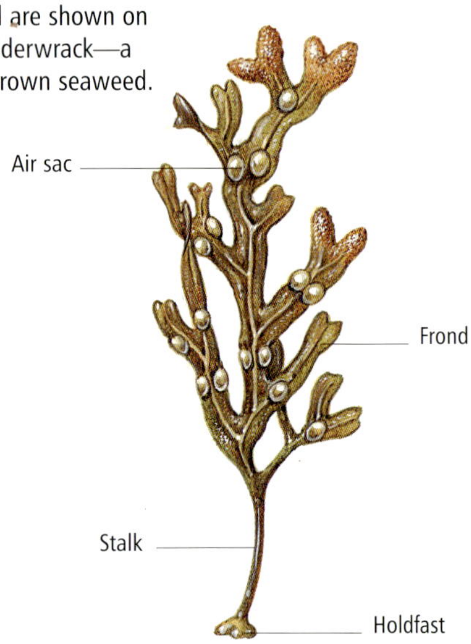

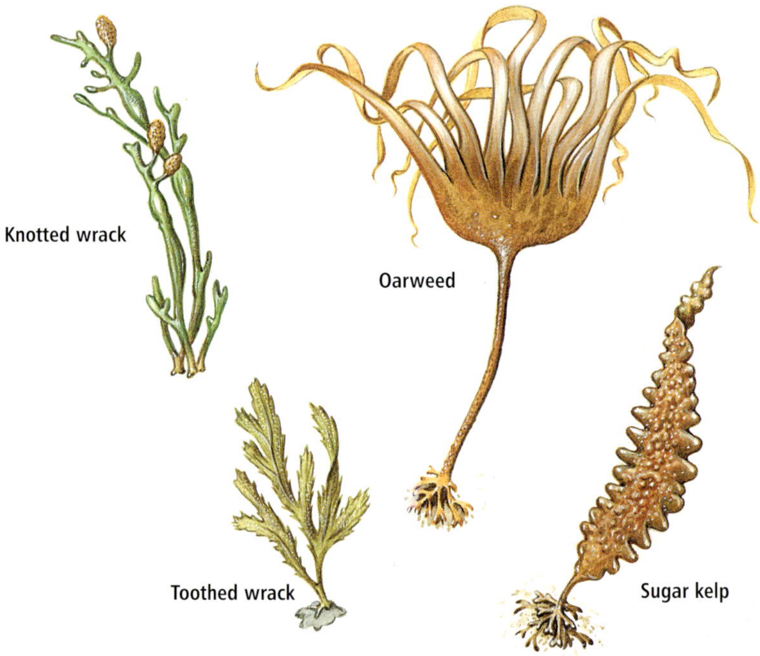

See also **vol 2:** 16–17, 44–45, 50, 51, 55; **vol 3:** 56

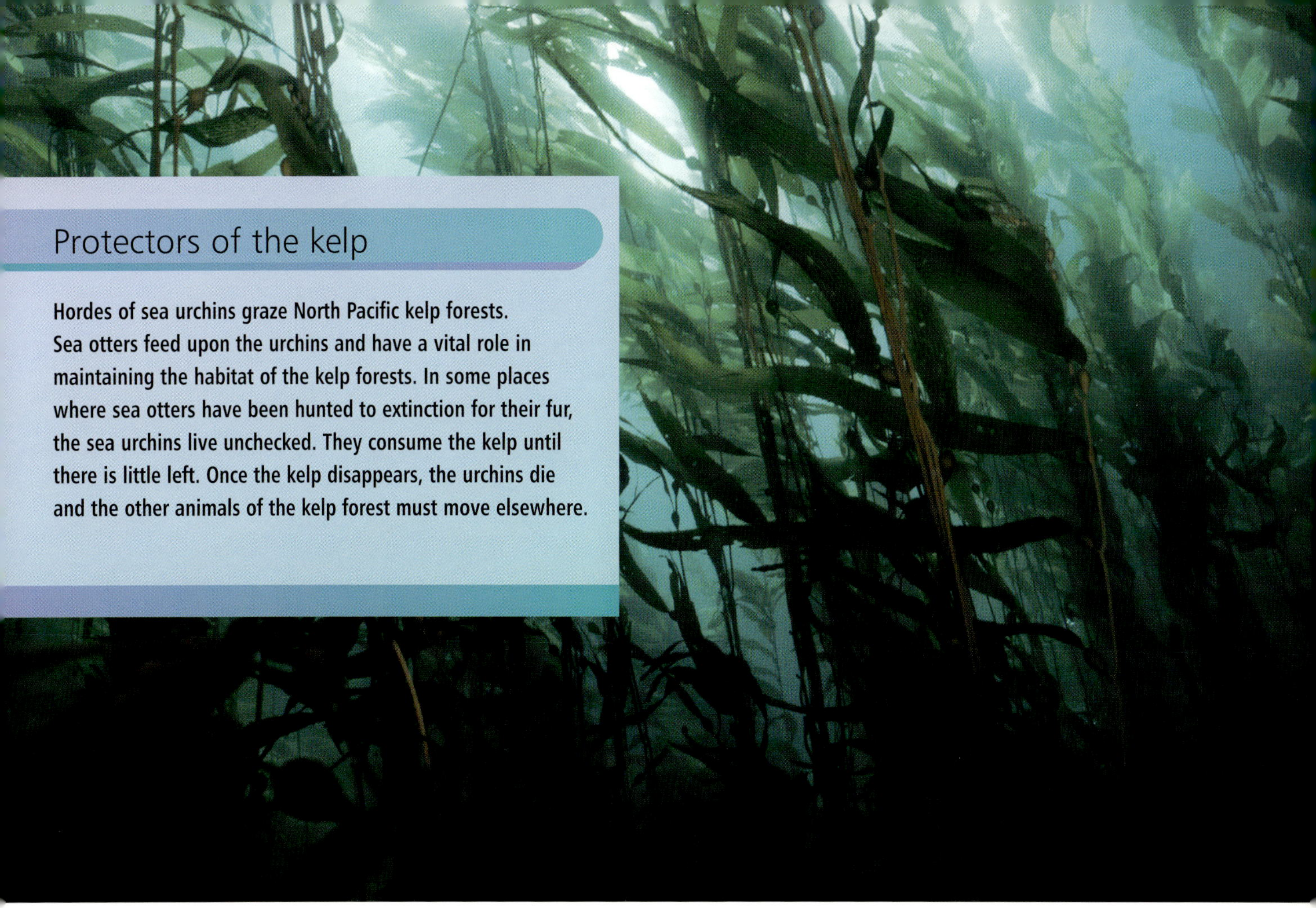

Protectors of the kelp

Hordes of sea urchins graze North Pacific kelp forests. Sea otters feed upon the urchins and have a vital role in maintaining the habitat of the kelp forests. In some places where sea otters have been hunted to extinction for their fur, the sea urchins live unchecked. They consume the kelp until there is little left. Once the kelp disappears, the urchins die and the other animals of the kelp forest must move elsewhere.

With their fronds floating to the surface and holdfasts anchored to the rocky sea floor, giant kelps form the biggest of the underwater forests. Like forests on land, kelps prevent much light penetrating beneath their canopy.

Callithamnion

Dulse

Seagrass

Sea lettuce

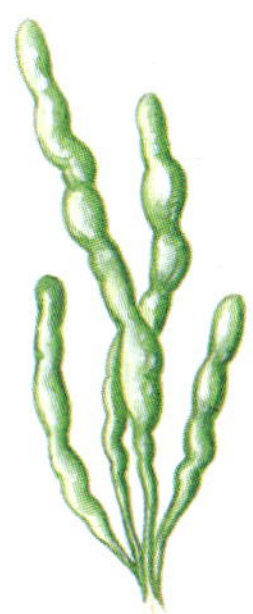

Enteromorpha

Seaweeds are grouped according to the coloration of the different light-trapping pigments they contain. There are green, brown, and red seaweeds. Also shown here is a seagrass, a flowering plant.

just like land plants. The roots of seagrasses bind together the mud and sand particles in which they grow, and their leaves help reduce the battering effects of waves and currents.

Like grass meadows on land, seagrass meadows provide food and refuge for many animals. Sea urchins, fish, mammals known as sea cows, and green turtles eat seagrasses. Microscopic algae grow on the leaves and provide food for various small grazers, especially sea snails. Many tiny animals attach themselves to seagrass leaves and filter the sea water for food. When seagrasses decay, their remains provide food for bottom-living worms, clams, and other creatures. Seagrass meadows are nurseries for the young of many species of fish, crustaceans, and mollusks.

In warmer waters, some of the inhabitants of coral reefs, such as sea urchins and lobsters, leave the safety of the reef at night to feed on nearby seagrass beds.

Food Chains and Webs

In the ocean, as on land, animals survive by eating plants or other animals. Energy is trapped by plants. Grazing animals eat the plants, and in turn are eaten by predators. At each stage, energy passes up the food chain.

Every living organism must have food to survive. Most plants make their own food by using sunlight to form substances such as carbohydrates (sugars and starches), fats, and proteins. Plants break these substances down to release the energy they need to carry out their living processes, such as growth and reproduction. Animals need carbohydrates, fats, and proteins for the same reasons, but they cannot make them. Instead, animals get them by eating plants or other animals.

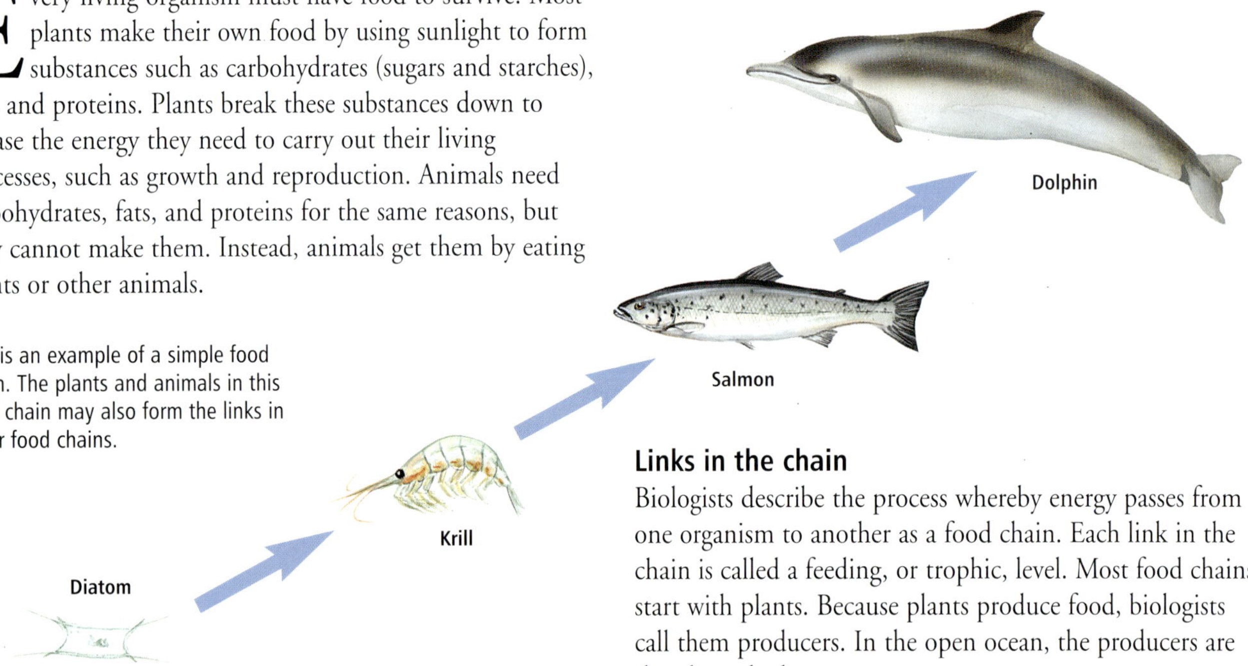

This is an example of a simple food chain. The plants and animals in this food chain may also form the links in other food chains.

Who eats what?

Food webs show what kinds of food different fish eat. This is especially helpful for researchers who study the effects of pollution or overfishing. For example, because of overfishing, the numbers of cod in the North Atlantic have fallen dramatically in recent years. Research shows that this decline has benefited dogfish, a kind of small shark. Dogfish have moved in to eat the cod's natural prey, and dogfish numbers have risen. Information like this helps the fishery experts who decide how many fish can be caught each year.

Links in the chain

Biologists describe the process whereby energy passes from one organism to another as a food chain. Each link in the chain is called a feeding, or trophic, level. Most food chains start with plants. Because plants produce food, biologists call them producers. In the open ocean, the producers are the phytoplankton.

Producers are eaten by plant-eaters (herbivores), which biologists call primary (first) consumers. In the ocean, most primary consumers are zooplankton—microscopic animals that drift in the sea among the phytoplankton. However, some fish—such as anchovies—also eat phytoplankton, and so do some other animals such as sponges.

The next link in the chain are called the secondary consumers. These animals eat the plant-eaters. In the ocean, secondary consumers include larger animal plankton, such as jellyfish, arrow worms (elongated, bristleheaded creatures), and comb jellies (jellyfishlike marine animals), as well as plankton-eating fish such as herring.

At the next link in the chain, other meat-eaters (carnivores) eat the secondary consumers. These larger meat-eaters are called tertiary (third) consumers.They include squids, larger predatory fish such as tuna and marlin, many

See also **vol 1:** 43; **vol 2:** 12–13, 14–15, 18–19, 20–21; **vol 3:** 40

Food chains are usually interconnected with other chains, forming a complex pattern known as a food web, like the one shown here. The food chain on the opposite page has been highlighted to show how it might fit into a typical marine food web.

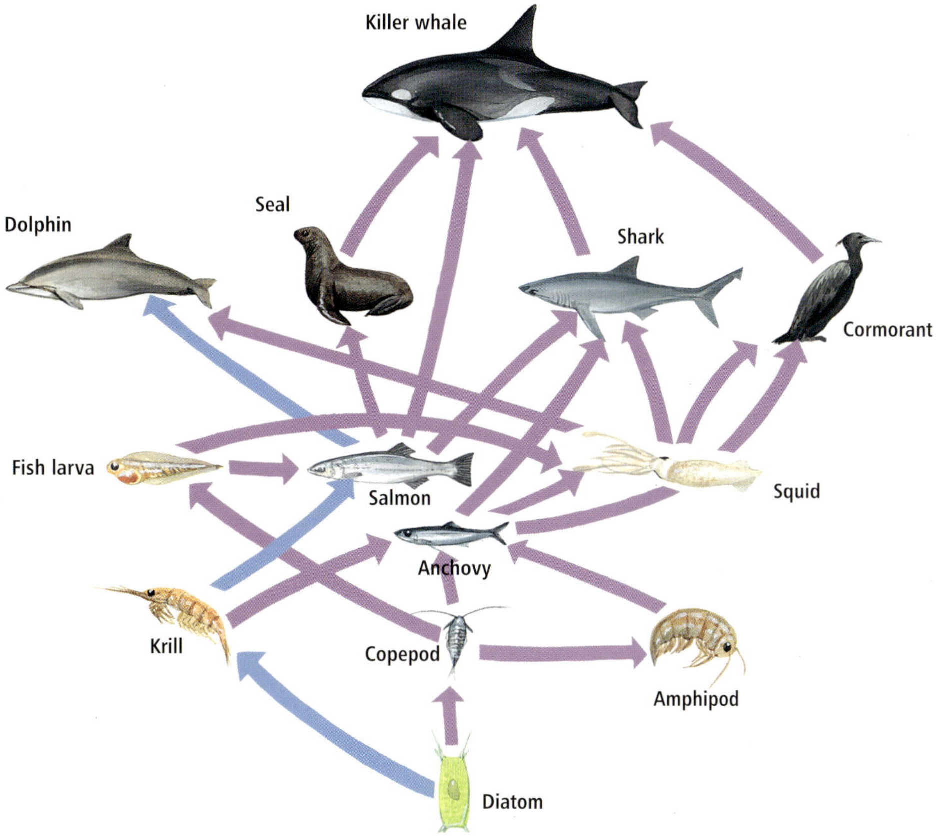

species of sea mammals, and seabirds. Also part of the food chain are the decomposers—such as bacteria and worms—that eat dead animal and plant remains on the sea bottom.

Food webs

Food chains are often very complicated. A plant or animal may be eaten by creatures at more than one feeding level. Creatures at a particular feeding level may eat many hundreds of different types of food—again, often from more than one level. For example, adult herring in the North Atlantic feed on arrow worms and sand eels, which are secondary consumers, as well as plant-eating copepods, which are primary consumers. To show feeding relationships accurately, biologists construct food webs (see above).

Losing energy

When one organism eats another, not all the energy "locked up" in the first organism is passed on. For example, parts of a phytoplankton are indigestible. Therefore, not all the food a phytoplankton makes is passed on to the primary consumer. In fact, only about 10 percent of the energy is passed from the first to the second feeding level. A similar situation occurs at each link in a food chain. The food energy passed onto the fourth feeding level is only about 0.1 percent of that at the first feeding level. The loss of food energy from one level to the next limits the number of feeding levels to six or less. There is simply not enough energy passed on to maintain a longer chain.

The killer whale is a fearsome predator that is not hunted for food by any other creature in the sea. It is at the top of the food chain.

Depth Zones

Marine scientists divide all the open ocean waters into layers, or zones, based on depth. The topmost layer is called the sunlit zone. Descending lower, the water becomes darker and cooler, and pressure increases.

The deepest zones are found in clear, tropical waters. There the sunlit zone (the ocean's topmost layer) extends down to about 660 ft (200 m). This is the greatest depth at which there is still enough light for phytoplankton to photosynthesize. At the lowest part of the sunlit zone the water is much cooler than at the surface, the pressure is 20 times more, and the light is filtered to a dull blue.

The sunlit zone is a place of great activity. Phytoplankton make food in this zone, and zooplankton graze upon them. The zooplankton, in turn, are eaten by larger animals. Most of the seafood humans eat comes from the sunlit zone.

The twilight zone

The next zone is the twilight zone (water below the sunlit zone). In clear, tropical water, a twilight world of dim light extends from the bottom of the sunlit zone to about 3,300 ft (1,000 m)—the depth at which all sunlight has been absorbed. This gray world is punctuated by flashes of blue and green light. The flashes are natural light, or bioluminescence, produced by some of the creatures living in this zone.

Bioluminescence has many purposes. Twilight-zone squid squirt bioluminescent ink to confuse predators while they make their escape. Hatchetfish (they get their name because

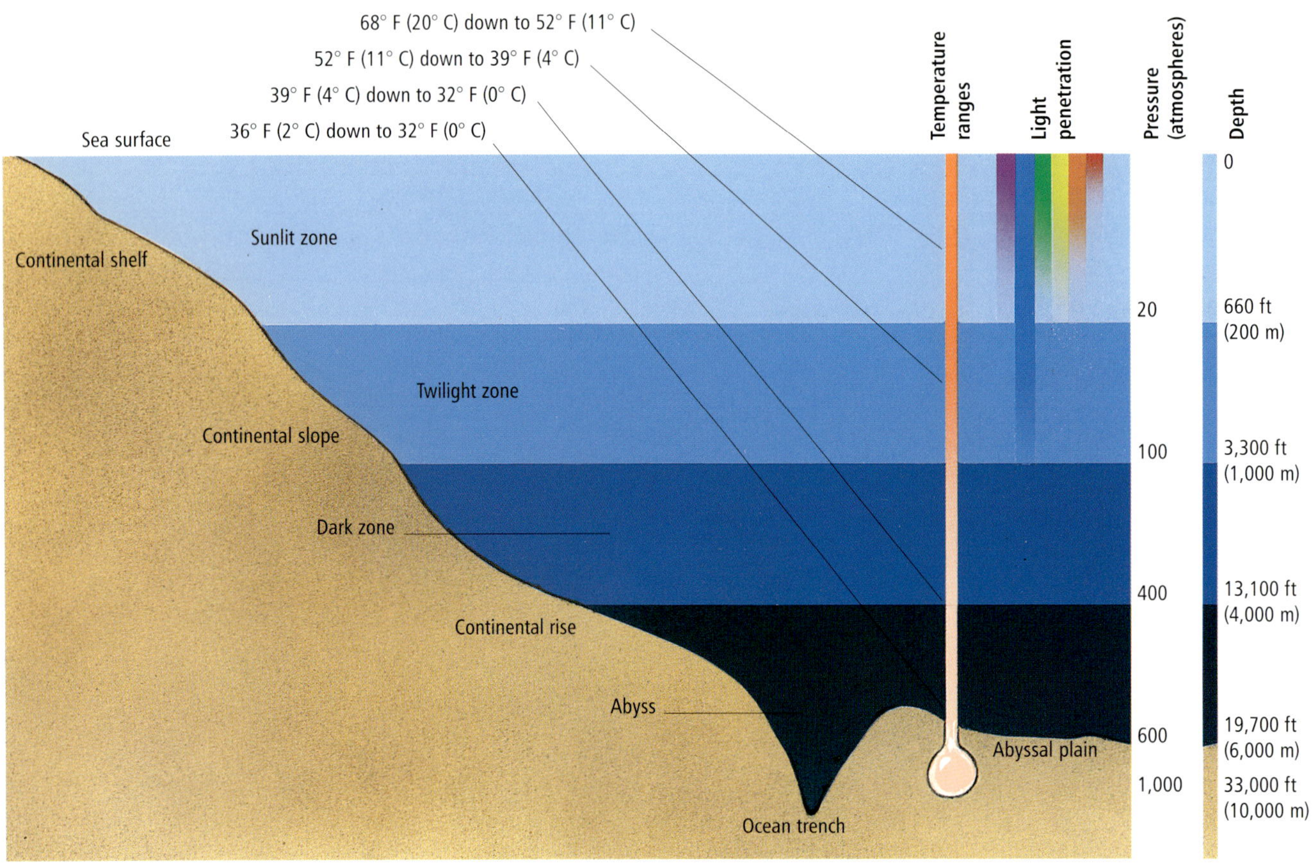

See also **vol 1:** 6–7, 43; **vol 2:** 16–17, 50–51, 56–57, 58–59; **vol 3:** 33

The sperm whale ranges through several of the ocean layers. It may bask at the surface but then plunge to depths of about 9,800 ft (3,000 m) into the dark zone to hunt for food.

their bodies are shaped like hatchet blades) use spots of light on their undersides to break up their outline so they are less visible to predators. Twilight-zone anglerfish dangle a luminous lure from their heads to attract prey. Lanternfish use patterns of light on their skin to attract mates.

The dark zone

Beyond depths of 3,300 ft (1,000 m) the ocean is a world of total darkness, illuminated only by the occasional flash of bioluminescent light. This dark zone extends to the plain that makes up most of the deep-ocean floor at depths to about 2.5 mi (4 km). Water pressure in the dark zone is more than 100 times greater than at the surface. The water is chillingly cold—between 32 and 39° F (0–4° C).

Most of the animals of the dark zone have bizarre appearances. They have shrunken bodies and huge heads with giant mouths armed with long teeth. In this realm of sparse food, these features enable the fish to concentrate all their energy on catching any prey that comes within range.

The depth zones in the ocean range from the sunlit surface waters down to the deepest ocean trenches. With ever-increasing depth, light decreases, the temperature falls, and pressure increases. Pressure is measured in "atmospheres." The pressure at sea level is 1 atmosphere. Below water, pressure increases by 1 atmosphere every 33 ft (10 m).

The abyss

Even deeper than the dark zone is the abyss found in the ocean trenches. In the abyss, the sea floor plunges to depths of 6 miles (10 km) or more. The sparse life that swims in these extreme depths is similar to that of the dark zone.

Up and down

Echo soundings sometimes pick up signals that seem to suggest the presence of seabeds in mid-water. Stranger still, these seabeds appear to rise and fall at different times of the day. In fact, the echo soundings do not come from a seabed, but are created by living organisms. When animals like shrimp, krill (shrimplike creatures), or small fish gather at a particular level in the water, they may form a dense layer that reflects sound waves. This appears as a sonar trace known as a deep scattering layer (DSL).

The DSL rises and falls as the organisms that produce it move up and down in the water. For example, a DSL produced by krill may rise near the surface at dusk and then descend to the dark zone at dawn. It seems that the krill wait until it is dark to feed on phytoplankton, thus reducing the chances of being eaten themselves. During the day, they sink out of the sunlit zone, probably to avoid predators and conserve energy by slowing their body activity in cooler water.

Zooplankton

Animals that drift or float in the ocean are known as zooplankton. Zooplankton eat bacteria, phytoplankton, and each other.

The smallest zooplankton are too tiny to see with the naked eye, but they are much larger than bacteria. They include foraminiferans (single-celled animals in a chalky shell) and radiolarians, which have intricate inner skeletons of silica. Both these creatures have sticky arms that project into the water to trap bacteria and phytoplankton.

Larger than these are copepods, tiny crustaceans that look like miniature shrimp. Together with their larger relatives, the krill, they are probably the most abundant animals on the planet. Some copepods have hairy limbs that beat at up to 2,000 times a minute to sweep small plankton toward

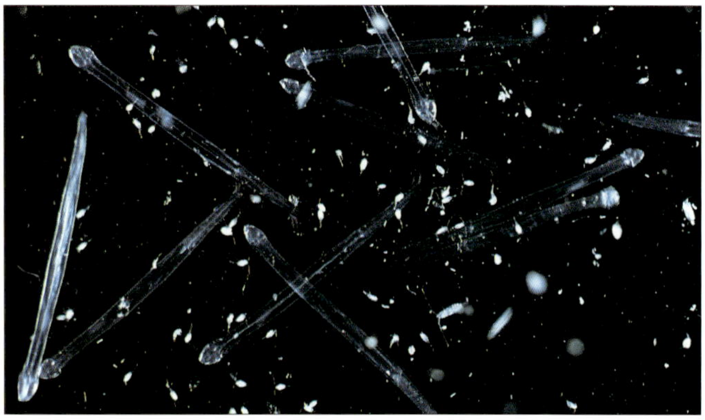

Arrow worms spend most of their time floating among the zooplankton, but they can dart rapidly through the water for short distances. Arrow worms grab their prey with grasping spines attached to the front of their heads. Most arrow worms are less than 0.8 in (2 cm) long.

Some common zooplankton include the *phyllosoma* larva, which is a stage in the life cycle of a crawfish. *Oikopleura* is a tadpolelike creature. Salps (related to sea squirts) join together to form chains. The salps shown here are about 0.25 in (0.6 cm) long. Shrimplike copepods are related to crabs. The biggest copepods are little more than 0.25 in (about 0.6 cm) long.

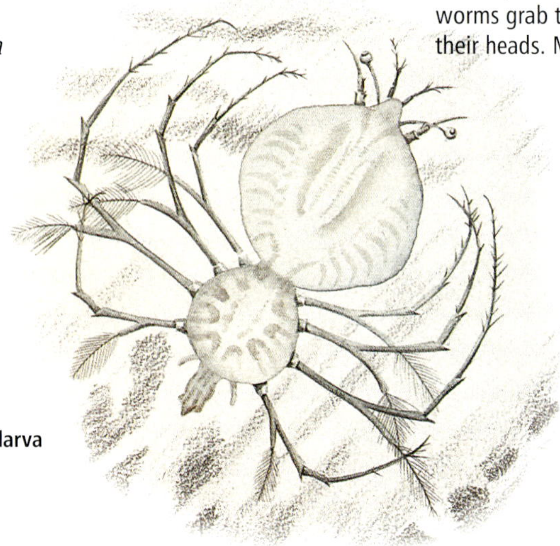

Phyllosoma larva

Copepod

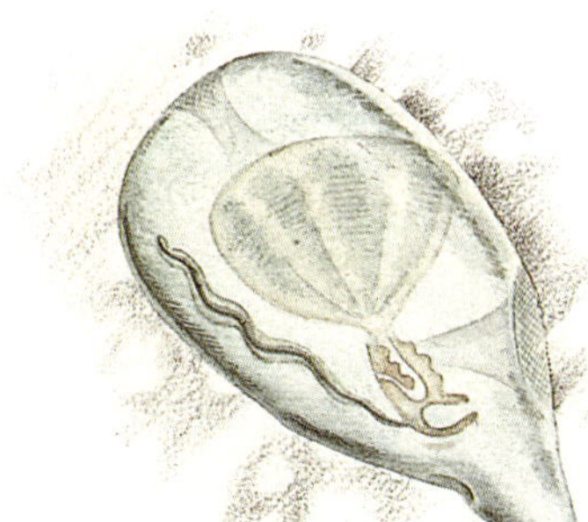

Oikopleura

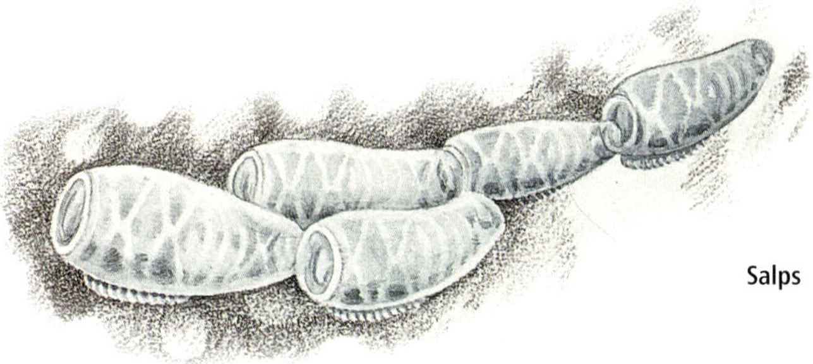

Salps

See also **vol 2:** 12–13, 16–17, 22–23, 40–41

The tiny larvae — immature forms— of many marine animals form part of the zooplankton. During the life cycle of this crab, for example, the zoea larva floats away from its parent and drifts close to the ocean surface. It passes through another planktonic stage, the megalops, when it becomes more crablike. Finally it turns into a crab, living on the sea floor.

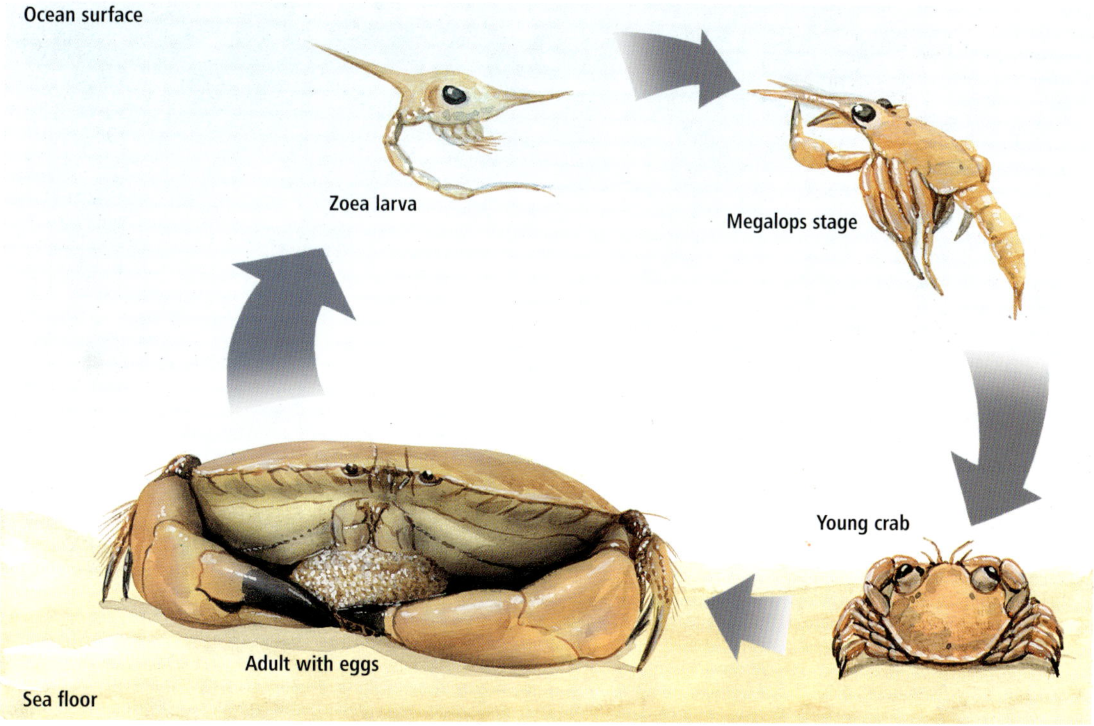

their mouths. Others stab and capture their prey with bristlelike projections on their legs or heads. Most copepods have long antennae, which they use to sense vibrations in the water, both to find food and to give early warning of approaching predators.

Salps are jellylike creatures with transparent, barrel-shaped bodies, while larvaceans look like miniature tadpoles. Larvaceans get their name because they resemble the tadpolelike larva of sea squirts. Both feed on phytoplankton by spreading nets of jelly into the water and then eating the nets and whatever is caught in them. Despite their simple and flimsy appearance, salps and larvaceans are quite closely related to vertebrates (animals with backbones).

Larger members of the floating community of zooplankton include arrow worms and jellyfish. Arrow worms float motionless in the water for much of the time. They are almost transparent and difficult to spot against the light streaming through the sea surface. When an unsuspecting small animal comes within range, the arrow worm darts forward and grabs it. Jellyfish float like transparent ghosts, trailing their tentacles through the water. Armed with stinging cells, the tentacles fire miniature poison darts into any creature that brushes up against them. Once the prey is immobilized, the tentacles draw it toward the jellyfish's mouth.

Most zooplankton spend their entire lives floating about in the ocean waters, but some animals become zooplankton for just a part of their lives. They include the larval (pre-adult) stages of certain animals. Some bottom-living animals such as barnacles, mussels, crabs, lobsters, sea urchins, and flatfish have a floating larval stage. After the breeding period, the ocean waters may teem with newly hatched planktonic larvae. The larvae are carried on ocean currents to colonize new habitats away from their parents. This is an important way for animals such as barnacles and mussels, whose adult stages are permanently fixed in one place, to spread out.

Their place in the food chain

Zooplankton form a vital link in the marine food chain. Small and medium-sized zooplankton feed on phytoplankton and are themselves eaten by larger zooplankton. The larger zooplankton, in turn, provide a meal for fish, squid, seabirds, and the largest marine mammals—the baleen whales. Zooplankton not only provide a vital link in the food chains of surface ocean waters, their feces (solid wastes) and the remains of their dead bodies sink down through the water. As the material descends, other sorts of creatures may swallow some of it. Whatever is left settles on the sea floor, providing food for bottom-dwelling creatures.

SPINELESS LIFE

Jellyfish, starfish, worms, crabs, sea anemones, and squid look very different from each other, but they have one feature in common—they lack a backbone. Spineless creatures make up 95 percent of sea-dwelling animal species.

Animals without a backbone are called invertebrates or spineless creatures. There are more than 30 different groups of marine invertebrates, and they live almost everywhere in the ocean. Most of the creatures that live in rockpools, in the sand, and under rocks on the beach are also invertebrates. The largest marine invertebrate is the giant squid. It can grow to 60 ft (18 m) in length, including its tentacles. At the other end of the scale are microscopic, single-celled creatures found almost everywhere, including the deep sea floor and floating in the surface waters.

Simple invertebrates

Sponges are the most primitive of the multicelled marine invertebrates. Some sponges are long and tubular, others are round, branching, or look like hard, colored coatings on rocks. The sponges we use in our homes are the leathery skeletons of one kind of sponge.

Sea anemones, corals, and jellyfish belong to a group of animals known as cnidarians (from the Greek word for "nettle," referring to the animals' stinging cells). Cnidarians capture their prey using tentacles armed with the stinging cells. Sea anemones are cnidarians that are fixed to rocks. At the top of the anemone is a mouth surrounded by stinging tentacles. These wave about in the water to catch small creatures. Corals are cnidarians that grow in clusters.

The jellyfish is a cnidarian that floats freely in the water. Trailing tentacles catch small fish and other creatures. Then the tentacles pull the prey toward the mouth of the jellyfish to be eaten. The 30-ft (9-m) long tentacles of the Portuguese man-of-war jellyfish carry enough venom to paralyze a human. *Obelia* is another type of cnidarian. It looks like a small, branching plant but it is an animal, like all cnidarians.

Thousands of different worms live in the sea. Most worms burrow into sand and sediment, but some crawl or swim about looking for food. Some have flattened, leaflike bodies, and others are long and thin. A few marine worms, known as ribbon worms, are nearly 100 ft (30 m) long.

Some of the many and varied species of marine invertebrates are shown here.

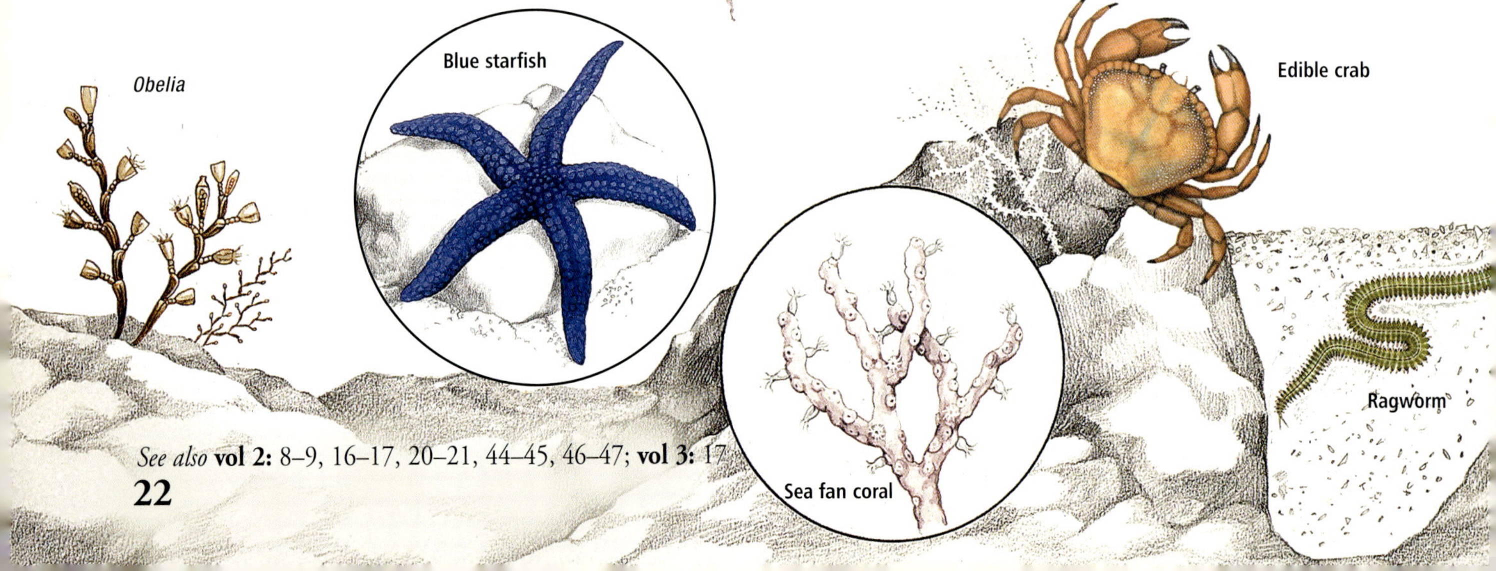

See also **vol 2:** 8–9, 16–17, 20–21, 44–45, 46–47; **vol 3:** 17

Shells and stars

The most numerous marine invertebrates are crustaceans. Crustaceans are joint-legged animals with a protective, chalky outer skeleton. They live by grazing, scavenging, or hunting. Many have powerful pincers for grabbing and tearing apart their prey. Crustaceans include the small, shrimp-like copepods and krill that make up part of the plankton, as well as about 10,000 species of shrimp, lobsters, and crabs. Barnacles are the most unusual crustaceans. Their bodies are hidden beneath chalky plates, and they remain fixed to rocks, piers, the bottoms of ships, and other places.

Most of the shells we see on the beach once belonged to creatures called mollusks. The shell provides a safe home for the soft-bodied mollusk that lives inside. Many mollusks glide about or dig into the sand with their muscular foot. Oysters, mussels, and clams are bivalve mollusks: this means their shells are in two parts joined by a hinge. Bivalve mollusks filter food particles from the water. Gastropod mollusks usually have a coiled shell. The word "gastropod" means "stomach-foot" because the animal's internal organs are on top of its foot. Their heads have tentacles for sensing the surroundings. Gastropods include sea snails and abalones. They are mostly scavengers and grazers.

Cephalopod mollusks ("cephalopod" means "head-foot" because their feet have become tentacles and are connected to their heads) include octopuses, squids, and cuttlefish. They are fast, predatory creatures. They have well-developed nervous systems and can change color if they need to hide.

Echinoderms—which means "spiny skinned"—include starfish, brittle stars, and sea urchins. They have a chalky skeleton and a body with a central mouth surrounded by arms. Most echinoderms walk and burrow using rows of water-filled stubby feet, called tube feet. Some starfish use their feet to pry open clam shells and eat the soft animal inside. Brittle stars get their name because their long, thin arms look so fragile. Many sea urchins look like round or flattened balls with long spines projecting from them. They use the spines for crawling about on the seabed while they search among the sediment for food fragments.

Sponges draw in sea water through tiny holes, or pores, in their body wall. As the water passes through the central cavity, food particles are filtered out of the water and eaten.

Jawless Fish

Jawless fish are the most primitive vertebrates (animals with backbones). As well as lacking jaws, jawless fish have no proper backbone. Instead, they have a strengthening rod called a notochord running the length of their backs.

The sucking mouth of a lamprey is ringed by horny teeth that scrape away the flesh of its victim.

Fresh water invasion

Between the 1890s and 1920s, marine lampreys entered the North American Great Lakes through newly deepened canal systems. When they arrived, the lampreys began devastating the local populations of trout and other food fish that local fishers depended upon. It was only in the 1950s that biologists began to turn the tide by poisoning the lampreys' spawning beds, and later releasing sterilized lampreys that mated with wild stock but produced no offspring.

About 500 million years ago, a small, fishlike creature swam clumsily along the bottom in the shallow waters of the ocean. Its body was covered in an armor of scaly, bony plates. The creature's mouth was a gaping oval, without jaws. Occasionally, the creature tilted its head down and sucked up small particles from the sea floor. This creature was a very early jawless fish known as a heterostracan ("heterostracan" means "differently shelled," and refers to the fish's body armor). Fish like these were the ancient ancestors of all the fish living in the oceans today. They breathed with gills—rows of feathery tissue that extract oxygen from water.

The closest living relatives of these early fish are the hagfish and lampreys. They are long and tubular like eels, but they lack the jaws and paired fins that eels and other more advanced fish have. Instead of jaws, hagfish and lampreys have horny teeth and tongues that scrape away prey's flesh.

Because they look quite similar, hagfish and lampreys are normally grouped together as "cyclostomes" (meaning "round mouths"). However, they are less closely related than their appearances suggest. Hagfish probably evolved more than 100 million years ago before lampreys. In fact, an examination of fossil—as well as living—lampreys, suggests they may be more closely related to modern, jawed fish.

Hagfish

If you place a hagfish in a bucket of sea water, the water soon turns to the consistency of wallpaper paste. This is because the fish produces enormous quantities of poisonous, sticky slime from special glands on its body as a deterrent against any predator that wants to make a meal of a hagfish.

Hagfish are the vultures of the underwater world. They are eyeless, but have tentacles that they use to scent rotting carcasses or dying fish. The hagfish gets right inside the dead or dying fish to eat the flesh. It does not have jaws to tear chunks of flesh away, so it anchors its head in the flesh and

See also **vol 2:** 9, 26–27, 28–29, 30–31

This kind of hagfish, *Myxine*, grows to 24 in (60 cm) in length.

Tentacles surround the mouth and nostrils of *Myxine.*

When a hagfish feeds, it twists its body into a knot to get extra leverage for pulling flesh from the prey.

then twists its body into a knot. It uses its tongue to scrape away at the flesh. As the hagfish attacks the flesh, it pulls itself against the knot it has made from its body to provide extra leverage.

This gruesome way of life has proved highly successful for hagfish. While many other kinds of fish—some much more advanced in evolutionary terms—have evolved and died out, the hagfish have held their own in the deep ocean for more than 400 million years.

Lampreys

Adult lampreys have a large mouth surrounded by a sucker and numerous teeth. The lamprey attaches itself to its live victim with the sucker, and then uses its teeth to penetrate the flesh. The lamprey releases bloodthinning chemicals into the wound, and consumes a "soup" of blood and damaged tissue. When it has taken its fill, the lamprey releases its prey, leaving a wound that continues to bleed and may become infected. A fish attacked by a large lamprey often dies.

Unlike hagfish, which live mostly in the deep sea, lampreys are found in fresh water and in shallow sea water. Some species of lamprey hatch in fresh water, migrate to the sea to mature, and then return to rivers and streams to lay their eggs. Other species of lamprey spend their entire lives in fresh water.

A sea lamprey builds a nest by moving stones with its sucker to make an oval, shallow depression on a gravelly bottom.

Some examples of the appearance and behavior of hagfish and lampreys are shown here.

FISH WITH JAWS

Between 500 and 410 million years ago an important evolutionary event occurred: a new type of fish appeared on the scene. It had a proper backbone instead of a notochord and, just as significantly, it had jaws.

There are far more species of jawed fish than there are jawless fish—a mark of their success. Jawed fish are divided into two main groups according to their type of skeleton. Bony fish have a skeleton made of bone and cartilage, a flexible material containing gristle (also found in a person's nose). In the other group, the skeleton is made entirely of cartilage. More than half of all species of jawed fish live in the sea, the fish with cartilage skeletons almost exclusively so. There is an enormous range of sizes, shapes, and life styles among jawed fish, from massive sharks that grow to 60 ft (18 m) in length, to fish that can glide through the air or crawl about on land, and to others that look more like a bunch of seaweed than a fish.

Jaws and fins

The development of jaws was a major evolutionary advance. During the course of evolution, parts of the skeleton that supported the gills of some primitive fish moved forward and formed structures to support the mouth and become jaws. (Gills are the breathing organs used by many aquatic animals.) We do not know exactly how this happened. However, jaws made fish more effective predators. They could now bite or chew instead of simply sucking up their food, a change that enabled them to prey on larger live animals. By 250 million years ago, 30-ft (9-m) sharks were roaming the oceans, attacking other creatures with their powerful teeth and jaws.

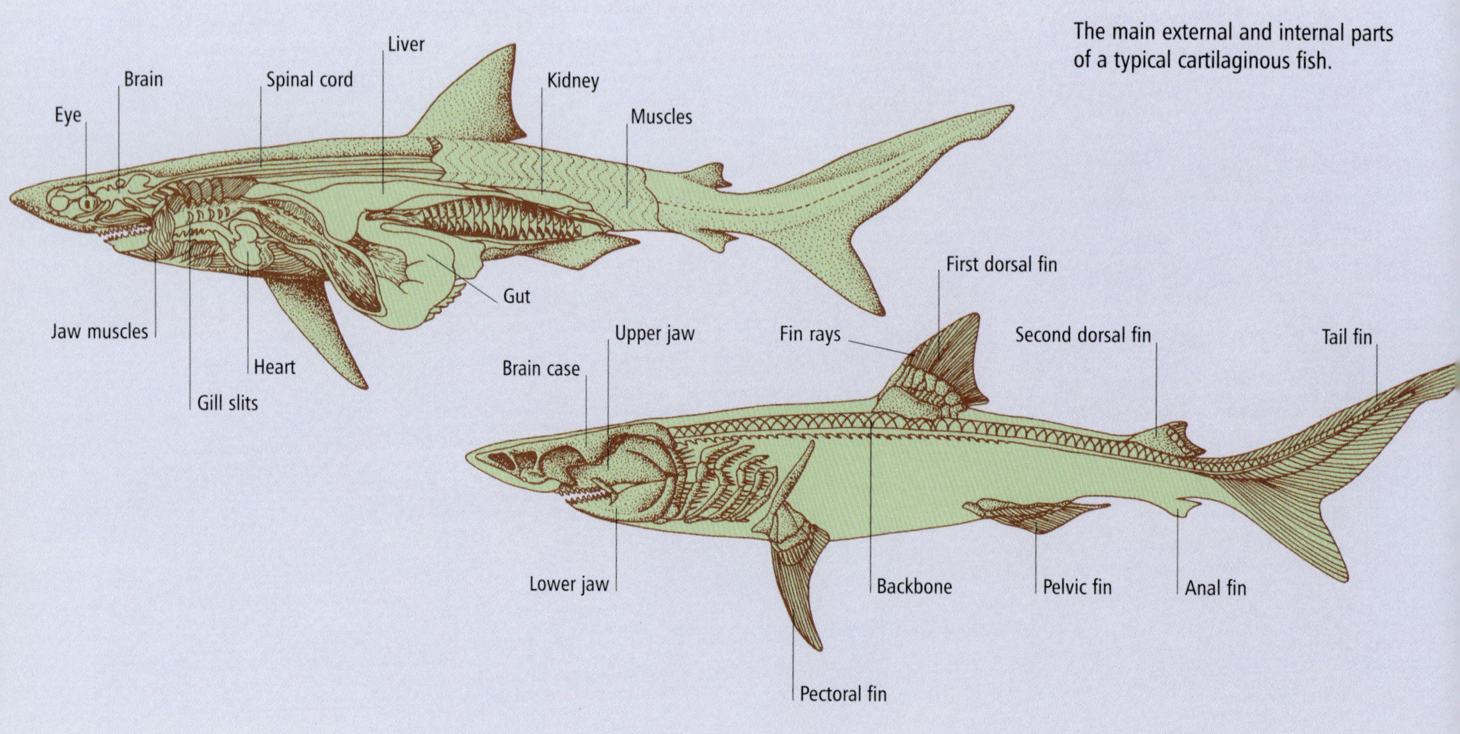

The main external and internal parts of a typical cartilaginous fish.

See also **vol 2:** 28–29, 30–31, 56, 58–59

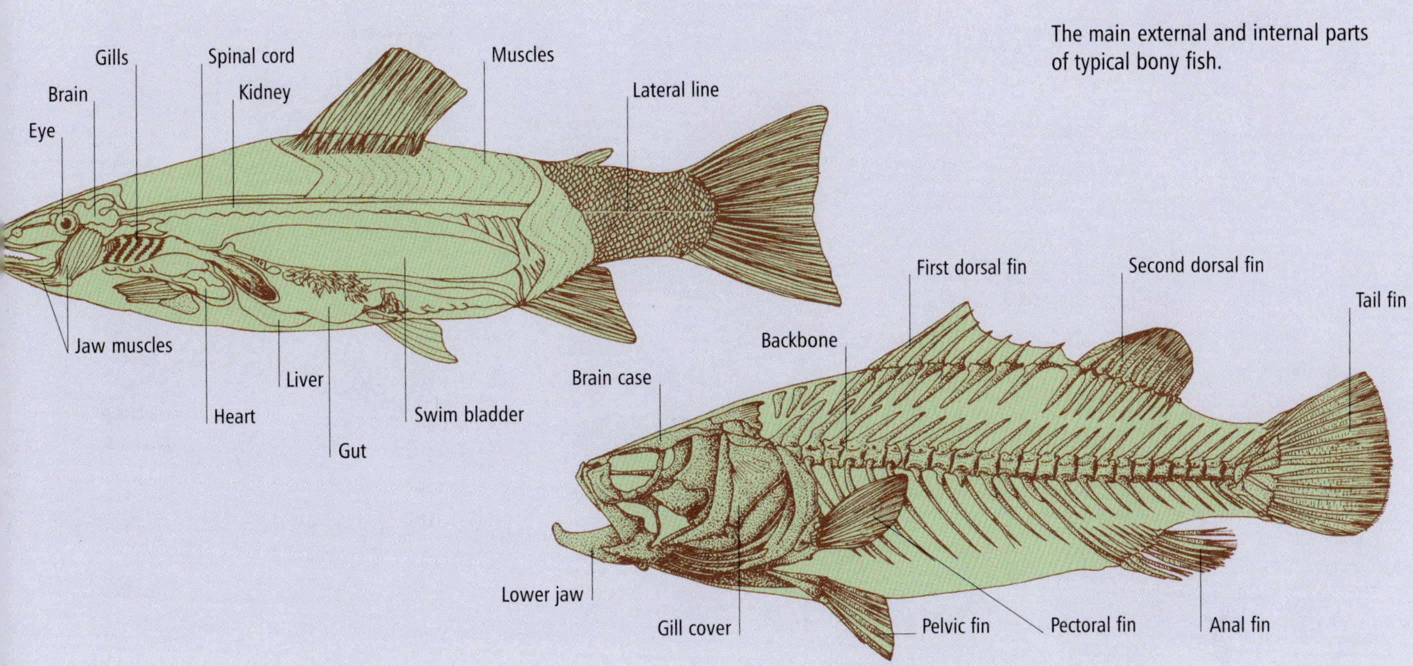

The main external and internal parts of typical bony fish.

The development of paired fins was another major advance. Almost all fish have a tail (caudal) fin and a back (dorsal) fin. The tail fin propels the fish forward. The dorsal fin helps the fish steer and prevents rolling. Sometimes there are other vertical fins, such as the anal fin. In addition, most modern fish have at least one extra pair of fins on each side of the body. Paired fins near the shoulder are called pectoral fins and a pair further back on the underside are called pelvic fins. These paired fins stabilize the fish when it swims. With paired fins, fish could swim better, hunt prey more easily, and escape their predators quicker.

Cartilaginous fish

The cartilaginous group of fish includes the sharks, skates, and rays. About 700 species of cartilaginous fish swim the oceans today. The ancestors of cartilaginous fish had a skeleton of bone, but cartilage is lighter than bone, and some fish developed a cartilage skeleton to help them float better. Unlike bony fish, cartilaginous fish do not have an air-filled chamber—a swim bladder—to help them float. Instead, they store oil in their liver to make them lighter. The absence of a swim bladder means that most cartilaginous fish sink when they stop swimming. But the fins of cartilaginous fish are shaped in such a way that when the fish swims forward, it lifts off the sea bottom automatically. The skin of most cartilaginous fish feels rough to the touch because the scales that cover the body are shaped like miniature teeth.

To breathe, fish take in water through the mouth or through nostril-like openings called spiracles. The water passes over the gills and oxygen is extracted before the water passes out of the fish through the gill slits. The gill slits of cartilaginous fish are visible on the side of sharks and on the underside of skates and rays.

Bony fish

Bony fish include the fastest swimming fish in the sea. There are about 24,000 types of bony fish, found all over the ocean from rockpools to the murky depths 5 miles (about 8 km) down. Most bony fish have fins supported by thin, bony structures called rays and a body covered in lightweight scales. The flexible fins of bony fish are a major reason for their success. The paired fins can move in almost any direction, enabling the fish to turn, hover, or even swim backward.

All bony fish have a swim bladder, a gas- or air-filled chamber below the backbone. By altering the amount of air in the swim bladder, bony fish can float at any level in the water, even if they are not swimming, so saving energy. The gills of bony fish are efficient at extracting oxygen from water. A good supply of oxygen is vital for releasing energy to power the muscles.

SHARKS, SKATES, AND RAYS

Sharks, skates, and rays have swum the oceans since before the start of the Age of Dinosaurs, 225 million years ago. There are about 400 species of shark and about 320 types of skate and ray.

Although sharks have a reputation as voracious predators, they do not all attack and eat large prey. The very largest sharks, the 33-ft (10-m) long basking shark and the 60-ft (18-m) whale shark, are harmless plankton feeders. The skates and rays generally feed only on small animals like crabs, scooped up from the seabed. Even the huge 22-ft (7-m) manta ray—the biggest of the rays—also feeds only on plankton.

Shark versus humans

Despite the fearsome reputation of some sharks, humans are a much greater threat to them than they are to us. In 2001 the International Shark Attack File (ISAF) recorded 76 unprovoked shark attacks on humans. Five people died as a result. In that same year, humans killed millions of sharks. Living sharks have their fins cut off to provide fins for shark-fin soup. The shark is then thrown back to die. In 2000 the U.S. House of Representatives banned shark definning in U.S. waters.

A manta ray swimming gracefully off the Great Barrier Reef, Australia. The manta ray takes in plankton as it moves.

Remarkable senses

Sharks have amazing senses. Some species can smell the equivalent of a drop of blood in a swimming pool of water. They also have a well-developed lateral line system (a line of sense organs along the side of a shark that detect vibrations). Sharks detect smells and vibrations from long distances and then use their eyesight and special pores that can detect electricity to hone in on prey at close range. These special pores on the shark's head pick up weak electrical fields produced by movements of their prey's muscles. Using this detector, sharks and rays can even uncover prey buried in sand.

Shark attacks

Sharks are greatly feared because of their attacks on humans, but their reputation as killers is exaggerated. Of about 400 shark species, 40 or so have been known to bite humans, but shark experts think that three species only—the great white shark, the tiger shark, and the bull shark—are responsible for the majority of attacks. In other species, most attacks only occur after a shark has been grabbed or prodded.

There is no simple explanation for why sharks attack people without provocation. Sometimes, as in some great white shark attacks, it may be a case of mistaken identity. Seen from below, the silhouette of a person on a surfboard can be mistaken for a seal or sea lion, the natural prey of the shark. In other instances, the shark is simply tasting a possible food item before deciding whether or not to swallow it.

Skates and rays

Skates and rays look quite different from most sharks, but they are related and have similar body designs. Skates and rays are basically sharks whose bodies are flattened. Their shape is an adaptation to living on or near the sea floor,

See also **vol 2:** 26–27, 30–31, 56; **vol 3:** 42–43, 54–55

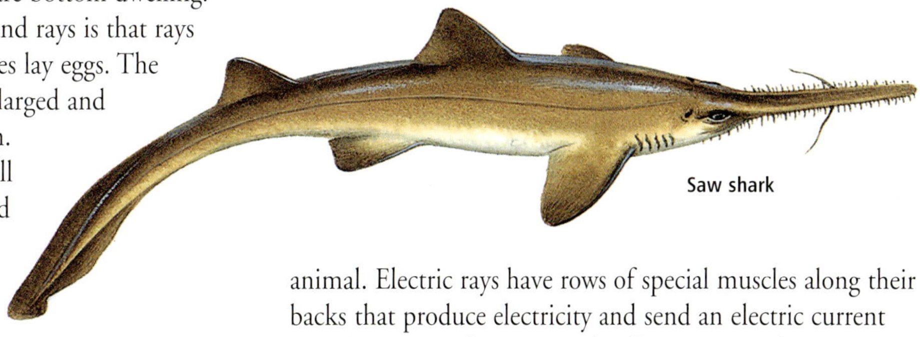

Some sharks, like the basking shark, have torpedo-shaped bodies with long fins. Other sharks, like the dogfish, are long and thin with narrow tails. The saw shark has feelers around its mouth for detecting mollusks and crustaceans hidden in sand, and a saw-edged snout for slashing its prey.

although not all species living today are bottom dwelling. The main difference between skates and rays is that rays give birth to live young, whereas skates lay eggs. The pectoral fins of skates and rays are enlarged and resemble wings that they flap to swim. Unlike sharks, skates and rays have gill slits on the underside of the body, and they take in water through holes on top of the head. In this way, they do not clog their gills with sediment when they breathe on the sea bottom.

Most skates and rays eat bottom-living crustaceans, mollusks, and echinoderms, which they crush with their flattened teeth. Some skates and rays have unusual, powerful defenses. Stingrays have one or more poisonous spines in their tail, which they lash forward to stab any threatening animal. Electric rays have rows of special muscles along their backs that produce electricity and send an electric current into the surrounding water of sufficient strength to stun small fish and deter predators.

Slow reproduction

Sharks, skates, and rays grow fairly slowly and do not produce many young. They invest more energy in their young than most bony fish and either lay a few large eggs or give birth to live young. This means that if a population of sharks or rays is heavily fished, it may take many years for it to recover. Some may never recover.

The 20-ft (6-m) long great white shark is the most feared of all sharks. This fierce predator, found in warm temperate and tropical coastal waters, feeds on fish, seals, and dolphins and also scavenges carcasses.

Bony Fish

Bony fish have developed many different ways of living and feeding in the ocean. Such variety and adaptability are key factors in their success.

On land, backboned animals have developed different types of jaws, teeth, or beaks so that they can eat every kind of natural food. Similarly, in the ocean, bony fish that use different methods of feeding have evolved. For example, barracudas are fast-cruising predators that seize their fish prey in daggerlike teeth. Parrotfish have a horny beak at the front of the jaws. They use it to bite off chunks of coral or nibble at coatings of green algae. At the back of the throat, the grinding teeth of parrotfish break down the coral or algae so it can be digested. Herring use parts of their gills, called gill rakers, to sieve zooplankton from the water.

Schooling

Many species of bony fish swim in tightly packed groups, called schools. Their movements are so closely coordinated that hundreds of individuals seem to move in unison, and then suddenly change direction simultaneously, like a single, giant creature. It is a fascinating sight. Fish are able to stay in tight formation, and yet they can change direction together at rapid speed by means of their lateral line system. A series of jelly-filled tubes along the fish's sides (the lateral line) are sensitive to vibrations in the water. As the fish swims, it creates a V-shaped wake under water, rather like the wave made when a ship moves along the water's surface. Nearby objects, including other fish in the school, sense the movement through their lateral line system. This sensory system provides the fish with a detailed "picture" of its surroundings and the movement of other creatures.

The main advantage of schooling is probably to reduce the risk of being hunted. Predators find it more difficult to locate a single school of fish in the ocean rather than lots of scattered individuals. When a predator attacks a school, the fish scatter in all directions, perhaps confusing the predator, so each individual has a better chance of survival.

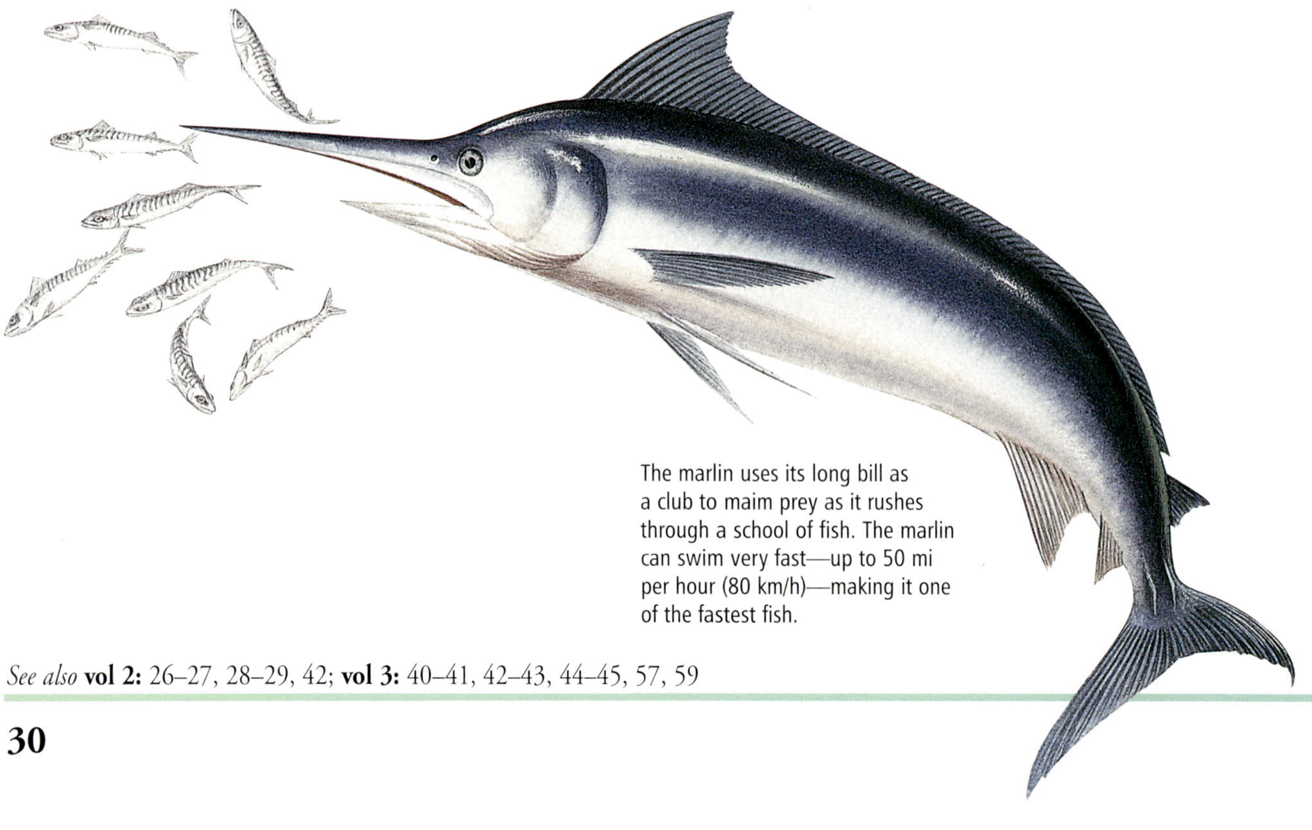

The marlin uses its long bill as a club to maim prey as it rushes through a school of fish. The marlin can swim very fast—up to 50 mi per hour (80 km/h)—making it one of the fastest fish.

See also **vol 2:** 26–27, 28–29, 42; **vol 3:** 40–41, 42–43, 44–45, 57, 59

Fish out of water

Around spring tides, during spring and summer, the sandy beaches of southern California and northern Baja California become a writhing mass of 6-in (15-cm) long silvery fish. These fish are grunion, one of two species of fish that breed on land (the other is the capelin, which lives in the Atlantic and breeds in Newfoundland). Waves on a rising tide wash female grunions up the beach, where they lay their eggs, and the male fertilizes them. The eggs hatch about two weeks later, when the next spring tides wash up the beach and carry the eggs back into the sea.

Millions of eggs

Bony fish produce young in a great many ways. In many species, the females release thousands of eggs at one time, and the male fertilizes the eggs in the water. The female of the North Atlantic cod typically produces 10 million eggs in a season. The fertilized eggs float up into the surface waters and form part of the plankton community. Thousands of the eggs, and the fish larvae they hatch into, are eaten by plankton-feeders, and few fish survive to adulthood.

Species of bony fish that live in coastal waters, on the sea floor, or among floating seaweed, are usually less wasteful. They produce fewer eggs and may spend more time and energy caring for them. Some coral reef fish are mouth-breeders. The males protect fertilized eggs by carrying them in their mouth until they hatch.

The seahorse gets its name because its body is shaped like a horse's head. The pregnant female seahorse places her eggs in a brood pouch on the belly of her male partner. He incubates the eggs, and when they hatch, miniature seahorses emerge form the pouch opening (below).

Flying fish

The four-winged flying fish (below) of tropical Atlantic waters feeds on zooplankton close to the sea surface. Fast-moving predators such as tuna hunt the flying fish. When threatened, the flying fish darts into the air, its tail propelling it out and over the water. The pectoral and pelvic fins snap open like wings, and the flying fish glides away from its pursuer. It can travel distances of more than 300 ft (90 m) in this way. Even so, a tuna can sometimes track the flying fish when it is above the water and grab it while it is in mid-air.

Marine Reptiles

Reptiles were the first vertebrates (backboned animals) to adapt fully to life on land. A few have returned to living in the sea. Among the 6,000 or so living species of reptile, about 80 are found in sea water.

Some 400 million years ago, complex life forms began to colonize the land. Within the space of 80 million years, mosses and ferns evolved from simple, multicelled aquatic algae. Then insects evolved from arthropods (invertebrates with hard, outer skeletons and jointed limbs) that lived in the ocean. Insects soon conquered the sky as well as the land. Amidst this "explosion" of new life, the amphibians also appeared. Amphibians are vertebrates such as frogs and toads. Many live on land and breathe air, but all must return to water to lay their eggs.

By 320 million years ago, reptiles, the first true land vertebrates, had evolved. Reptiles lay leathery eggs that retain moisture and can survive outside water. They also have a hard, waterproof skin made of scales or bony plates. Despite the fact that reptiles are really more suited to life on land, some species have returned to the water. All reptiles breathe air using lungs, so even those that swim under water must surface to take in air. Nearly all come ashore to lay eggs.

Sea turtles

Sea turtles live in nearly all the world's oceans, although they are absent from very cold waters. They swim with flippers, which are modified legs. They are chiefly carnivorous, feeding on fish and invertebrates, but the green turtle grazes on seagrasses and seaweeds. The sea turtle's hard shell helps to protect the animal from predators, but turtles cannot entirely withdraw their large, flat flippers under the shell like tortoises can, and sometimes a limb is bitten off by a shark.

Male sea turtles spend their entire lives at sea, but females come ashore to lay eggs. After mating at sea, a female turtle comes out of the sea at a carefully selected sandy beach—the same one where she herself hatched. She hauls herself laboriously up the shore and digs a hole in which she lays 100 or so eggs. She then covers the eggs in sand and drags herself back to sea.

Inside the nest, temperature determines the sex of the baby turtles. Typically, warm eggs (above 87° F/30.5° C) yield females, while cooler eggs (below 82° F/28° C) hatch into males. When the eggs hatch, the hatchlings return to the sea, guided by a combination of reflected light, the

Baby loggerhead turtles make their journey to the sea after hatching on the beach. Many fall victim to predators such as crabs and seabirds.

See also **vol 2:** 42, 54

sound of the surf, and high air humidity—all of which tell them which direction to go. On the way, many are picked off by birds and crabs. Even if they reach the sea, sharks or crocodiles may be waiting for them. Fewer than one hatchling in a thousand reaches adulthood.

All seven species of sea turtle are threatened or endangered because of the demand for their shells and their meat. All are protected by international law.

The banded sea snake lives in the coastal waters of the Persian Gulf, Indian Ocean, and parts of the Pacific Ocean. It can remain submerged for up to two hours at a time. If accidentally washed ashore, the banded sea snake is helpless because its muscles are too weak for use on land.

Saltwater crocodiles

Among crocodiles, only the Indo-Pacific saltwater crocodile lives in the open sea, although a small population of the American crocodile lives in slightly salty (brackish) water at the southern edge of the Everglades in south Florida. Both species are voracious hunters. They eat invertebrates and fish, and will attack mammals on land, pulling them under water and drowning them before eating them.

Sea snakes

In many parts of the world, a few land-based snakes hunt in shallow sea water. True sea snakes (those that spend most or all their lives at sea) are largely restricted to the warmer parts of the Indian and Pacific oceans. Many evolved from cobras.

Some sea snakes have highly potent venom that paralyzes their fish prey almost instantaneously. The venom can also be fatal to humans, and fishers are sometimes fatally bitten when they try to remove sea snakes from their nets. One species, the yellow-bellied sea snake, is very widespread, and is probably the most abundant reptile on Earth. Like other well-adapted sea snakes, its tail is flattened into a paddle for swimming. Unlike other marine reptiles, most sea snakes give birth to live young at sea.

The marine iguana is the only lizard well adapted to a life in the sea. It lives on the Galápagos Islands where it dives for seaweeds. The marine iguana uses its flattened tail for swimming.

Seabirds

Soaring over the ocean, bobbing like corks on the waves, or diving arrowlike into the water in search of fish, seabirds are superbly adapted for a marine life. Some simply feed in the water, but others spend almost their whole lives at sea.

Seabirds have developed special features to cope with a life at sea, including webbed feet for swimming and special glands that empty into the nostrils to get rid of excess salt. All seabird species breed on land.

Penguins and auks

Penguins are the seabirds best adapted for diving and swimming. They no longer use their wings for flight, but as flippers for "flying" under water in search of food. The Antarctic emperor penguin (the world's largest penguin) dives deeper than any other bird. It regularly plunges to depths of 820 ft (250 m) or more in search of fish and squid. On land, penguins waddle about with a comical, almost human gait or slide on their bellies, propelling themselves with their flippers.

All penguins live in the Southern Hemisphere, and most dive in cold water. For protection against the cold, they have an insulating layer of fat beneath the skin and dense, waterproof feathers that trap warm air near the body. Male emperor penguins incubate a single egg in a pouch on top of their feet to keep them from freezing on the ground.

Puffins have thick, parrotlike beaks. They can catch several small fish in a single dive, holding each one securely while they return to the nest. Puffins breed on cliff tops, often laying their eggs in a burrow in the ground. Many species of puffins spend the winter at sea.

A big catch

Many seabirds prey on fish, and large numbers of birds can seriously affect other marine life populations. A study of a colony of 5 million Adélie penguins showed that the birds ate about 9,900 tons (9,000 tonnes) of krill and young fish every day. Along the coast of Peru, cormorants and other seabirds catch about 2.8 million tons (2.5 million tonnes) of fish in most years—nearly one-third of the annual catch netted by fishing boats and trawlers fishing in the same waters.

See also **vol 1:** 30–31, 51; **vol 2:** 16–17, 43, 48–49; **vol 3:** 50, 52

Auks, a group of birds that includes puffins, are found in oceans throughout the world. Auks can fly, but they catch their food by diving. The razorbill, another auk, is found in many parts of the Northern Hemisphere. It looks rather like a penguin and has a similar lifestyle. Like penguins, razorbills swim under water using their wings.

Superb fliers

One group of seabirds are called "tubenoses" because their nostrils join to form a single tube running along the top of the bill. They expel excess salt through this tube. Tubenoses include petrels, shearwaters, and albatrosses—some of the greatest fliers of the oceans. Most impressive of all is the wandering albatross, which has a wingspan of 11 ft (3.4 m)—the largest of any bird. The wandering albatross makes nonstop flights around the Southern Hemisphere before settling to breed on islands in the Southern Ocean. The flights often last two years or more.

The smaller tubenoses include petrels and shearwaters. Many hover and fly just above the water surface. Storm-petrels patter their feet on the water surface and pluck out plankton and small fish. Other petrels, and most shearwaters, are superb divers, able to catch small fish at high speed.

Some of the many different feeding methods used by seabirds are shown here.

Like other albatrosses, the black-browed albatross shown here has long, narrow wings. The shape is perfect for soaring and gliding on updrafts of air, making it possible for albatrosses to spend long periods aloft.

Gulls and terns

Best known of all the seabirds, and with the biggest number of species, are the terns and gulls. Terns are plunge-divers, plummeting into the sea and using their slender beaks to snatch small fish. Gulls will scavenge for almost any kind of food. They gather in squabbling hordes or pounce on discarded fish and offal thrown into the water by commercial fishers, or sort through litter left by humans.

The gull-like jaegers and skuas are the pirates of the seabird world. These aggressive birds are found in most oceans of the world, especially cooler ones. They steal fish caught by other seabirds, and they nest near the breeding colonies of other species, consuming their eggs and young.

Webbed feet and pouches

Members of one seabird group, spread throughout the temperate and tropical oceans, have webbing between all four toes to help them swim more efficiently. They include the brown pelican, which plunges onto the sea surface to scoop up fish using the pouch that hangs below its bill. Related to pelicans are cormorants, which swim on the surface and then dive and pursue fish. After coming out of the water, cormorants often stand on rocks with their wings open to allow them to dry. Frigatebirds snatch prey from the beaks of other seabirds or harass them so that they regurgitate their food. Male frigatebirds inflate their bright red chests like a balloon in the breeding season to court females.

MARINE MAMMALS

Mammals are warm-blooded creatures that breathe air and suckle their young with milk. Most mammals live on land, but some have adapted to life in the sea.

Mammals evolved on land, but several groups have returned to the sea. Ancestors of whales, dolphins, and porpoises were the first mammals to do so. Whales, dolphins, and porpoises are called cetaceans (from the Latin word "cetus" meaning "large sea creature"). Cetaceans have streamlined, fishlike bodies. Since returning to the sea their front limbs have altered to become paddle-like flippers for steering, and their hind limbs have all but disappeared. Cetaceans have a long, tapering tail which is flattened horizontally into two wide blades called flukes. Up and down movements of the tail drive the cetacean forward.

The blowholes on top of cetaceans' heads are their nostrils, through which they breathe. The animals come up to the surface and blow out stale air before taking in fresh air and resubmerging. Some whales can stay submerged for more than an hour. Cetaceans spend their entire life in the sea, and even give birth and suckle their young under water.

Sea cows

Only one other group of mammals gives birth in water, the sea cows (sirenians). Sea cows look like a cross between a whale and a walrus and may have been responsible for the "mermaid" myth. Apparently, sailors mistook suckling sea cows for creatures that were half women and half fish. They live in shallow, warm water and graze on seagrasses and seaweeds. Like whales, sea cows have almost no hair.

Today, only four species survive—three kinds of manatee and the dugong. The tail of the dugong is similar to a whale's, while that of the manatee looks more like a beaver's paddle. Sea cows are less streamlined, and less well adapted to diving, than whales. Sea cows can hold their breath under water for only about 10 minutes.

Seals, sea lions, and walruses

The seals, sea lions, and walruses are known as pinnipeds ("pinniped" means "feather-footed" in Latin). Their closest land-based relatives are bears, cats, and dogs. To live in the sea, pinnipeds have developed a dense layer of short, oily,

A dugong grazing seagrass on the seabed. The scars on its body are probably the result of fights with other males.

See also **vol 2:** 38–39, 40–41; **vol 3:** 38–39, 54–55

Most seals (above) are true or earless seals. Sea lions (right) and fur seals are also known as eared seals.

waterproof fur, and a thick layer of fat (blubber) beneath their skin, to keep out the cold.

Seals spend much of their time in the water chasing after prey. They have sleek, streamlined bodies and are superb swimmers and divers. The southern elephant seal can hold its breath for two hours while diving to a depth of more than 5,500 ft (1,675 m). Most seals are clumsy on land, however, pulling themselves along laboriously with their flippers. Most seals eat fish, but the 12-ft (3.5-m) long leopard seal (so-called because it is aggressive and has spotted skin like a leopard) eats penguins and other seals. The crabeater seal has special teeth for straining krill from the water. Most baby seals are ready to take to the water in a matter of hours, or at most a few days, after being born.

Sea lions and fur seals also have streamlined bodies for moving easily through the water. They can swivel their rear flippers forward and use all four limbs to walk or run on land. Their young spend weeks or months on land before venturing into water. The Californian sea lion lives on rocky islands off the Pacific coast.

Walruses live in arctic seas. Their two tusks can grow to a length of up to 3 ft (1 m). The walrus uses the tusks for a variety of purposes—breaking air holes in the ice for breathing, hauling itself out of the sea, and for fighting. Biologists once thought that walruses used their tusks to dig out clams. However, this is not the case. They locate food on the seabed with their whiskers and then dig it out with their snouts or uncover it by blowing away the sand with jets of water.

The North Pacific sea otter is a marine mammal related to land-dwelling skunks. It spends most of its life at sea but can stay under water for only a few minutes. It catches a sea urchin, clam, or crab before returning to the surface. Floating on its back, it uses a stone to smash open its prey. The sea otter is one of only a few animals to use tools.

Toothed Whales

There are about 70 species of toothed whale in the ocean—nearly 90 percent of all cetaceans. They include agile, leaping dolphins and porpoises, as well as sperm whales, the largest of the toothed whales.

As their name suggests, all toothed whales have teeth—usually arranged in prominent, beaklike snouts. Some types of dolphin have more than 100 teeth. Porpoises have shovel-shaped teeth and blunt snouts, but otherwise resemble dolphins. One of the strangest of the toothed whales is the arctic-dwelling narwhal. The narwhal has only two teeth, and in the male the left-hand tooth grows into a long tusk. Males fight each other with their tusks to decide who will mate with females. The sperm whale's large, square forehead makes up one-third to one-quarter of the animal's length of 65 ft (20 m). It contains a waxy oil called spermaceti, which keeps the animal afloat. Humans once hunted sperm whales for this oil, which was prized as a lamp fuel. The sperm whale dives deeper than any other creature in search of prey. Most toothed whales hunt fish or squid, although some search for crabs, sea urchins, and other bottom-dwellers. Killer whales, orcas, also eat other mammals.

Seeing with sound

Toothed whales find their way about and locate food using sound. They produce a series of loud, high-frequency clicks and analyze the returning echoes bounced back from objects in the water to create a "sound picture" of their environment. This way of "seeing" with sound is called echolocation. A dolphin's echolocation system is highly sensitive. Dolphins in aquariums can distinguish between a kernel of corn and a lead shot using echolocation.

Strong social ties

Many species of toothed whale live in tightly knit groups. Schools of dolphins or porpoises leaping through the waves are familiar sights to seafarers. As they move rapidly through the water, dolphins keep in touch with each other by clicking and whistling. Each dolphin has its own call, recognized by the others when it is separated from the rest of the group.

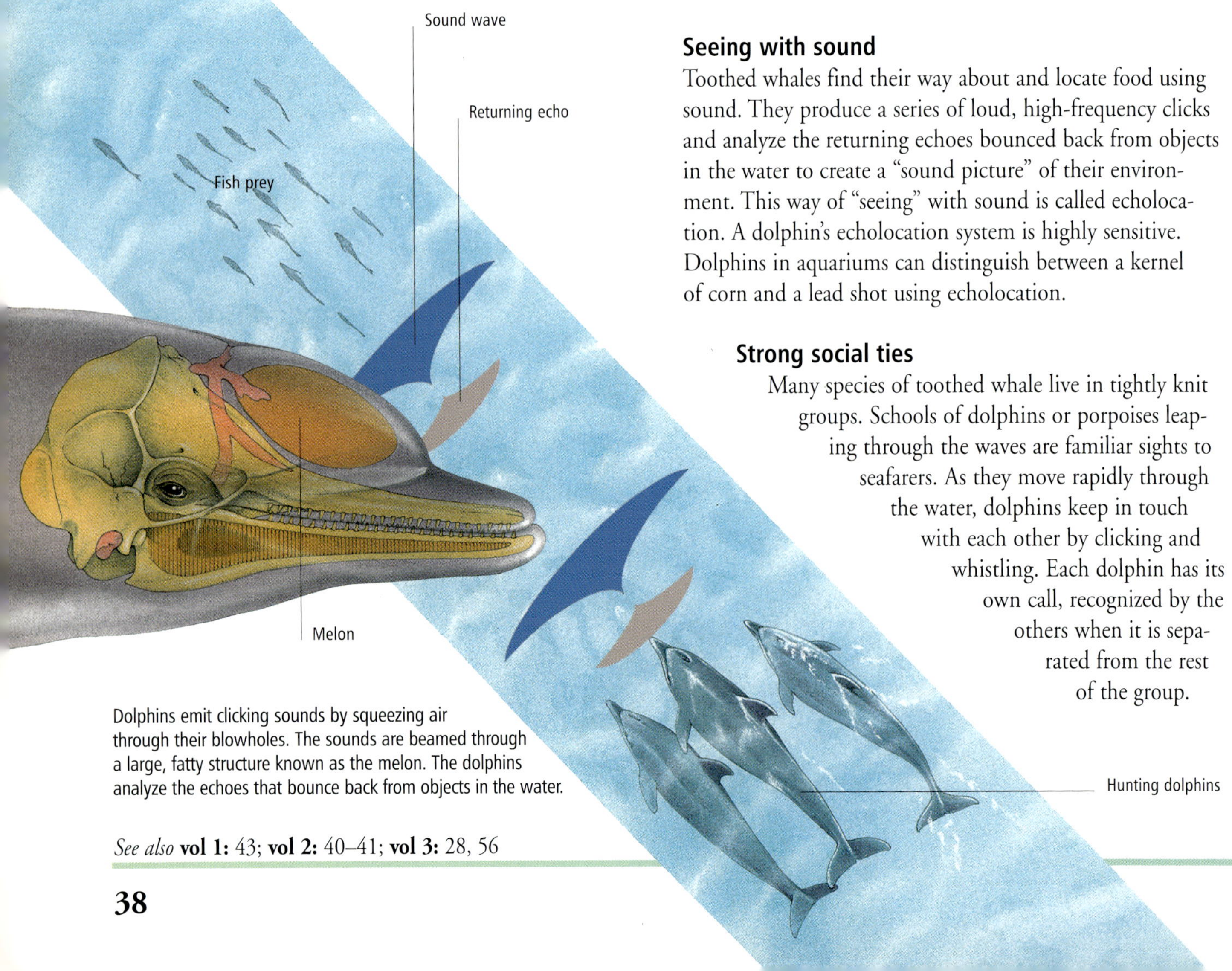

Dolphins emit clicking sounds by squeezing air through their blowholes. The sounds are beamed through a large, fatty structure known as the melon. The dolphins analyze the echoes that bounce back from objects in the water.

See also **vol 1:** 43; **vol 2:** 40–41; **vol 3:** 28, 56

Long-finned pilot whales stranded on Ocean Beach in Tasmania, Australia. Despite the efforts of the rescue team, none of these whales was successfully returned to the sea.

Working together

Killer whales or orcas, the largest members of the dolphin family, live in groups called pods. They often stay together for life. As well as living together, some pods cooperate when feeding. For example, some herd schools of fish into a small bay before eating them (below). At sea, killer whales may surround a school of fish and then stun them with slaps of their tails. Members of a killer whale pod will try to separate a mother gray whale from her calf so they can attack the infant.

Sometimes large groups of toothed whales become stranded on the shore. They will almost certainly die unless human rescuers guide them back to sea. Whales may strand for a variety of reasons. Those species that strand most often, such as pilot whales and sperm whales, are deep-water species that live in social groups. They may be less skilled at navigating in shallow water. When a whale strands itself, other individuals remain close by because of strong social ties, and they may also become stranded.

Scientists think that certain whales navigate by following natural magnetic fields in the seascape, and that strandings are more common where regions of low magnetism cross coastlines. Individuals may also run aground because their navigational system is not working properly as a result of injury or disease. Studies made of pilot whales that became stranded on the Massachusetts coastline showed that some individuals were suffering from a harmful viral infection.

Intelligence

Dolphins and some other toothed whales have large brains in comparison with their body size. Some experts believe that certain types of toothed whale are as intelligent as apes. However, this is difficult to prove because the way a dolphin behaves in experimental conditions may not be the same as how it behaves in the wild. But there are plenty of examples of natural dolphin behavior that point to the animal's intelligence. In Shark Bay, Western Australia, some dolphins carry sponges on their snouts when probing for food in the sand. The sponge may offer the dolphins protection against being scratched by the sand or being injured by the spines of prey buried on the seabed.

Baleen Whales

The blue whale is 110 ft (33 m) in length and weighs up to 200 tons (180 tonnes). It is probably the largest animal that has ever lived, yet feeds mainly on small zooplankton, as do most of the other 10 species of baleen whale.

The sievelike baleen plates are clearly visible in the top jaw of this Californian gray whale (left). A baleen whale feeds by opening its mouth to gulp in sea water containing tiny creatures (below). Then it partly closes its mouth and raises its tongue, forcing water out but trapping food on the baleen (bottom).

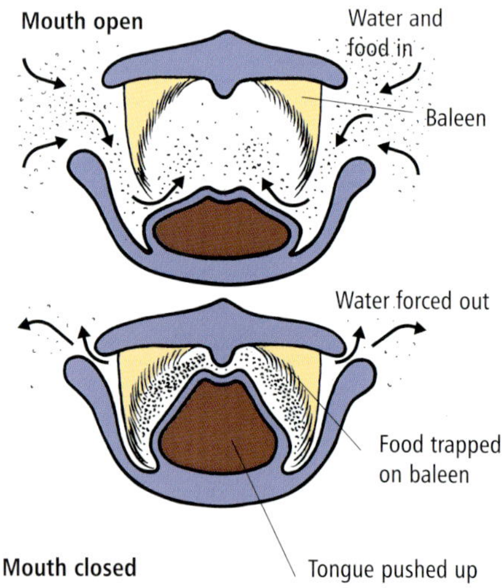

Baleen whales do not have teeth. Instead, stiff, hairlike material called baleen hangs down from either side of their upper jaw like a giant, many-layered comb. The stiff baleen hairs are sometimes called whalebone.

The whales use baleen to filter food from the sea water. Most baleen whale species feed in a similar way. First they take huge gulps of sea water containing zooplankton and other small creatures. They partly close their mouth, raise their tongue, and squeeze the water out through the baleen. Thousands of tiny creatures are strained off by the baleen, to make the whale's meal. Right whales (so-called because these slow-moving, coastal species were the "right" whale for whalers to catch) use a variation on this method. They simply swim along with their mouths open, trapping and skimming off food more or less continuously. Only the gray whale has a markedly different diet from the other baleen whales. It churns up sand on the shallow sea bottom, exposing swarms of crustaceans, crabs, mollusks, and worms, which it strains in through its baleen.

Rorquals are baleen whales that have grooves, or pleats, along the throat and belly. The grooves enable the mouth cavity to expand, so a rorqual can take in a vast volume of water. There are six species of rorqual, including the humpback whale and the blue whale. A blue whale can hold more than 10 tons (9 tonnes) of water in a single mouthful.

Baleen whales have two blowholes. Toothed whales have one. Whale experts identify a species of whale at sea by the size, shape, and direction of its "blow" or waterspout—the spray of air and water the whale breathes out from its lungs. The blow condenses in the air, forming water vapor.

See also **vol 1:** 43; **vol 2:** 36, 38–39, 43; **vol 3:** 54, 56

Blue whale facts

- The heart of a blue whale weighs about 1 ton (0.9 tonne).
- The blue whale's largest blood vessels are big enough for a human to swim along.
- A blue whale's tongue weighs about 3.3 tons (3 tonnes).
- There is room for at least six elephants to stand head to tail along the back of a blue whale.

The song of the whale

Like toothed whales, baleen whales communicate with each other using a wide range of sounds, from deep grunts and wails to high-pitched chirps and whistles. During the breeding season, adult male humpbacks float motionless under water, with their bodies tilted head up, and "sing" to attract females and warn off other males. The male's complicated song can last more than 15 minutes at a time. Each male has a different song, although males from the same population show regional similarities—rather like a local dialect.

The songs may change from one season to the next. Occasionally, whales that have lost their way arrive from distant populations and bring their new songs with them. Influenced by the new tunes, the local males often quickly change their songs.

The deepest notes made by baleen whales can travel thousands of miles across the ocean. The sound waves are carried in a layer 1,000 to 5,000 ft (300–1,500 m) below the surface that has the right combination of temperature, saltiness, and pressure to transmit sound waves best. Submariners also use this layer as a secret communication channel. Whales have already been using it for millions of years.

Blowing bubbles

Humpback whales are found in all seas from the polar regions to the tropics. They have an unusual method of catching prey. It is called bubblenet feeding, and it is one of the best-coordinated hunting strategies of any whale species. In cold waters, two or more humpback whales sometimes work together to blow streams of bubbles to form a curtain around a school of fish or a swarm of krill. The bubbles act like a net, keeping the prey tightly packed inside. Eventually, the circling whales and rising bubbles guide the fish or krill toward the surface, and the whales then rise up together and gulp down as many of their prey as they can.

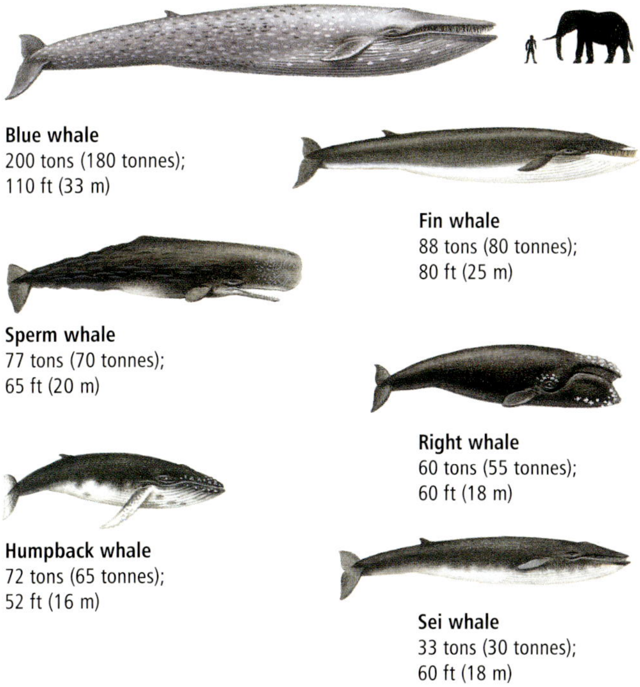

The sizes and weights of some baleen whales. The sperm whale, the largest of the toothed whales, is also shown for comparison. An African elephant and a human are shown drawn to the same scale.

Telltale surface bubbles, photographed in Alaska, show where humpback whales are bubblenet feeding.

Marine Migrations

At certain times of the year, some marine animals journey across the oceans to feed or breed in another part of the world, following the same route every time. Any movement that is seasonal or is part of an animal's life cycle is called a migration.

Many of the creatures in the ocean—including some fish, crustaceans, and squid—regularly migrate from shallow water to deeper water, and back again, to seek shelter and to find food. It is also quite common for some marine animals, such as Atlantic cod and crabs, to live in different parts of the ocean during different stages of their life cycle. The breeding grounds of some animals may be different from their usual feeding grounds, and they move between the two, sometimes covering huge distances in the process. Environmental changes, such as increasing day length in spring or a change in temperature, may act as the trigger for migration. Once migration is under way, animals use a variety of different methods to help them navigate.

The arctic tern, shown here with a map of its migration routes (in blue), migrates farther than any other animal. Breeding birds travel up to 20,000 miles (32,000 km) in the course of a year.

Fishy journeys

Some kinds of fish spend part of their life cycle in fresh water and part in sea water. For example, European and American freshwater eels breed in the Sargasso Sea in the North Atlantic. The newly hatched young travel on ocean currents to rivers in Europe and North and Central America. After about 10 years, when the adult eels have fully grown, they migrate downstream and return to the deep waters of the Sargasso, where they breed and die.

Atlantic and Pacific Ocean salmon make the opposite journey. Salmon spawn in rivers but grow to maturity in the sea. As adults, salmon return to spawn in the very same river in which they hatched. In some species, but not all, the exhausted adults die after spawning.

Turtle voyagers

Sea turtles, some of the most determined marine travelers, perform amazing feats of navigation. One species of sea turtle, the green turtle, spends its adult life grazing on seagrass meadows off the coast of Brazil. Every year it migrates to breed on Ascension Island in the middle of the Atlantic Ocean, some 1,300 mi (2,100 km) away. Ascension Island is only 5 mi (8 km) wide—a tiny target in a big ocean.

Scientists think that green turtles, like some other migratory marine animals, follow ocean currents and navigate by local magnetic fields. They recognize the scent and other telltale signs of their intended destination when they get close. These turtles get their name because their fat has a green color.

See also **vol 2:** 30–31, 34–35, 40–41

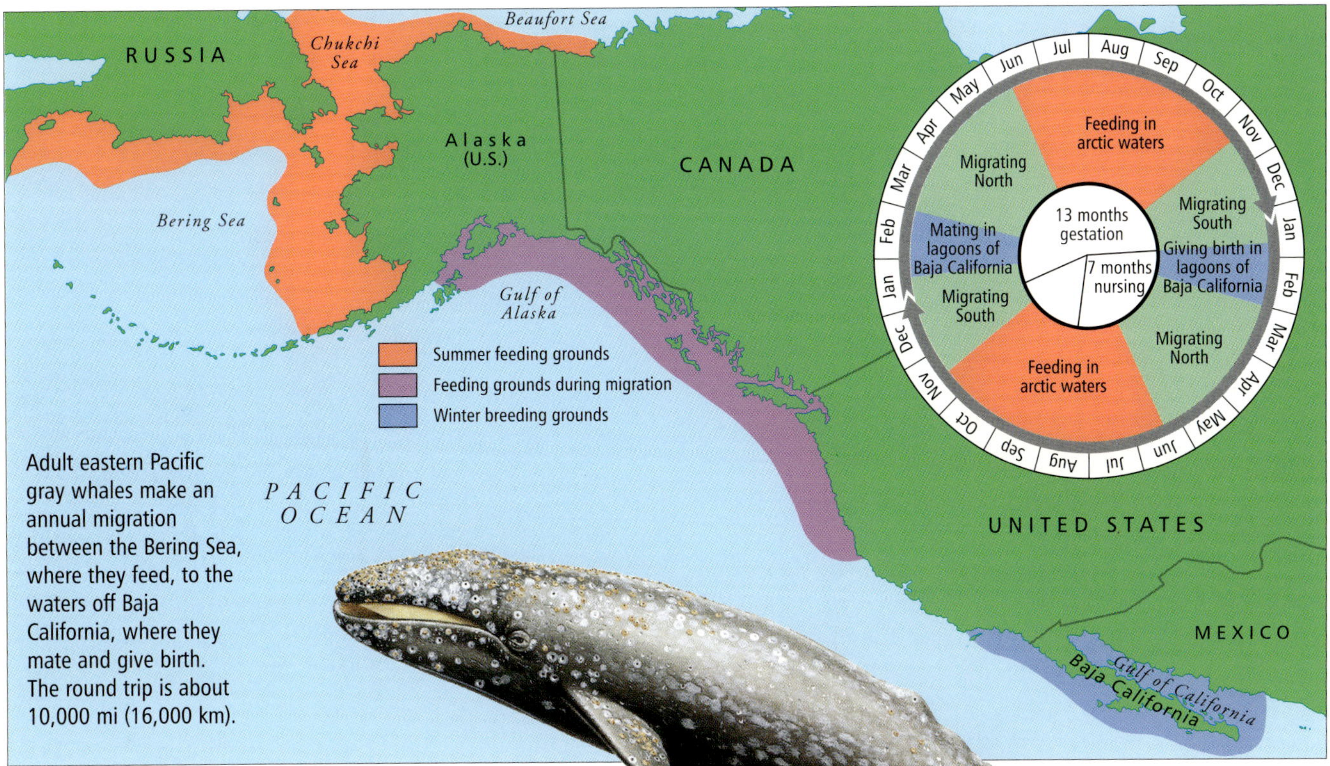

Adult eastern Pacific gray whales make an annual migration between the Bering Sea, where they feed, to the waters off Baja California, where they mate and give birth. The round trip is about 10,000 mi (16,000 km).

Whale migration

Among marine mammals, gray whales and humpback whales migrate long distances following regular routes. They breed in tropical and subtropical waters in winter and migrate to polar or subpolar waters in summer to feed. Scientists are unsure why they make this journey. It may be because the newborn whale calves, which do not yet have a thick insulating layer of blubber, survive better in warmer water. It is also likely that adults need to find less food to help keep them warm in mild seas than in cold polar seas.

Because these large creatures travel the same route at the same time each year, they were easy prey for the whaling ships of the past. The whalers, who hunted them for their baleen, oil, and meat, simply waited until the whales began their migration runs and then intercepted them.

Airborne travelers

Arctic terns (relatives of gulls) make the longest migration of any animal. They breed along the northern coasts of Europe and North America. During the fall in the Northern Hemisphere they fly south and cross the equator. They arrive at feeding grounds in the Southern Ocean and the far south of the Pacific, Indian, and Atlantic Oceans where concentrations of plankton and small fish are at their highest during the southern summer, which is from December to March. In the late southern summer the terns begin their return journey to the Northern Hemisphere, arriving back at their northern breeding grounds in the northern spring.

In the skies above the oceans, some species of seabird make long aerial migrations using a variety of clearly visible sea and landscape features, such as wave patterns and coastline shapes, and fly toward the Sun's path as it travels east to west. At night, some species navigate by the stars or by using the Earth's local magnetic field as a guide.

Life on a Rocky Shore

Rocky shores are harsh environments for animals and plants. They are especially challenging for creatures living in the intertidal zone, the part of the shore that is alternately covered and then exposed by the rising and falling tide.

Animals and plants have a range of special adaptations to help them survive on rocky shores. Many rocky shores are exposed to the full force of the sea. Storm waves smash against the shore, and anything that lives there must hide in crevices or anchor itself firmly to the rock to stop being swept away.

Although seaweeds have tough, flexible fronds and strong holdfasts, they are sometimes torn away from the rocks by wave action. The shells of some of the invertebrate animals that cling to rocks, such as barnacles and limpets, are shaped like cones or pyramids. The force of the waves crashing down on their shells presses them even more firmly against the rock, instead of levering them off.

Between land and sea

Once or twice a day the tide comes up to cover the intertidal zone—the region of the shore normally covered by the highest tides and exposed by the lowest tides. Then the tide recedes, leaving the plants and animals living at the top of the shore exposed to the air for many hours until the rising tide resubmerges them. Lower down, on the middle shore, the periods of exposure and submersion are more equal in length, while the lower part of the shore is submerged for most of the time. Each part of the intertidal zone presents particular challenges for the plants and animals that live there, and each has its own community of creatures.

The upper shore

The upper part of the intertidal zone is also called the splash zone. It is hardly ever covered by tides, but it is often splashed with spray from crashing waves. The splash zone is the boundary between land and sea, and it contains both land and marine creatures, which must be able to live exposed to the air without drying out.

In the splash zone, lichens (plants consisting of an alga and a fungus) hug the rocks and form a mosaic of colors.

When the tide goes out, many rocky shore animals are still visible. Some cling to the rocks, and others hide among the seaweeds or lie in rock crevices.

See also **vol 1:** 26–27, 54–55; **vol 2:** 14–15, 22–23, 50–51; **vol 3:** 52

Lichens grow very slowly and can live on bare rock. Farther down, slippery cyanobacteria cover some rocks, interspersed by patches of green seaweeds. Small snails called periwinkles graze on the seaweed but retreat to moist crevices in the rocks when conditions become hot and dry. Limpets (mollusks with conical-shaped shells) also graze on the algae when the tide comes in, clamping themselves back on the rocks before they are uncovered by the falling tide. Oystercatchers and other shorebirds feed at low tide on mollusks, and crabs come to scavenge and to graze algae.

The middle shore

The middle shore is the widest part of the intertidal zone. In most temperate regions of the world, it has two clearly distinct bands. The upper band, farthest from the sea, is crowded with barnacles, which can survive here because they are able to close their shells and retain moisture when the tide goes out. Marine animals that feed on barnacles can only reach this part of the shore when it is covered in sea water. The barnacles themselves can only feed when the tide comes in because they filter plankton from the water.

The lower band of the middle shore is mainly colonized by mussels and brown seaweeds. Both live anchored permanently to rocks, as do the sea anemones that are scattered among them. Gastropod mollusks like the brightly colored topshells graze the algae. At high tide, marine predators such as crabs and starfish emerge to scavenge carcasses or eat mussels and barnacles. At low tide, these predators disappear under seaweed or into the safety of crevices and rockpools.

The lower shore

The lower part of the intertidal zone is submerged most of the time. It is dominated by red, green, and brown seaweeds that compete with each other for sunlight. Spiny bodied sea urchins graze the seaweeds, and predatory starfish and dog whelk mollusks hunt here, too. When the tide is out, the seaweeds trap plenty of moisture under their fronds, providing a safe haven for sea anemones, crabs, worms, and small fish, marooned temporarily by the receding tide.

As soon as the tide returns, rocky shore animals resume their feeding. A giant intertidal barnacle opens its shell, extends its feet into the water, and uses them to waft food into its mouth. Also seen here are some purple sea urchins. Nearby, some chitons (mollusks with platelike shells) graze pink encrusting algae from the rocks.

This diagram shows the different zones of animal and plant life on a rocky shore.

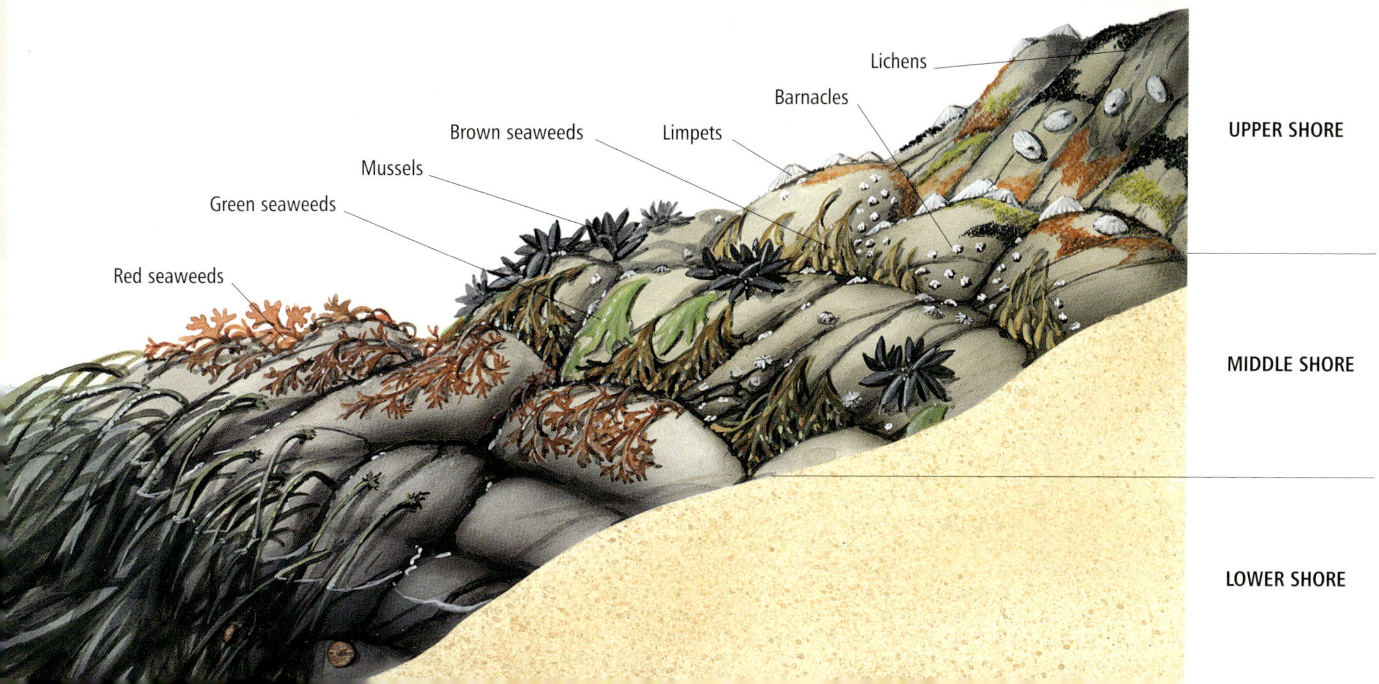

Life on a Sandy Shore

A sandy beach may seem empty of life at low tide, but this is because most of the animals are hiding beneath the surface. When the tide rises to cover the sand, the sand-dwellers spring into action.

There are two quite different communities of animals on a sandy shore. When the tide is out, the sandy shore becomes a fairly dry habitat. In summer, the Sun bakes the surface; in winter, cold winds chill it and blow about the abrasive sand particles. At low tide, predators such as shorebirds and foxes scour the beach for food. Scuttling crabs and leaping beach hoppers (beach fleas) gather among seaweed fragments and other debris washed up on the beach. But apart from these creatures, the beach seems curiously uninhabited. Even seaweeds are rare on sandy beaches because they cannot anchor themselves securely in the sand.

Do not be fooled by appearances, however. Plenty of animals live on the beach, but they burrow into the sand at low tide to hide from predators and to avoid drying out. As the tide comes in, the sand-dwellers become active. Many of them feed on the phytoplankton and zooplankton floating in the incoming ocean waters. The tide also washes the carcasses of dead animals and pieces of seaweed onto the sand, and these are devoured by sand-dwelling scavengers.

Large animals of the sand

The largest animals on a sandy beach are segmented worms, bivalve (two-shelled) mollusks such as clams, crustaceans such as crabs, and echinoderms such as starfish. Many of the creatures feed by filtering plankton from sea water. *Chaetopterus*, a segmented worm, strains plankton from the water using a mucus net, which it sets at the mouth of its burrow. When the net is full, *Chaetopterus* eats the contents.

The sand mason worm hides under the sand at low tide. When the tide comes in, the worm extends its feeding tentacles into the water to trap small animals. The worm coats its tentacles with tiny grains of sand.

See also **vol 1:** 28–29, 54–55; **vol 2:** 48–49, 50–51; **vol 3:** 52

Some worms are active predators: they roam the sand surface looking for food. Bivalve mollusks usually stay buried in the sand. They have tiny, hairlike structures (cilia) under their shells. The cilia beat to draw in water through a tube called a siphon, the mollusk's gills filter the plankton out of the sea water, and the water is expelled through another siphon.

Clams have a muscular foot for burrowing into the sand. They push the foot into the sand and then use it to drag the rest of the body downward. Bivalve mollusks such as razor clams (so-called because their long, straight shells resemble old-fashioned straight razors) can dig themselves down into the sand quicker than a human being can dig them out!

Echinoderms of the sandy shore, such as heart urchins and sand dollars, move their spines and tube feet to plow through the sand. They feed on the organic matter that coats sand grains. Sea cucumbers (cucumber-shaped echinoderms) feed while they burrow. They swallow sand, digest what is useful, and expel the rest as pellets. Starfish are voracious predators of crustaceans, mollusks, and other echinoderms.

Starfish are voracious predators. They use their arms and tube feet to burrow through sand to find bivalve mollusks such as clams. Then they pull open the shells to reach the mollusk inside. They extend their stomach into the shell, digesting the unfortunate victim while it is still alive.

Worlds among grains of sand

Sand grains—and the tiny spaces between them—are home to a community of miniature animals, ranging in size from 0.004 to 0.08 in (0.1–2 mm). They are extremely varied. Most of them move about by squeezing between sand grains, and so they often have wormlike or flattened shapes. Some are tiny versions of larger animals found on the sand, such as worms and bivalve mollusks. Others are only found living in this tiny world. They include polyps with snakelike tentacles, single-celled organisms that constantly flow into different shapes, and worms that have unusual suction devices to attach themselves to sand grains, suck up tiny food particles, or grab other animals.

Like the larger animals of the sandy shore, the creatures of the sand-grain world include plant-eaters, meat-eaters, and omnivores (organisms that eat both plants and animals). These tiny animals themselves become food for the larger burrowers and scavengers of the sandy shore such as sand mason worms and crabs, but also for shrimp and young fish that arrive to hunt and forage on the sandy surface below the water when the tide comes in.

Low tide on a sandy beach. The only clues to life beneath the surface come from the numerous lugworm casts (mud expelled from burrows). Each one indicates the position of a lugworm lying in its burrow.

LIFE IN AN ESTUARY

Creatures that live in an estuary, where a river joins the sea, must deal with rapid changes in the saltiness of the water when the tide ebbs and flows.

When the rising tide flows into an estuary—the place where a river joins the sea—water is very salty because sea water predominates. However, when the tide goes out, the water gradually gets less salty because most of the water in the estuary is now river (fresh) water. Depending on the state of the tide, the saltiness of estuary water varies from about 35 parts of salt per 1,000 parts of water (near to that of normal sea water) to about 5 parts of salt per 1,000 parts of water, or less. Water that is a mixture of sea water and fresh water is described as brackish.

Living in the mud

At low tide, the estuary becomes a place of mudflats and shallow creeks. Because drainage is poor, water remains trapped between the particles of mud and becomes stagnant. Beneath the top inch or so, the mud is often black and evil-smelling—a sure sign that oxygen is lacking.

Some of the organisms that live in this mud can survive without oxygen. However, oxygen-breathing creatures must dig tubes or burrows in the mud that allow fresh, oxygen-rich sea water to penetrate when the tide comes in. Many worms and crustaceans beat their cilia to create currents that draw water into their burrows, while some bivalve mollusks extend their siphons to the top of the burrow to obtain fresh supplies of sea water.

A flock of curlews swirl in to land on mudflats. At low tide, the rich feeding grounds attract many kinds of shorebirds as well as seabirds such as gulls and terns.

See also **vol 1:** 30–31, 32–33, 54–55; **vol 2:** 34–35, 54–55; **vol 3:** 52

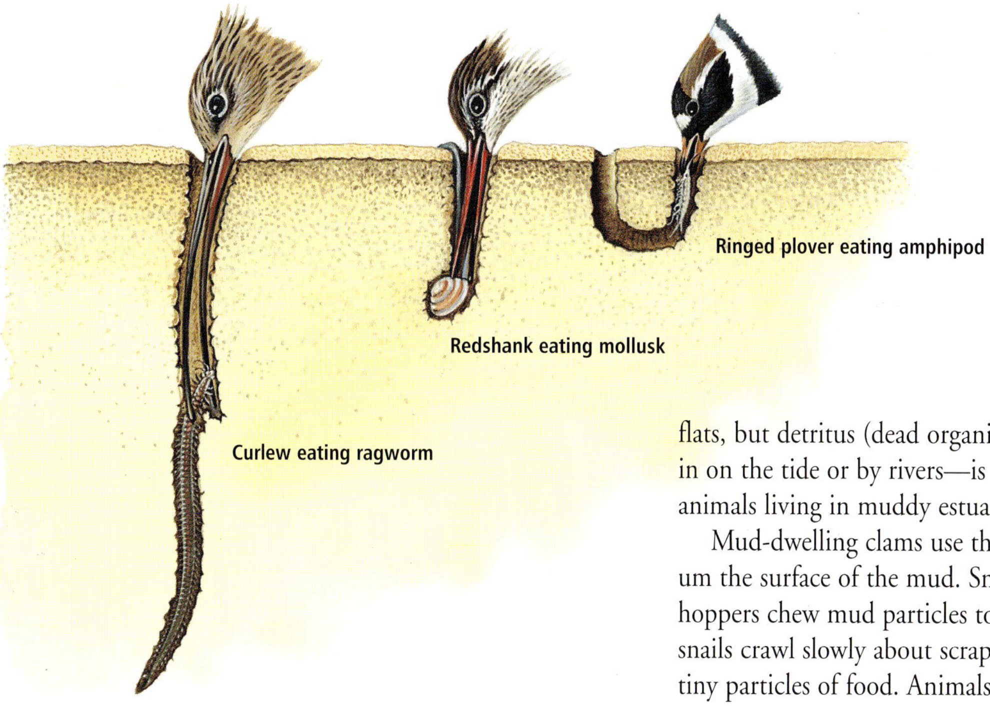

This diagram shows how the beaks of different birds are adapted to reach food buried at various depths in the mud.

Getting food

There is little plant life to provide food or shelter for animals beneath the waters on an estuary. Mud particles make the water very murky, so sunlight does not penetrate far. Therefore, phytoplankton growth is limited. Nor are there many seaweeds in an estuary because they cannot anchor themselves easily in the shifting mud and sand. Cyanobacteria (primitive organisms) and single-celled algae like diatoms and dinoflagellates grow on the surface of mudflats, but detritus (dead organic matter)—most of it brought in on the tide or by rivers—is the main source of food for animals living in muddy estuaries.

Mud-dwelling clams use their elongated siphons to vacuum the surface of the mud. Small crustaceans called mudhoppers chew mud particles to obtain their food, while mud snails crawl slowly about scraping the mud surface to gather tiny particles of food. Animals that feed off the muddy bottom in this way are called deposit feeders.

Fish and birds are the biggest predators of the mudflat community. Fish such as mullet swim in at high tide to find their prey. Mullet suck up sediment containing tiny creatures, filter it through their gills, and then swallow the larger food particles. Shorebirds gather at low tide. Different bird species are specialized for eating prey found at various depths in the mud. This specialization reduces competition for food. Some shorebirds, for example, have fairly short bills and probe the surface of the mud for snails, tiny crustaceans, and shallow-burrowing clams. Other shorebirds have longer beaks and search for deeper-burrowing clams or worms. Herons and egrets have long, sharp beaks and long legs. Instead of probing in the mud for prey, they wade into the water and grab fish, shrimp, and other animals with a lightning fast stab of their beaks.

Nursery grounds

The waters of estuaries—including mudflats, salt marshes, and mangrove swamps—provide food and relative safety for the young of many kinds of ocean fish, as well as juvenile shrimp and prawns. In North American waters, for example, the immature forms of fish such as mullet and species of flatfish such as sole live in estuaries. Salmon and eels pass through estuaries both as juveniles and as spawning adults. If estuaries become polluted or drained for farming, fish nurseries can be wiped out, causing a drop in commercial fish numbers.

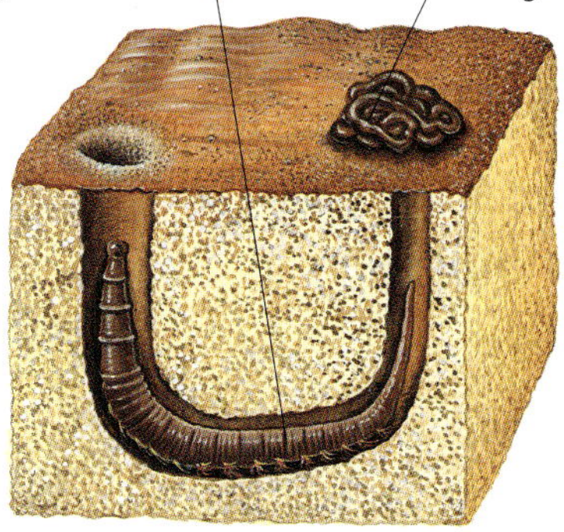

Lugworms live by swallowing mud and digesting dead organic matter and microbes as they pass through the gut. Small piles of expelled mud, called casts, pile up outside the lugworms' burrows.

The Shallow Sea Floor

Colorful and varied creatures inhabit the shallow parts of the ocean, where there is sunlight, warmth, and plenty of nutrients for life to flourish.

The shallow sea floor extends from the lower end of the intertidal zone to the edge of the continental shelf. The maximum depth of these waters is usually about 650 ft (200 m). They are the richest part of the ocean for marine life. Waves, currents, and tides stir the water and ensure that nutrients are well distributed. Rivers add more nutrients from the land, and sunlight penetrates easily below the surface of the water. Phytoplankton thrive, providing food for the plentiful zooplankton. Larger invertebrates and fish graze on the plankton.

The soft seabed

Where large amounts of sediment settle in shallow sea water, the seabed is usually sandy or muddy. In many parts of the world seagrass grows on the soft seabeds of sheltered, shallow waters. The seagrass forms underwater meadows that are a haven for marine life. Seaweeds do not grow on soft seabeds, however, unless there are any firm surfaces to which they can attach themselves.

Many of the invertebrates living on sandy seabeds are similar to those found on nearby sandy shores. However, there is usually a greater variety of animals in the shallows than on the seashore because the seabed is always covered by sea water and conditions are less demanding. Among the animals that filter plankton and drifting plant and animal remains from sea water are bivalve mollusks such as the soft-shelled clam and the razor clam. Sea pens (so-called because they look like an old-fashioned quill pen) are related to sea anemones. They have clusters of feeding polyps attached to a stalk, giving them the appearance of delicate plants. The sea

This anglerfish is perfectly designed for a life on the shallow sea floor. The flattened body helps the fish remain concealed, and the huge mouth can swallow large animals such as crabs.

See also **vol 1:** 16; **vol 2:** 44–45, 46–47; **vol 3:** 50–51, 52

The octopus is a fierce predator. Lying concealed within a rocky crevice, it waits until its prey comes close before darting out to grab it with powerful, suckered tentacles. The octopus then scuttles back to the safety of its hideaway to consume the catch.

pen unfurls its fan of feeding polyps into the water at night to catch zooplankton with its tiny stinging tentacles.

A busy community of tiny animals lives in the soft sand and mud, feeding on the organic matter that drifts through the waters or settles on the seabed. Tube-dwelling worms known as terebellids have a crown or fan of tentacles on their head that they use to gather food particles from the surface of the sediment. Some deposit feeders roam over the seabed looking for food. They include the seamouse, a type of worm that gets its name from the fine layer of silky bristles that covers its body, and crustaceans such as small crabs, and shrimp. Brittle stars—echinoderms closely related to the starfish—feed on detritus and tiny food particles. They get their name because of their long, fragile-looking arms.

Many predators actively hunt the creatures that feed on particles floating in the water and on the seabed. Starfish roam the seabed looking for mollusks to devour. Crabs and lobsters skulk in crevices waiting for an unsuspecting victim. A marine snail called the dogwhelk can bore into the shells of other mollusks with its special tongue and then eat the victim. Some species of fish, such as skates and rays, and flatfish like flounders and plaice, have flattened bodies that help them remain concealed in the sand until they spot their prey. Then with a sudden movement they swallow it whole.

The hard seabed

On rocky shores, the shallow seabed is hard. In cooler waters, wherever sunlight reaches the bottom, seaweeds grow in profusion. In seaweed communities grazers include sea urchins and mollusks such as chitons and limpets. They eat the more fragile seaweeds as well as the fine turf of green algae that grows on seaweeds and rocks. Some seaweeds have defenses to resist grazing. Kelps are usually tough and leathery, while some red and green algae, known as coralline algae, contain chalky deposits to deter plant-eaters.

Barnacles, sponges, sea squirts (saclike creatures that live in colonies), and tube-dwelling worms all find a home on the seaweed fronds. They filter food particles from the sea water. Many of the same predators of soft seabeds are also found on hard seabeds. Octopuses favor hard seabeds, where there are plenty of crevices for them to hide in while they wait to grab crabs and other creatures with their tentacles.

The speckled sandperch remains motionless on the seabed, blending in with the sand and pebbles. It ambushes unwary fish and invertebrates, sucking them into its cavernous mouth with lightning speed. It grows to about 10 in (25 cm) in length.

THE CROWDED REEF

In the clear, shallow waters of warm, tropical seas, coral reefs provide shelter and food for thousands of species of plants and animals.

Coral reefs are formed by tiny, anemonelike creatures known as hard coral polyps. The polyps secrete a protective, calcium carbonate covering around themselves. When the polyps die, new polyps grow on their remains, and the reef steadily enlarges. Coral reefs cover less than 10 percent of the area occupied by rainforests on land, yet they probably contain at least half as many types of animals and plants. Almost all of the 30 or so major groups of living invertebrates has members that inhabit the coral reef. A tropical rainforest has a far smaller range of invertebrates. Coral reefs occupy just 0.2 percent of the world's total ocean area, but they contain a quarter of all marine fish species. Reefs are full of chambers, nooks and crannies, and channels where creatures of all sizes can find food and shelter.

The most common creatures on the reef are the coral polyps, but there are also many animals such as worms, starfish, shrimp, and crabs. Moray eels lurk in dark crevices, ready to spring out and catch unsuspecting prey. Sharks and other predatory fish such as barracudas seek out the schools of smaller fish that swim among the reef. Sea snakes glide through the coral looking for unwary fish.

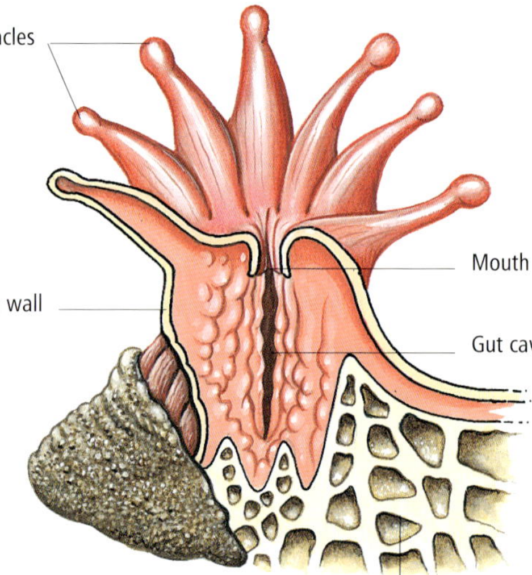

A hard coral polyp secretes a protective, calcium carbonate cup around itself. When threatened, the polyp can withdraw into the cup for safety.

Various kinds of algae also grow on the reef. Some form clumps, and others coat the coral in a thin, colorful layer. The rich plant life helps to make the reef an oasis for all kinds of animals.

Toxic attack

The competition for space is so intense on a coral reef, that some corals, as well as many creatures that live attached to coral, have evolved ways to prevent themselves being smothered by their neighbors. Many do this by producing poisons that prevent them from being eaten and also repel other creatures that grow too close. Scientists have studied some of these chemicals and found them to be medicinally useful. For example, drug developers recently extracted a chemical from an Australian soft coral called *Eleutherobia*. The substance was found to be a powerful anticancer drug, now dubbed Eleutherobin.

Coral spawning

Several days after a full Moon, thousands of sexually reproducing corals release pale bundles of eggs or sperm. The water becomes full of tiny spheres, giving the appearance of an upside-down snowstorm, drifting from the reef toward the surface. By methods not clear, corals on a reef all produce their spawn at the same time in order to maximize the chances of successful fertilization and, by overwhelming predators with the sheer abundance of food, increase the chances of fertilized eggs surviving to become larvae.

See also **vol 1:** 22–23, 43; **vol 2:** 22–23, 55, 60–61; **vol 3:** 51, 52–53, 54, 57

Coral grows in a variety of forms. The pinkish, branching coral seen at the top is antler coral. To its left and slightly below is brain coral. The large, flat yellowish coral is called plate coral.

Coral reefs teem with brightly colored fish. Some of the most common species are shown here.

Colorful fish

Because the reef is so crowded with predators, many animals have evolved ways of keeping out of danger. Some hide away in crevices. Many animals, including some reef fish, have special colors to warn that they are poisonous and so help prevent them from being eaten. These are called warning colors, and they are often in the form of bold contrasting stripes. The bizarre-looking lionfish, for example, is vividly striped in red and white to advertise the fact that its spines are venomous. Other fish, such as the Moorish idol and butterfly fish, have stripes that break up their outline, making them less visible. Bands and blotches on their bodies conceal the position of their eyes. Butterfly fish also have a false "eye" pattern on the dorsal fin. It is possible that these features divert attacks away from the fish's vulnerable head.

Many reef fish live in schools. Their bright colors and patterns help individuals identify themselves to others of the same species, but it is unclear why this does not attract predators as well. Perhaps the wealth of hiding places on a coral reef provides sufficient defense against a possible attack. Or maybe the way predatory fish see light, shade, and color on a coral reef helps prey species avoid being seen by their enemies. A dazzling school of highly colored or patterned fish may confuse predators, making it difficult for them to single out an individual to attack.

Threatened Communities

The richest marine habitats include those found on the coastline, the continental shelf, and the coral reef. However, it is precisely these communities that are most at risk from the activities of humans.

Less than 10 percent of the world's oceans cover the continental shelves and their shore and shallow-water communities, yet about one-third of all living matter produced in the ocean each year occurs there. Most commercial fishing is done in these waters, together with commercial farming of species such as oysters, mussels, and salmon. Many coral reefs, the ocean's most diverse biological communities, are also found in this region of the sea. Unfortunately, coral reefs, and estuaries and coastal lagoons—together with their associated salt marshes and seagrass meadows—are among the most threatened of all marine environments.

Mangrove swamps

The mangrove swamps of tropical and subtropical coasts are extremely important natural habitats. The roots of mangrove trees are nurseries for the young of many commercially important species of fish and shellfish. Mangrove roots also help trap and bind coastal sediments. Where mangroves are removed, coastal erosion and flooding become more likely.

For thousands of years, local people left mangrove swamps more or less undisturbed. But in the last few hundred years, mangrove trees and shrubs have been overharvested for timber or fuel. More than one-half of all coastal mangroves have been removed to make way for farms and coastal towns.

More recently, mangroves have been cleared to allow fish and shellfish farming to take place. In Ecuador, during the 1970s and 1980s, shrimp farmers cleared mangroves to create ponds for rearing shrimp. But the waters around mangrove roots were the natural breeding ground for shrimp. Removal of the mangroves destroyed the supply of juvenile shrimp that was needed for stocking these farms. Many of the shrimp farms have long-since been abandoned.

Mangrove swamps are very important habitats. In addition to stabilizing coastal sediments and preventing erosion, they support a wide range of animal species. Some of the animals that live in mangroves are shown here.

See also **vol 2:** 14–15; **vol 3:** 50–51, 52–53, 54, 57

Seagrasses in retreat

Seagrass meadows the world over are homes to endangered species such as seahorses and sea cows. In the United States, seagrass meadows provide nursery grounds for commercially important fish and crustaceans. But seagrass communities are threatened. Some of the damage to seagrass meadows in Europe and North America has been caused by a wasting disease spread by a tiny, single-celled organism living in the water. Other factors, such as chemical pollution and rising sea water temperatures, have probably weakened seagrass plants, making them more susceptible to disease.

Some coastal developments, such as dredging deep-water channels for shipping, cause cloudiness in the water. This blocks the sunlight that seagrasses need for photosynthesis. Sewage discharges from towns enrich coastal water, encouraging the growth of algae that smother seagrasses. Oil and heavy metal pollution kill seagrasses. Plants are also damaged by boats' propellers and anchors and are uprooted by fishing nets dragged through the water. Since the 1950s, more than three-quarters of the seagrass in Galveston Bay, Texas, has been lost.

An underwater view of a seagrass meadow. Seagrass beds reduce wave action and slow currents, so reducing coastal erosion.

Coral crisis

Coral reefs are very delicate habitats. They are vulnerable to damage from a wide range of human activities. The coral polyps that build coral reefs are often killed by pollutants such as sewage or metals such as tin from boat paint. Coastal developments may produce sediment-laden water that runs into the sea, smothering corals. In the Indian Ocean, some marine tourist developments are built using limestone quarried from local coral reefs.

There is still a worldwide trade in animal shells and corals taken from reefs. Some people use chemicals to stun tropical fish so they can be gathered for the aquarium trade, but they kill many smaller reef organisms in the process. Sometimes, fishers blast reefs with dynamite to catch fish for food.

Ships' anchors damage reefs, and many people kill coral polyps when they snorkel or dive, simply by touching the coral. The 1997–98 El Niño (a change in wind patterns that affects ocean currents) raised sea water temperatures in parts of the Indian Ocean. This resulted in the death of huge areas of coral. The future for coral reefs is uncertain.

In 1997 Reef Check—a worldwide survey by divers—estimated that 90 percent of the world's coral reefs showed visible signs of damage. In 2000 the Global Coral Reef Monitoring Network reported that one-quarter of the world's coral reefs have already been destroyed.

Deep-Sea Life

Most of the deep-sea floor is featureless, soft sediment. But scavengers roam across its surface, and a thriving microworld exists within the mud. Oases of life occur where hot waters spew out of rocks or gases seep onto the ocean floor.

Beyond the continental slope, much of the deep-ocean floor is a muddy plain. Little light penetrates there, and we see only a tiny area at a time in the searchlight of a submersible. Mounds, holes, and tracks suggest that there is life on the bottom, but it is usually buried from view. Occasionally, a strange, semitransparent sea cucumber (a cucumber-shaped, spiny skinned animal) bearing a "sail" or spines lumbers past. The bottom sediment also seethes with microscopic life—like single-celled, chalky shelled foraminiferans that feed on bacteria. The foraminiferans in turn are eaten by animals such as small roundworms.

Seasons in the deep

Until the 1980s, most biologists thought the ocean floor changed little from season to season. We now know this is not the case. In the North Atlantic, for example, the spring bloom of phytoplankton results in a "snowfall" of detritus that arrives on the deep-sea floor weeks later. This detritus consists of dead phytoplankton and zooplankton, and fecal pellets (the zooplankton's solid waste). The mixture forms a green-brown sludge that provides welcome seasonal food for bottom-living scavengers and deposit feeders.

Baited camera

One of the best ways to find out what lives on the deep ocean floor is to leave a carcass alongside a remotely operated camera and see what arrives to take the bait. Within hours, scavengers in the form of large, shrimplike crustaceans, ghostly pale crabs, giant red prawns, hagfish, primitive sharks, and rattail fish arrive and begin to demolish the carcass. Later, slower-moving creatures like sea snails arrive to join the feast. Within weeks, even the carcass of an animal the size of a whale is reduced to a pile of bones. The deep-sea scavenging community wastes no time in exploiting any sources of food.

Some deep-sea fish. *Xenophthalmichthys* feeds on small crustaceans and worms. Catsharks live on the ocean floor and feed on mollusks and small fish. The goblinshark grows to 14 ft (4.3 m) in length and lives in waters over 1,000 ft (300 m) deep.

See also **vol 1:** 15; **vol 2:** 18–19, 58–59; **vol 3:** 32–33

Sea spiders get their name because they look like true spiders, although they are related to crabs. Sea spiders are found from intertidal waters to the deep sea. Deep-sea species are usually red in color, like the one shown here. They feed on colonial creatures such as sponges.

Unique communities

In 1977 scientists in the submersible *Alvin* explored the hot-water vents—places where hot water spews from cracks in the deep ocean floor—along the Galápagos Rift off the coast of Ecuador. At a depth of 8,200 ft (2,500 m), they discovered an astonishingly rich community of unusual animals. For example, they encountered giant 3-ft (1-m) long tubeworms (a kind of segmented worm that builds a protective tube around its body) and mussels and clams about 10 in (25 cm) long. Since then, similar animal communities have been discovered at more than 100 other hot-water vent sites.

Without sunlight, no photosynthesizing plants can grow here to produce energy. It was a mystery where the energy was coming from to sustain such vibrant animal communities. The answer lay in the water that seeped from the vents. Bacteria use hydrogen sulfide dissolved in the water to make their food by chemosynthesis, just as the earliest life forms did 3.6 billion years ago. These bacteria sustain the rest of the community. While many of the bacteria are free-living, some live inside the animals themselves. The tubeworms, for example, used their plumelike gills to trap hydrogen sulfide, which they deliver to bacteria growing inside a modified cavity in the body. The bacteria make up about half of the tubeworm's weight.

Iceworms

In 1997, on the floor of the Gulf of Mexico, scientists discovered some marine worms burrowing in mounds of methane hydrate—crystals of methane and water, sometimes called "methane ice." No one yet knows exactly what the worms feed on or how they are able to survive in this toxic, methane-rich environment.

Cold seeps

In 1984 biologists studying the sea floor in the Gulf of Mexico found animal communities as rich and varied as those living at deep-sea hot-water vents. But there was no hot-water vent nearby. Instead, the scientists discovered that cold methane gas and hydrogen sulfide seeping from the sea floor provides fuel for chemosynthetic bacteria living inside mussels. Giant tubeworms also live close to the methane-gas community. They are a different species from the tubeworms found at hot-water vents, but like them they also absorb hydrogen sulfide for chemosynthetic bacteria—but this time by means of rootlike structures on their bodies.

Sea cucumbers, like this leopard sea cucumber, suck up sediment on the deep-sea ocean floor and extract tiny food particles.

The Dark Zone

Few creatures from the upper water layers ever enter the dark zone that extends below 3,300 ft (1,000 m). This is a world of strange and misshapen monsters.

The dark zone stretches from the bottom of the twilight zone (the part of the ocean still lit by the Sun's rays) to the deep-sea plains several miles down. It is a vast and unvarying habitat holding three-quarters of the ocean's water.

Food is scarce in the dark zone. Only about 5 percent of the food produced in sunlit and twilight zones finds its way into the dark zone. Most dark-zone animals are predators, cruising slowly about looking for prey or floating motionless until a likely victim swims near. As long as they can swallow it, predators eat whatever comes within range.

Viperfish

Anglerfish

Window on the deep

Studying dark-zone animals poses problems for scientists. It is very difficult to bring up live specimens from the depths, since the pressure changes usually kill them. In recent years, scientists have succeeded in observing deep-sea creatures from submersibles or by mounting a camera on a robot vehicle. This has thrown new light on many of these creatures. For example, scientists who had only been able to study the remains of the deep-sea squid *Vampyroteuthis* (its name means "vampire squid from hell") thought it was a poor swimmer. When they observed one alive in the deep sea, they discovered it moves at high speed, using a looping motion.

Anglerfish

Gulper eel

These deep-sea fish all show adaptations for catching prey in the dark world they inhabit. All have huge teeth, and large mouths for swallowing anything that comes their way. Some also have lures to attract prey.

See also **vol 1:** 42–43; **vol 2:** 18–19; **vol 3:** 33

Turning on the light

Most deep-sea creatures use bioluminescence to identify each other, to attract prey, or to startle predators. They create "cold" light by mixing together two chemicals—one called luciferin and another called luciferase. Most bioluminescent animals produce their own luciferin and luciferase. Using their nervous system, they can switch the light on and off at will.

Deep-zone adaptations

Creatures of the dark zone have weak skeletons and muscles. They rely on their cavernous mouths and long, curved teeth to overcome prey. They also have stomachs that can expand to accommodate an enormous meal if necessary. Anglerfish have an ingenious method of capturing food. They take their name from the lure they dangle on the end of a rodlike stalk attached to their noses. The lure flashes with a special light created by bacteria in their bodies. When prey approaches to investigate the lure, the anglerfish snaps it up. There are more than 100 types of anglerfish in the dark zone.

In the darkness, senses other than vision take on greater importance. Many species have small eyes or even lack eyes altogether. Most dark-zone fish have highly developed lateral line sensory systems, and many have sensory hairs or antennae that detect vibrations in the water. Some blood-red shrimps have enormously long, touch-sensitive antennae that detect prey or predators venturing near. Most dark-zone creatures have a highly developed sense of smell for detecting prey, the odor of rotting food, or even a potential mate.

Finding a mate

Finding a mate in the deep ocean can be even harder than finding food. Most deep-water fish species probably locate their potential mates by smell. They release pheromones—chemicals that attract the opposite sex.

Some species use unusual strategies to help ensure breeding success. A few deep-sea crustaceans and fish are hermaphroditic. This means that individuals have both male and female sex organs. Any adult can mate with any other adult of the same species. They fertilize each other's eggs.

In some species of anglerfish, the male lives as a parasite of the female. When a male finds a female, he fuses with her body, and he gains nutrients from her blood. The male is always available to fertilize the female's eggs, whenever she releases them. Fertilized anglerfish eggs float to the surface waters. Young anglerfish feed on plankton in the sunlit and twilight zones, then descend into the dark zone later in life.

A glass jellyfish swimming in arctic waters illuminates the gloom with bioluminescent light, hoping to attract prey to its stinging tentacles.

Living Together

Many marine creatures are found living in close association with other species. Sometimes different species live together for their mutual benefit, but on other occasions only one partner gains from the association.

Scientists describe close associations between two species of organisms as symbiosis, or symbiotic relationships. Some symbiotic relationships benefit both creatures, but others can be harmful to one of the individuals concerned. An example of an association that is of benefit to both partners is the one that exists between a coral polyp (the host) and the algae that live inside it. The coral polyp receives food and oxygen from the algae to help it grow. The algae are protected inside the coral and receive carbon dioxide and nutrients to help them photosynthesize. This form of association is known as mutualism, and the partners are called symbionts.

There is a second type of symbiosis, called commensalism, in which only one member of the association—called the commensal—clearly benefits from the arrangement. The other, the host, does not seem to benefit but is not harmed. For example, remoras live in association with sharks and other large fish. Remoras are small fish with suckers on their heads that they use to attach themselves to larger fish. The remora hitches a ride, gains protection, and may get food scraps from a host's meal but does no real harm to its host.

In the third form of symbiosis, known as parasitism, the host is clearly harmed in some way, but the parasite benefits. The parasite usually feeds on the host's body tissues and also uses the host's body as a home and a form of protection. The fish louse, for example, is a parasitic crustacean related to crabs that attaches itself to the surface of a fish and consumes its body fluids and tissues.

Sharing house space

On the Pacific coast of North America, a worm, aptly named the fat innkeeper, shares its U-shaped burrow with several guests: a type of segmented worm known as a scale worm, a fish, and at least one species of crab. The guests all use the burrow as a refuge. The fat innkeeper feeds by pumping water through a funnel-shaped net of mucus that filters out food, and the resident scale worm snatches scraps. It is possible that the crabs might help defend the burrow, or provide an early warning of danger, which could help "repay" the innkeeper's apparent generosity.

Cleaning stations

At certain places on coral reefs, medium and large fish sometimes gather to be "groomed" by shrimp or smaller fish. The places where they gather are called "cleaning stations." The

A yellow-edged moray eel waits patiently while a cleaner shrimp attends to it at a cleaning station. This a form of mutualism.

See also **vol 1:** 22–23; **vol 2:** 13, 22–23, 52–53; **vol 3:** 40

smaller creatures feed by removing harmful parasites from the skin of their "clients" and by cleaning their wounds to help heal damaged or diseased tissue. The cleaner wrasse fish (a small, striped fish) even enters the mouth and gills of its client.

This cleaning arrangement benefits both parties, but what stops the cleaner fish or shrimp damaging their clients or being eaten by them? The answer is that cleaner and client send signals to each other to allay suspicions. A cleaner wrasse performs a zigzag "dance" to attract a client. The client, in return, moves forward, spreads its fins, opens its mouth, and flares its gills, inviting the fish to groom it. This arrangement usually works well. However, the appearance and behavior of the cleaner wrasse is copied by at least one species of blenny, another kind of small fish. Instead of attending a client, the impostor bites a chunk out of the unsuspecting fish!

The small striped fish accompanying this great white shark are called pilot fish. They stay close to the shark and so gain protection from their enemies. They may also eat scraps of food left after the shark has eaten. This is an example of commensalism.

The soft lining at the edges of this giant clam's shell contain algae. As the algae photosynthesize, they provide the clam with oxygen. In return, the algae receive carbon dioxide from the giant clam, as well as nutrients such as nitrates. This is another example of mutualism.

Parasites everywhere

Of the 34 major groups of animals, at least 19 of them have members that are parasites. A scientist studying marine parasites on the Great Barrier Reef in Australia estimated that 1,000 local species of fish were probably infected by at least 2,000 species of monogenean—a type of flatworm—and by about 20,000 types of parasite in all.

Although parasites feed on the body tissues of their host, they do not usually kill their host. Killing the host would destroy the parasite's food supply and home. However, parasites can cause harm to their hosts. For example, North Atlantic cod, whiting, and haddock lose up to one-third of their weight when heavily infested by parasitic crustaceans. This greatly reduces their economic value to fishermen.

There are two different sorts of parasites. Endoparasites live inside their host. Many endoparasites are tiny, single-celled organisms, although a few—such as some parasitic worms—can be enormous. For example, a parasitic roundworm that lives inside the blood vessels of large whales is 30 ft (9 m) long, but it is very thin and weighs only a tiny fraction of its host. Ectoparasites live on the surface of their host. One parasitic crustacean that attaches itself temporarily to small fish can grow to one-half the volume of its unfortunate host.

INDEX

Page numbers in *italics* refer to illustrations

abalones 23
abyss *18*, 19
abyssal plain *18*
acid rain 10
Alaska *41*
albatrosses 35
  black-browed *35*
  wandering 35
algae *7*, 14, 15, 44, 45, *45*, 49, 51, 55
  brown 14
  of coral reef 52, 60
  coralline 51
  evolution 32
  green 51
  red 51
  symbiotic relationship 60, *61*
algal bloom 13, 56
Alvin (submersible) 57
amphibians 32
amphipod, in food web *17*
anchors, damage from 55
anchovy *5*, 16
  in food web *17*
anemones *see* sea anemones
angelfish, regal *53*
anglerfish 19, *50*, *58*, 59
animals
  of dark zone 19
  in food chains 16–17, *16*, *17*
  largest 4
  marine, extinctions 10, 11
  soft-bodied, fossils 8
  species 4
Anomalocaris 9, *9*
archeobacteria *7*
archer fish *55*
arrow worms 16, 17, *20*, 21
arthropods 8–9, 32
Atlantic 17
  animal species in 4, 56
atmosphere 6
auks 35
Aysheaia 9

bacteria 4, 5, 6–7, *7*, 17, 20, 57
  chemosynthetic 57
  energy source 6–7
Baja California *43*
baleen 40, *40*
barnacles 21, 23, 44, 45, *45*, 51
barracudas 30, 52
beach, sandy, life on 46–7, *46–7*
beach hoppers 46
beadlet anemone *23*
benthic environment *4*, 5
Bering Sea *43*
bioluminescence 18, 19, 59
birds
  beaks *49*
  in estuaries *48*, 49
  sea 17, 34–5
    feeding methods *34–5*
    food 33, 34
    in food web *17*
    migrations *42*, 43
  shore 45, 46, 49
bivalve mollusks 23, 46, 47, 48
bladderwrack *14*
boobies *34*
brittle stars 14, 23, 51
bubblenet feeding 41, *41*
Burgess Shale 8–9
butterfly fish 53, *53*

calcium carbonate 52, *52*
Callithamnion 15
Cambrian explosion 9
Canada, Rockies 8–9
cancer, drugs against 52
capelin 31
carbohydrates 7, 16
carbon dioxide 4, 10
catshark *56*
cephalopod mollusks 23
cetaceans 36, 38
Chaetopterus 46
chemical pollution 55
chemosynthesis 7
chitons *45*, 51
chordates 9
cilia 47, 48
clams 15, 23, 46, 47, 49, 57
  razor 47, 50
  soft-shelled 50
  symbiotic relationship *61*
cleaning stations 60–1
climate, changes 10
clownfish *53*
cnidarians 22
coastline, threat to life on 54
cobras 33
cod 16
  Atlantic, migrations 42
  North Atlantic 31, 61
  reproduction 31
coelacanth 11, *11*
cold seeps 57
colors, warning 53
comb jellies 16
commensalism 60, *61*
communities
  deep-sea 57
  under threat 54–5
continental rise *18*
continental shelf, threat to life on 54
copepods 20–1, *20*, 23
  in food web *17*
coral 9, 22
  as food 30
  poisons produced by 52
  spawning 52
  species
    antler *53*
    brain *53*
    plate *53*
    sea fan *22*
coral polyp *52*
coral reefs 15, 52–3
  animal life on 52, 53, *53*
  threat to 54, 55
cormorants 34, 35, *35*
crabs *4*, 14, 21, 22, 23, 46, 51, 52, 56
  feeding 33, 45
  as food 45
  life cycle *21*
  sharing home 60
  species
    edible *22*
    soldier *55*
crawfish, larva *20*
crocodiles,
  American 33
  food 33
  Indo-Pacific saltwater 33
  saltwater *54*
crustaceans 8, 9, 14, 15, 23
  migrations 42
  in mud 48
  parasitic 61
curlews *48*, *49*
currents, ocean, change of direction 11
cuttlefish 23
cyanobacteria 4, *5*, 6, *6*, 7, 12–13, 45, 49
cyclostomes 24

dark zone *18*, 19, 58
  animals of 58–9
darter, American *54*
dead man's fingers *23*
dead zone 13
decomposers 17
deep scattering layer 19
deep-sea life 56–7
deposit feeders 49, 51
depth zones 18–19, *18*
diatoms *12–13*, 13, 49
  in food chain *16*
  in food web *17*
dinoflagellates 12, 13, 49
dinosaurs 10, 11
DNA 7, 12
dog whelks 45, 51
dogfish 16, *29*
dolphins 36, 38
  echolocation 38, *38*
  in food chain *16*
  in food web *17*
  intelligence 39
  schools 38
  social ties 38
dredging 55
drugs, from corals 52
dugong 36, *36*
dulse *15*

echinoderms 23, 46, 47, 51
echo soundings 19
echolocation 38, *38*
ectoparasites 61
Ediacara Hills, Australia 8
eels 24, 49
  migrations of 42
  species
    gulper *58*
    moray 52, *60*
egrets 49
El Niño 55
electricity, fish producing 29
Eleutherobia 52
endoparasites 61
energy, from food 16, 17
Enteromorpha 15
estuary
  life in 48–9
  saltiness 48
  threat to 54
eukaryotes 7
evolution 4, 7
extinctions, mass 10–11

fat innkeeper 60
fats 7, 16
fertilizers 13, 14
fins *26*, 27, *27*
  anal *26*, 27, *27*
  dorsal (back) *26*, 27, *27*
  pectoral *26*, 27, 31
  pelvic *26*, 27, *27*, 31
  tail (caudal) 26, *26*, *27*
fish 5, 14, 15
  ancestors 9
  attracting mate 59
  bony 26, 27, *27*, 30–1
  breeding on land 31
  cartilaginous 26, *26*
  of coral reefs 52, 53, *53*
  deep-sea *56*
  feeding methods 30
  fins *26*, 27, *27*, 31
  gills 24, 26, *27*
  grooming 60–1
  hermaphroditic 59
  internal and external parts *26*, *27*
  jawed 26–7, *26*, *27*
  jawless 24–5, *24*, *25*
  lateral line *27*, 30
  migrations 42
  mouth-breeders 31
  of mudflats 49
  nurseries 49
  reproduction 31
  on sandy shore 47
  schooling 30–1, 53
  warning colors 53
fish farming 54
fish larva, in food web *17*
fish louse 60
fishing, commercial 54
flagella 13
flatfish 21
flatworms 61
flippers 36
flounders 51
flukes (tail) 36
flying fish 31, *31*
food chains 16–17, *16*, *17*, 21
food webs 16, 17, *17*
foraminiferans 20, 56
forests, underwater 14, *15*
fossils 6, 8–9, *9*, 10

frigatebirds 35, *35*
frogs 32
fronds 14, *14*
fungi *7*, 44

Galápagos Islands *33*
gastropod mollusks 23, 45
gill rakers 30
gills 7, 24, 26
Global Coral Reef
  Monitoring Network 55
goblinshark *56*
grasses 12
Great Barrier Reef 61
Great Lakes 24
grooming 60–1
grunion 31
gulls 35

habitats, destruction 10
haddock 61
hagfish 24–5, *25*, 56
Hallucigenia 8, *8*
hatchetfish 18–19
heart urchins 47
herbivores 16
herons 49
herrings 16, 17
  feeding 30
heterostracan 24
holdfasts 14, *14*, *15*
hydrogen sulfide 7, 57

ibis, scarlet *55*
iceworms 57
iguana, marine *33*
Indian Ocean 11, 33, 55
industrial waste 13
insects 32
International Shark Attack
  File (ISAF) 28
intertidal zone 44–5
invertebrates 22–3, *22*

jaegers *34*, 35
jawless fish 24–5, *24*, *25*
jaws, of fish 26
jellyfish 5, *5*, 8, 9, 16, 21,
  22, *59*
  common *22*

kelps 14, *15*, 51
  sugar *14*
knotted wrack *14*
krill 20, 23
  in food chain *16*
  in food web *17*
  reflecting sound waves 19

lagoons, threat to life on 54
lampreys 24, *24*, 25, *25*
landmasses, collision 11
lanternfish 19
larvaceans 21
larvae 21
lateral line system *27*, 30
lichens 44–5
light, bioluminescent 18,
  19, *59*
limpets 44, 45, 51
lionfish 53
living hotspots 5
lobsters 15, 21, 23
louse 60
luciferase 59
luciferin 59
lugworms 46, *49*
  casts *46–7*, *49*

mackerel *4*
mammals, marine 5, 17,
  21, 36–7
manatees 36
mangrove swamps 49
  clearing 54
  life in *54*
  threat to 54
manta ray 28, *28*
marine life, studying 4–5
marlin 16, *30*
mass extinctions 10–11
megalops stage *21*
merlin *5*
mermaids 36
metal (heavy) pollution 55
meteorites 11
  crater *10–11*
methane 7, 57
methane hydrate (methane
  ice) 57
Mexico, Gulf of 13, 57
microbes, species 4
migrations 42–3
mites 9
mollusks 15, 23, 45
  bivalve 23, 46, 47, 48
  cephalopod 23
  gastropod 23, 45
monkey, proboscis *54*
monogenean 61
Moorish idol 53, *53*
moray eels 52, *60*
mosses 32
mudflats 48, *48*, 49
mudhoppers 49
mudskipper *54*
mullet 49
mussels *4*, 21, 23, 45, 57
  common *23*
  farming 54
mutualism 60, *61*
Myxine 25
narwhal 38
navigation 39
  for migrations 42, 43
nekton 5
nitrogen
  sources 13
  trapping 12–13
notochord 24
nucelic acids 7
nucleus 7
nutrients, nitrogen-based 12

oarweed *14*
Obelia 22, *22*
ocean, depth zones 18–19,
  18
ocean trenches *18*, 19
octopuses 23, 51, *51*
Odaraia 9
Odontogryphus 9
Oikopleura 20
oil pollution 55
Olenoides 9
orcas (killer whales) *17*, 38,
  39
  in food web *17*
organisms
  earliest 6–7
  groups 5
  multicelled 7, *7*, 8
  single-celled 4, *5*, 22, 47,
    49, 61
  smallest 4
otters, sea 15, *37*
overharvesting 10
oxygen
  in atmosphere 7
  levels 6
  organisms surviving
    without 48
  production 4, 7, 12
oystercatchers 45
oysters 23
  farming 54

Pacific, sea snakes 33
parasitism 60, 61
parrotfish 30
pelagic environment *4*, 5
pelicans *34*, 35
penguins 34, *34*
  Adélie 34
  emperor 34
perch, sand *51*
periwinkles 45
Permian period 10
petrels 35
  diving *34*
  storm- *34*, 35
Pfiesteria 12
pheromones 59
photosynthesis 4, 7, 11, 12,
  13, 18, 55, 60
phyllosoma larva *20*
phytoplankton 4, *4*, 5,
  12–13, 18, 46, 49, 50, 56
  in food chain 12, 16, 17
  types 12–13
pilot fish *61*
pinnipeds 36–7
plaice 51
plankton 5, 12, 23, 43, 46
  see also phytoplankton;
    zooplankton
  colors 13
  creating dead zone 13
plankton bloom 13, 56
plants
  in food chains 16
  species 4
plover, ringed *49*
poisons
  produced by corals 52
  in red tides 12
pollution 13, 55
polyps 47, 50–1
  coral 52, *52*, 55, 60
porpoises 36, 38
Portuguese man-of-war 22
prawns 56
primary consumers 16
proteins 7, 12, 16
puffins *34*, 35

radiation, ultraviolet 6
radiolarians 20
ragworm *22*
rainforest, tropical 52
rattail fish 56
rays 27, 28–9, 51
  fins 29
  food 28, 29
  reproduction 29
  species
    electric 29
    manta 28, *28*
    sting- 29
razor clams 47, 50
razorbills 35
red tides 12
redshanks *49*
Reef Check 55
remoras 60
reptiles 32–3
  breathing 32
  reproduction 32
ribbon worms 22
rockpools, creatures in 22
rocks, formation 8
  acids eating away 7
rocky shores 44–5, *44*,
  45
Rodinia 9
rorquals 40
roundworms 56, 61

salmon 49
  farming 54
  in food chain *16*
  in food web *17*
  migrations 42
salps *20*, 21
saltmarshes 49
sand, creatures in 22
sand dollars 47
sand eels 17
sand mason worms *46*, 47
sand-perch *51*
Sargasso Sea 42
scallops *4*
scavengers, deep-sea 56
schooling 30–1, 53
sea anemones 22, 45, *53*
  beadlet *23*
  jewel *23*
sea cows 15, 36
  threat to 55
sea cucumbers 47, 56
  leopard *57*
sea floor
  deep 56
  shallow 50–1
sea lettuce *15*
sea lions 28, 36, 37
  Californian 37, *37*
sea otters 15, *37*
sea pens 50–1
sea slugs 14
sea snakes *see* snakes
sea spiders *57*
sea squirts 51
  larva 21
sea urchins *4*, 8, 14, 15,
  21, 23, *23*, 45, *45*, 51
seabirds *see* birds
seagrasses *4*, 14–15, *15*, *36*
  meadows 15, 55, *55*
seahorses *31*, 55
seals 14, 28, 36, 37
  crabeater 37, *37*
  elephant 37
  fur 37
  in food web *17*
seamouse 51
seaweeds 5, *5*, 14, *14–15*,
  44, *44*, 45, 46, 49, 50
  brown *14*, 45
  colors *15*
  grazers among 51
  green 45
  red 45
secondary consumers 16
sediment 10

sewage 13, 55
sharks *4*, 14, 26, 27, 28, 52, 56
attacks on humans 28
fish living on 60
food 28, 33
in food web *17*
legislation to protect 28
reproduction 29
senses 28
species
basking 28, *29*
bull 28
catshark *56*
dogfish 16, *29*
goblin *56*
great white 28, *29*, *61*
saw *29*
tiger 28
whale 28
shearwaters 35
shores
rocky, life on 44–5, *44*, *45*
sandy, life on 46–7, *46–7*
shrimp 14, 23, 47, 51, 52, 59
farming 54
reflecting sound waves 19
silica 20
silicate 13
single-celled organisms 4, *5*, 22, 47, 49, 61
sirenians 36
skates 27, 28–9, 51
fins 29
food 28, 29
reproduction 29
skuas 35
snails 14, 23, 56
snakes
banded sea *33*
sea 33, 52
yellow-bellied sea 33
sole 49
species
mass extinction 10
numbers 4
spermaceti 38
spiders 9
spineless creatures *see* invertebrates
spiracles 27
splash zone 44–5
sponges 9, 16, 22, *23*, 51
dead man's fingers *23*
squid 5, *5*, 16, 22, 23, 58
in food web *17*
giant 22
ink produced by 18
migrations 42
starches 16
starfish *4*, 22, 23, 45, 46, 47, *47*, 51, 52
blue *22*
feeding 45
stinging cells 22
stingrays 29
storm-petrel *34*, 35
stromatolites 6, *6*
sugar kelp *14*
sugars 16
sulfur dioxide 10
sulfur springs 6
sunlight
deprivation of 11
trapping 7
sunlit zone 18, *18*
surgeon fish *53*
swim bladder 27, *27*
Sydneyia 9
symbionts 60
symbiosis 60–1

terebellids 51
terns 35
arctic, migration *42*, 43
tertiary consumers 16–17
tides, red 12
toads 32
tool users *37*
toothed wrack *14*
topshells 45
tortoises 32
trees 12
trenches, ocean *18*, 19
trilobites 8–9, *9*
trophic levels 16
tube feet 23
tubenoses 35
tubeworms 51, 57
tuna 16, 31
turtles *5*, 32–3
migrations 42
reproduction 32–3
species
green 15, 32, 42, *42*
loggerhead *32*
twilight zone 18–19, *18*, 58

ultraviolet (U.V.) radiation 6
upwelling 5

Vampyroteuthis 58
vents, hydrothermal 6, 57
viper fish *58*
volcanoes, eruptions 10, 11

walruses 36, 37
water, in body fluids 6
whalebone 40
whales 36
baleen 21, 40–1, *40*, *41*
blowholes 40
breathing 36
communication 41
feeding 40, *40*, 41
intelligence 39
migrations 43, *43*
navigation 39
parasites 61
schools 38–9
social ties 38–9
song of 41
species
blue 4, *4*, 40, 41, *41*
fin *41*
gray 39, 40, *40*, *43*
humpback 40, 41, *41*
killer *17*, 38, 39, *39*
in food web *17*
narwhal 38
pilot 39, *39*
right 40, *41*
rorquals 40
sei *41*
sperm *19*, 38, *41*
toothed 38–9
stranding 39, *39*
whaling 43
whelks, dog 45, 51
whiting 61
winds, change of direction 11
winkle, mangrove *54*
worms 8, 9, 14, 15, 17, 22, 45, 46–7, 52
arrow 16, 17, *20*, 21
flat 61
iceworms 57
in mud 48
parasitic 61
predators 46–7
ragworm *22*
ribbon 22
roundworms 56, 61
sand mason *46*, 47
scale 60
seamouse 51
segmented 46
terebellids 51
tubeworms 51, 57
wrack
bladderwrack *14*
knotted *14*
toothed *14*
wrasse, cleaner 61

Xenophthalmichthys 56

zoea larva *21*
zones, depth 18–19, *18*
zooplankton *4*, 5, 18, 20–1, *20*, *21*, 30, 40, 46, 50, 56
in food chain 16, 21

Picture Credits

Photographs

Abbreviations

A Ardea; C Corbis; NHPA Natural History Photographic Agency; OSF Oxford Scientific Films; SPL Science Photo Library

1 Max Gibbs/OSF; 3 Georgette Douwma/SPL; 5 Microfield Scientific Ltd./SPL; 6 Roger Garwood & Trish Ainslie/C; 9 Kevin Schafer/NHPA; 10–11 Jean-Paul Ferrero/A; 11 Peter Scoones/SPL; 12 Pete Atkinson/NHPA; 12–13 Jan Hinsch/SPL; 15 Brandon D. Cole/C; 17 T. Kitchin & V. Hurst/NHPA; 19 Howard Hall/OSF; 20 OSF; 23 Kurt Amsler/A; 24 Ken Lucas-Photo/A; 28 Aquavisuals/OSF; 29 A.N.T./NHPA; 31 Rudi Kuiter/OSF; 32 Lynda Richardson/C; 33t Brandon D. Cole/C; 33b Galen Rowell/C; 34 Mark Hamblin/OSF; 35 Jean-Paul Ferrero/A; 36 A.N.T./NHPA; 37 François Gohier/A; 39 Dave Watts/NHPA; 40 François Gohier/A; 41 Norbert Wu/NHPA; 42 Allan G. Potts/Bruce Coleman Collection; 45 Anne Wertheim/Animals Animals/OSF; 46 Roy Waller/NHPA; 46–47 G.J. Cambridge/NHPA; 47 Fred Winner/Jacana/SPL; 48 R.J.C. Blewitt/A; 50 Lawson Wood/C; 51 David Fleetham/OSF; 53 Nancy Sefton/SPL; 54–55 Frank Lane Picture Agency; 55 William Boyce/C; 57l Agence Nature/NHPA; 57r Mark Spencer/A; 59 Norbert Wu/NHPA; 60 Matthew Oldfield, Scubazoo/SPL; 61 Rudiger Lehnen/SPL; cover photograph: Powerstock.

Illustrations

Andromeda Oxford Ltd 8, 9, 20, 22–23, 25, 26–27, 29, 30, 31, 37, 38, 39, 40, 41, 42b, 43, 50, 53, 54–55, 56, 58, 61; Bernard Thornton Artists 4–5, 7, 13, 14–15, 16, 17, 18, 21, 34–35, 42t, 44, 45, 49, 52.

EXPLORING THE OCEAN

VOLUME 1

THE PHYSICAL OCEAN

Trevor Day

OXFORD
UNIVERSITY PRESS

Consulting Editor Professor Dorrik Stow
University of Southampton Oceanography Centre, UK

Published in the United States of America by
Oxford University Press, Inc.
198 Madison Avenue
New York, NY 10016
www.oup.com

Oxford is a registered trademark of
Oxford University Press

Planned and produced by
Andromeda Oxford Limited
11-13 The Vineyard
Abingdon
Oxon OX14 3PX
UK
www.andromeda.co.uk

Library of Congress Cataloging-in-Publication Data
Day, Trevor.
Exploring the ocean / Trevor Day.
p . cm.
Summary: Explores various aspects of the world's oceans including currents, tides, weather patterns, the sea floor, marine life, uses by humans, and ecological issues.
Contents: v. 1. The physical ocean -- v. 2. Life in the ocean -- v. 3. Uses of the ocean -- v. 4. Reference and set index.
ISBN 0-19-515738-9 (set : alk. paper)
1. Oceanography--Juvenile literature. 2. Marine organisms--Juvenile literature. 3. Marine biology--Juvenile literature. 4. Ocean [1. Oceanography. 2. Marine biology. 3. Ocean.] I. Title.

GC21.5.D385 2003
551.46--dc21 2002042583

Project Director Susan Kennedy
Project Manager Derek Hall
Design Tony Truscott
Editors Marion Dent, Deborah Fox, Suzanne Juby
Cartographers Tim Williams, Richard Watts
Picture Manager Claire Turner
Picture Research Alison Floyd
Production Clive Sparling

Printed in the U.A.E. by Oriental Press Limited, Dubai

Using this set

This four-volume set describes the fascinating world of the ocean through informative text and colorful illustrations and maps. Each volume explores a different theme. **Volume 1** is concerned with the physical ocean: its origin and structure, and associated features such as tides, weather systems, islands, and coral reefs; **volume 2** describes the varied natural life in the ocean: evolution, marine animals and plants, food webs, ocean zones, communities of creatures, and conservation; **volume 3** looks at the way humans use the ocean: exploration from earliest times to the present day, exploitation of ocean resources, pollution, and ocean management.

In each volume, feature boxes highlight items of special interest. Cross-references at the bottom of each left-hand page point the reader to further information on related topics in other sections and volumes: the relevant volume number is given in bold type, followed by the page numbers. Each volume also has its own subject index.

Volume 4 contains a comprehensive series index, together with further reference resources, including maps, datafiles, full descriptions of the five oceans, biographies of men and women who have contributed to our knowledge of the oceans, timelines, places to visit, websites, further reading, and a glossary.

Throughout the set the following abbreviations are used for measurements:–
Length in (inches), cm (centimeters), ft (feet), m (meters), mi (miles), km (kilometers)
Area sq in (square inches), sq cm (square centimeters), sq ft (square feet), sq m (square meters), sq mi (square miles), sq km (square kilometers)
Volume cu mi (cubic miles), cu km (cubic kilometers)
Temperature F (Fahrenheit), C (Celsius)
Weight Given in U.S. tons, followed by metric tonnes

Previous page: An ocean wave breaking.

Right: Part of the group of sea stacks called the Twelve Apostles, on the coastline at Victoria, Australia.

Contents

The Physical Ocean 4
Blue Planet 6
Birth of the Oceans 8
The Changing Oceans 10
Global Jigsaw Puzzle 12
New Sea Floor 14
Anatomy of an Ocean 16
Recycling Rock 18
Islands 20
Coral Reefs 22
Attack and Defense 24
Rocky Shores 26
Sandy Shores 28
Estuaries and Deltas 30
Coastal Wetlands 32
Changing Sea Level 34
The Ocean Waters 36
Recycling Chemicals 38
Buried Treasure 40
Heat, Light, and Sound 42
Frozen Seas 44
Global Winds 46
Ocean Currents 48
Subsurface Currents 50
El Niño 52
Rhythm of the Tides 54
Making Waves 56
Ocean Weather 58
Climate Change 60
Index 62
Picture Credits 64

The Physical Ocean

Without the ocean, humans would not survive on Earth. The ocean controls the mechanisms that support life and regulates climate. Yet scientists are only just beginning to understand its physical structure.

As far as we know, life on our planet began in water, and for much of the Earth's history the ocean has been the birthplace of new species, from microscopic, single-celled creatures to huge, marine mammals. For a long time dangerous radiation from space bombarded the land, and the land was therefore unable to support life. But some 500 million years ago conditions changed, and life forms from the sea and fresh water began moving onto land. Among the pioneers were fishlike amphibians that most scientists believe eventually gave rise to the two- and four-legged creatures we see on land today, including ourselves.

Life-sustaining oceans

The ocean acts like a giant air-conditioner. It picks up heat in the tropics and dumps it at the poles, cooling the tropics and warming the poles. The ocean dampens down temperature changes both on the land and in the air and thus helps to keep them within a range that is suitable for life to exist on the planet. The ocean also regulates climate and weather. As water evaporates from the ocean, it forms clouds that carry rain. Clouds of moist, rising air swirl across the globe, creating weather systems. The weather systems determine where rain falls. Life depends on water, so the ocean helps establish where life can exist on land.

Microscopic, plantlike organisms called phytoplankton float in the surface water of the ocean. "Phyto" is from the Greek word for "plant," and "plankton" from the Greek for "drifting." These tiny marine organisms trap the energy from sunlight and use it to combine simple substances like water and carbon dioxide (a gas in the air) to make their own food substances, such as fats and carbohydrates, in a process called photosynthesis. As they do this, they enrich the Earth's air by taking in carbon dioxide and giving off oxygen, which is vital for most life on Earth.

Exploring beneath the ocean

Scientists have only begun to explore the sea systematically in the last 150 years or so. The earliest investigations

Earth or ocean?

- **Our planet should really be called "Ocean." In the 1960s, when astronauts first looked at Earth from space, they saw a cloud-shrouded blue planet, not a brown or a green one.**
- **Sea water covers 71 percent of our planet.**
- **More than 90 percent of the Earth's living space, by volume, is in the ocean or below the seabed. The remainder is on or below the land surface and in the air.**
- **About three-quarters of the world's ocean is at least 3 mi (5 km) deep. This is deep enough to contain 13 Empire State Buildings stacked one on top of the other.**

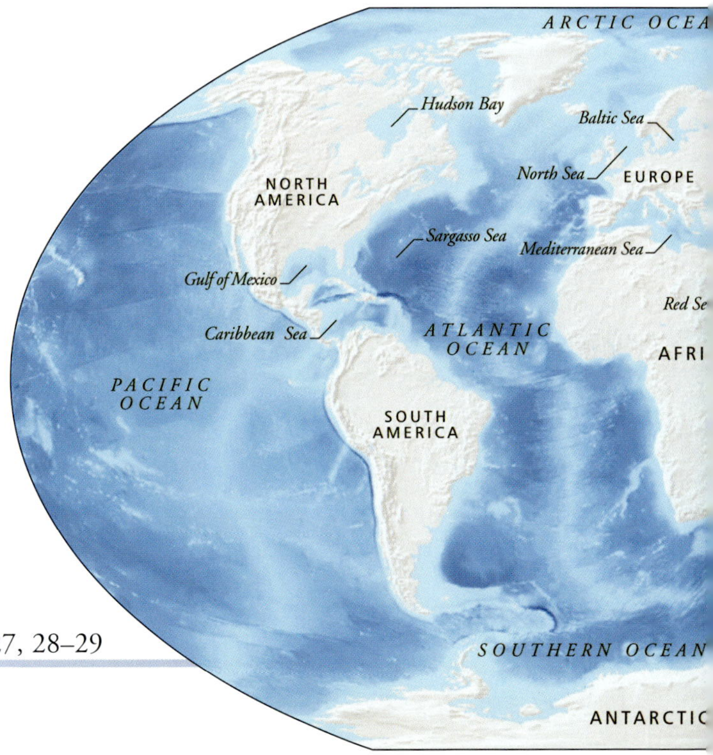

See also **vol 1:** 15, 46; **vol 2:** 12–13, 32–33, 57; **vol 3:** 26–27, 28–29

Oceans and seas

People now divide the world ocean—the single expanse of sea water that covers the Earth—into five oceans. They are the Pacific Ocean, the Atlantic Ocean, the Indian Ocean, the Southern or Antarctic Ocean, and the Arctic Ocean. Until recently, most marine scientists did not classify the Southern Ocean as a true ocean because, unlike other oceans, it is not surrounded by landmasses. It is the most southerly parts of the Pacific, Atlantic, and Indian Oceans.

"Sea" is a general name for the ocean water. But a sea is also a geographic region within an ocean. For example, the Sargasso Sea is a large, oval patch of the Atlantic Ocean, off the coast of southeastern United States. The Caribbean Sea and Mediterranean Sea are extensions of the Atlantic Ocean. "Gulf" and "bay" are other names for a sea partly enclosed by land—for example, the Gulf of Mexico and the Bay of Bengal.

A true-color satellite image of North America. Remote-sensing satellites such as the U.S. Nimbus-7 and the European ERS-2 monitor the oceans and land from space.

involved simply dragging nets through the water or pulling up huge buckets of mud from the sea floor. These crude methods gave scientists only a hint of what really lay beneath the sea. Within the last 50 years other methods have revolutionized the study of the ocean. They include bouncing echoes off the seabed to find out its shape and the water depth (a method known as sonar), taking photographs using satellites in space, and the use of cameras fixed to unmanned craft that can dive down thousands of feet to the deepest parts of the sea. For example, in the 1950s sonar surveys of the ocean floor revealed that by far the biggest mountain range lay below the sea, not on land.

Five oceans fill massive hollows on the Earth's surface. Seas, gulfs, and bays may be parts of oceans, too, or they may be land-bound areas of sea water connected to each other by oceans.

Scientists who study the physical ocean and the life within it are known as oceanographers. In 1977 a team of oceanographers in a submersible (a research vessel that can go under water) explored cracks known as hydrothermal (hot-water) vents in the ocean floor. They discovered amazing new life forms—for example, giant 3-ft (1-m) worms—living close to the warm water heated by the Earth's core that spewed from these cracks. Despite such developments, scientists have explored less than 15 percent of the sea floor in any detail. We know less about the deep-ocean floor than we know about the dark side of the Moon.

Blue Planet

The world's oceans are vastly bigger than the landmasses they surround. Some parts of the ocean also have much higher mountains and much deeper valleys than any of those to be found on land.

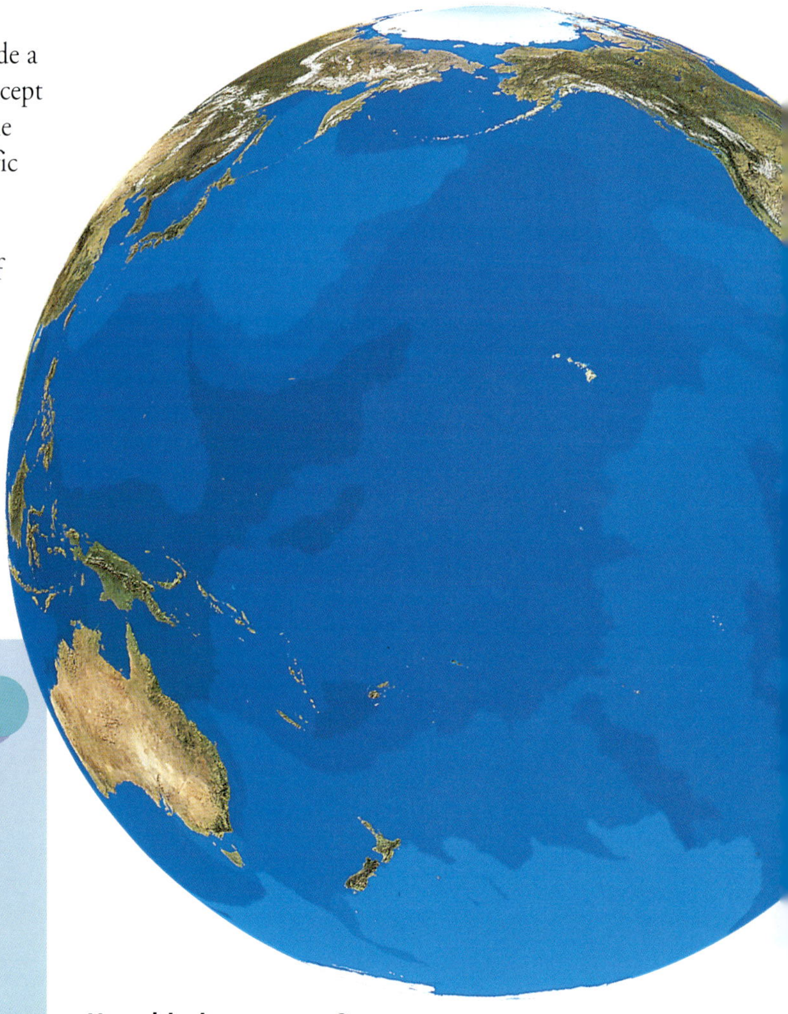

Imagine you are an astronaut floating, weightless, inside a space shuttle. You are looking down on the Earth. Except for swirling clouds, and edges of green and brown, the planet looks almost entirely blue. You are above the Pacific Ocean, the biggest volume of water on the planet. This single ocean covers nearly one-third of the globe.

The world's major landmasses mark the boundaries of the five oceans: the Pacific, Atlantic, Indian, Southern, and Arctic. Each ocean is contained within a gigantic hollow called a basin. The basins are interconnected. Water flows from one ocean to another. Together, the oceans form a huge "superocean."

The world's smallest ocean, the Arctic, covers an area only about 14 percent that of the Pacific. The Arctic is also by far the shallowest ocean, and the Pacific is the deepest.

Why are oceans blue?

Sunlight contains all the colors of the spectrum (the range of colors seen in a rainbow). But when sunlight strikes clear sea water, the colors at the red end of the spectrum tend to be absorbed by the water, so we do not see them. Mostly colors at the blue end of the spectrum are reflected by the water. This makes the sea look blue to us. Water in the air has a similar effect. On a clear day, the sky looks blue.

Of course, the sea does not always look blue. Near river estuaries, for example, where fresh water enters the sea, sea water often looks brown because soil is washed into it. Chemicals discharged into the sea may also stain it. The sea surface also reflects light from the sky. A red sunset can make the sea look ablaze.

How big is an ocean?

Experts do not always agree on the size of each ocean. When calculating ocean sizes some scientists include the seas that are attached to them, while others do not. An estimate of the size of the Atlantic Ocean that includes the Caribbean Sea, the Gulf of Mexico, and the Mediterranean Sea, is clearly bigger than one that does not!

Scientists sound the depths to measure the volume and determine the shape of ocean basins. Sailors used to take a

See also **vol 1:** 4–5, 42; **vol 2:** 18–19; **vol 3:** 28–29

sounding by lowering a weight attached to a long rope (called a sounding line) from a ship. They divided the line into 6-ft (1.8-m) lengths known as "fathoms." As the line ran out, a crew member called out the number of fathoms until the weight touched the bottom. The final reading gave the depth of the ocean at that point, in fathoms. To create a sea floor map, scientists took soundings at regular intervals and charted the different depths to show peaks and valleys, also known as troughs.

From the mid-20th century "sonar" was regularly used to take ocean soundings. Sonar involves bouncing sound waves off the seabed and timing how long it takes for the echoes to return. The longer it takes, the greater the depth. In the 1970s scientists added a technique called "radar altimetry." This method involves bouncing radio waves between a satellite and the sea surface. This way, scientists map the bumps and hollows on the sea surface that follow the contours of the sea floor. Scientists revise their estimates of the shape of the sea floor as more information becomes available.

The world's tallest mountain is Mauna Kea on Hawaii, but most of it lies beneath the sea. From base to summit Mauna Kea is 31,601 ft (9,632 m) tall—more than 2,500 ft (762 m) taller than Mt. Everest. If the two mountains could be placed side by side, this is how they would compare in height.

Tallest mountains and deepest valleys

A trough in the Pacific, the Mariana Trench, is the deepest place on Earth. It plunges to nearly 7 mi (about 11 km) beneath the sea surface. If Earth's tallest peak, Mt. Everest (29,035 ft/8,850 m high), could be sunk into this trench, there would still be about 1.2 mi (2 km) of water between the mountain top and the ocean surface. In 1960 French undersea explorer Jacques Piccard and American engineer Don Walsh traveled to the bottom of the Mariana Trench in the deep-diving vessel *Trieste*. They are the only people to have gone there.

The Pacific Ocean is by far the biggest of the five oceans. It contains nearly as much water as the other four oceans combined. If all the world's land was scraped off at sea level, it could be dumped into the Pacific Ocean basin with room to spare.

The table below shows a comparison of the dimensions of the world's five oceans. Areas are given in square miles (sq mi) and square kilometers (sq km). Depths are given in feet (ft) and meters (m). Volumes are given in cubic miles (cu mi) and cubic kilometers (cu km).

Ocean (without seas)	Area in sq mi (sq km)	Average depth in ft (m)	Volume in cu mi (cu km)	Greatest depth in ft (m)
Pacific Ocean	58,957,258 (152,617,160)	13,874 (4,229)	154,960,672 (645,375,552)	35,840 (10,924)
Atlantic Ocean	31,563,463 (81,527,400)	12,391 (3,777)	73,902,416 (307,905,776)	28,232 (8,605)
Indian Ocean	26,064,036 (67,469,536)	12,720 (3,877)	62,780,380 (261,590,400)	23,376 (7,125)
Southern Ocean	8,102,165 (20,973,318)	11,188 (3,410)	17,165,720 (71,518,408)	23,737 (7,235)
Arctic Ocean	3,351,811 (8,676,520)	6,349 (1,935)	4,028,950 (16,790,582)	18,400 (5,680)

Birth of the Oceans

Where did the water come from that fills our oceans? Most of the water formed on this planet, but it is probable that some of it also came from space.

Like other planets in our solar system, it is likely that the Earth formed from space debris that gathered together into a large clump. As the clump grew, its gravity increased and it attracted even more space debris. In this way, the Earth and other planets got bigger. When space debris collides at high speed it releases energy, and so the planets grew very hot, forming a glowing sea of molten rock. As they mopped up the debris in space, collisions became fewer, and the planets eventually cooled.

About 4 billion years ago the Earth's surface hardened. The hot rock released steam into the atmosphere (the layer of gas that surrounds the Earth), and the Earth became covered in a thick layer of water vapor. As the atmosphere cooled, clouds formed, the water vapor turned to liquid, and the liquid fell as rain. This downpour must have lasted thousands of years. As the water filled the hollows (basins) on the Earth's surface, the first true oceans were born. However, some of the Earth's water came from other sources.

Comets are made of ice, gas, and dust. Some of the Earth's water may have come from comets that collided with the planet. Other comets have sprinkled water into the Earth's atmosphere as they passed.

Volcanoes and dirty snowballs

Since the first watery oceans formed, rocks from space containing ice (known as meteorites) have continued to strike the planet regularly, adding water. So, too, have the occasional giant, dirty snowballs hurtling through space. These snowballs are known as comets. Both meteorites and comets are left-over debris from the birth of the solar system. And volcanoes erupting on the Earth's surface release steam and add to the water vapor in the air.

Are there oceans—and is there life—elsewhere?

As far as we know, Earth is the only planet in our solar system that has watery oceans. But it might not always have been so. The surface of the planet Mars is etched with features that look like ocean basins, shorelines, and river valleys. It is likely that Mars once held liquid water. If so, it is possible that some form of Martian life—probably microbes (microscopic organisms)—may have evolved. In 1996 NASA scientists believed they had found fossilized microbes in a meteorite originating from Mars and recovered from Antarctica. Others disagreed. These tantalizing, bacterialike traces could be the result of non-living, chemical processes.

Some scientists believe there may still be liquid water below the surface of Mars. Simple life forms might one day be discovered there. In the last 30 years scientists have uncovered life on Earth in unlikely places. The discoveries

See also **vol 1:** 6; **vol 2:** 6–7

include "rock-eating" bacteria from nearly a mile (1.6 km) below the Earth's surface and bacteria that live at incredibly high pressures and temperatures close to hot-water vents in the deep ocean. These strange creatures raise the possibility that Mars may also sustain life.

Life and water

Is there anywhere else where life might exist? All life as we know it needs water. Find liquid water, and there is a chance of finding similar life forms. Europa—one of the moons of the planet Jupiter—may have a surface layer of ice about 6 mi (10 km) thick. NASA scientists plan to send probes to Europa to drill through the ice to reach any liquid water below. There, they will search for signs of life. Many people are convinced that we will eventually find life somewhere in the Milky Way, our galaxy consisting of billions of stars.

When the Earth was born, some 4.6 billion years ago, it was very unlike the planet we know today. It was blazingly hot, and it had no watery oceans. Instead, it was covered in an ocean of liquid rock. This may have looked like the runny, red lava that flows from some volcanoes today, such as the one shown here.

The Changing Oceans

Over millions of years oceans grow and shrink. New ones are born, and old ones die. The shapes of the continents are gradually changing, too.

About 220 million years ago, soon after the beginning of the Age of Dinosaurs, the Earth consisted of just one single landmass and one ocean. Both were huge. The ocean, called Panthalassa (from the Greek word "pan" meaning "all", and the Greek word "thalassa" meaning "sea"), covered about two-thirds of the planet. The landmass, called Pangaea (from the Greek word "pan" meaning "all", and the Greek word "gaea" meaning "Earth"), filled the rest of the planet's surface. Pangaea was spread north-to-south across the globe, almost from pole to pole. Its two giant bays enclosed massive seas: the Sinus Borealis near the North Pole and the Tethys Sea across the equator.

1

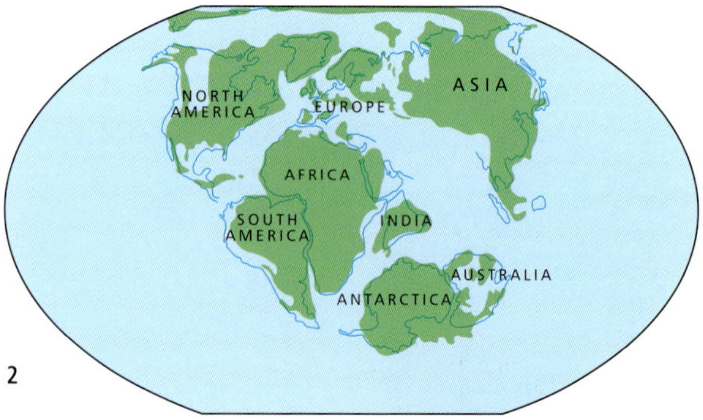

2

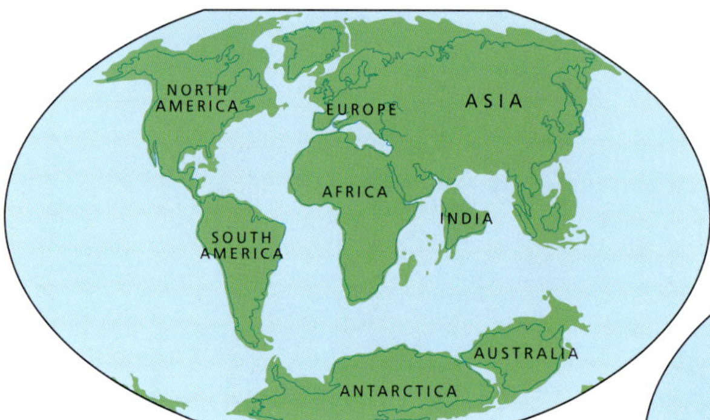

3

4

The two giant continents of Laurasia and Gondwana formed 200 million years ago (1). Over millions of years they broke into smaller landmasses (2). Australia broke away from Antarctica and India moved toward Eurasia (3). The continents took on their present shape within the last 5 million years (4).

The breakup of Pangaea

By 200 million years ago, Pangaea had begun to split in two along a rift that divided it into two supercontinents: Laurasia to the north and Gondwana to the south. Laurasia contained the present-day continents of North America and Eurasia (without India). Gondwana contained South America, Africa, India, Australia, and Antarctica. Water from the Tethys Sea flowed into the rift between the two supercontinents. The flooding of this rift marked the birth of the northern Atlantic Ocean.

By 130 million years ago, a rift had formed between Africa and Antarctica, triggering the breakup of Gondwana. A landmass that was to become India split away and moved northward, in the process establishing the Indian Ocean. By 80 million years ago—toward the end of the Age of Dinosaurs—the Sinus Borealis had moved north to become the Arctic Ocean. A rift between South America and Africa was giving birth to the southern Atlantic Ocean. As the Atlantic Ocean began to spread, the Pacific Ocean (the remains of Panthalassa) began to

See also **vol 1:** 4–5, 14; **vol 2:** 11

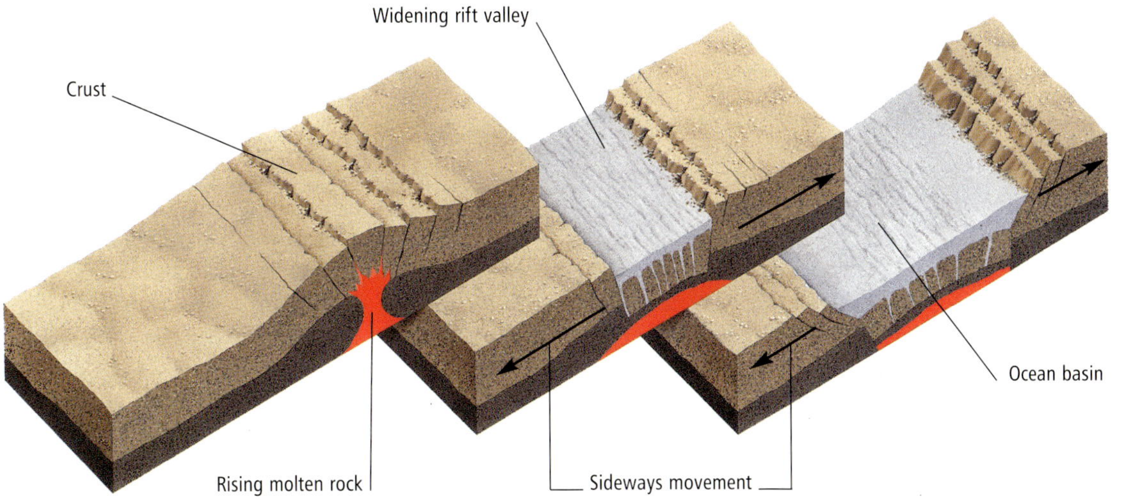

Some oceans formed after sea water flooded into a widening rift in the Earth's crust. Rising molten rock and new sea floor first spread into the rift valley and caused the landmasses on either side to move apart. Eventually, the valley formed an ocean basin.

shrink. The Pacific Ocean has been shrinking ever since.

In the last 80 million years the Indian Ocean has extended, and Antarctica has separated from Australia. The Atlantic Ocean has grown into a mature ocean and, in the process, North America has become separated from Europe. India has moved northward and collided with Eurasia. And an extension of the Indian Ocean has created the Red Sea, which is still widening.

Within the last 5 million years the Isthmus of Panama has joined together North and South America. The Mediterranean Sea is probably all that is left of the once-great Tethys Sea. About 5 million years ago the Atlantic Ocean broke through to the Mediterranean Sea at the Straits of Gibraltar that lies between Spain and Morocco. Sea water from the Atlantic Ocean then poured into the Mediterranean basin.

In 100 million years time, as sea floor spreading continues and the continents jostle their way across the Earth's surface, the arrangement of the oceans and continents will look quite different from the way they look today.

The 3,100-mi (about 5,000-km) long Great Rift Valley, seen here at Hell's Gate, Kenya, shows what a rift might look like before it floods with sea water and gives birth to an ocean. Although this particular rift is unlikely to flood soon, farther north the rift system has flooded with sea water and widened to become the Red Sea.

Global Jigsaw Puzzle

The sea floor is a landscape of vast, flat plains, volcanoes, mountains, and ridges. This landscape is slowly changing, as the plates of the Earth's crust drift on a layer of hot rock beneath the Earth's surface.

If you examine a world map, you will notice that the east coast of South America and the west coast of Africa look as though they could fit together, like pieces of a jigsaw puzzle. You can shuffle around other landmasses on a world map and fit them together in similar ways. Evidence such as this convinced German scientist Alfred Wegener that the continents of the world are not fixed in place but drift slowly across the surface of the globe over long periods of time. Over millions of years some continents might split apart, and others might crash together. In 1915, when Wegener suggested this idea, most scientists thought his notion was crazy. It was not until the 1960s that scientists who study the Earth's rocks (geologists) gathered together enough evidence to realize that Wegener's theory of "continental drift" was more or less correct.

Moving plates

The outer layer of the Earth's surface is called the "crust," and it rests on top of a hard layer known as the "upper mantle." The crust and upper mantle together form the lithosphere. "Litho" is from the Greek word for "rock", and "sphere" from the Greek for "ball." The Earth's crust is thin and heavy beneath the oceans, thus creating the gigantic basins that hold sea water. The crust is thick and slightly lighter on the continents. There it bulges upward to form land. The lithosphere is not a continuous layer over the Earth's surface. Instead, it is divided into seven gigantic slabs, and several smaller ones, called plates. They fit together like a giant jigsaw puzzle across the surface of the globe. The plates drift very slowly—at about the same rate as your fingernails grow. As the plates move, they carry the oceans and continents with them.

What is it that makes the plates move? Liquids and gases expand and become lighter as they warm up. They shrink and become heavier as they cool. Heat creates rising currents, while cooling produces downward movements. These temperature-produced currents are called convection currents.

The Earth's plates. Arrows show the direction of plate movement. At ridges (shown in blue), new material is added to plate boundaries. Where plates collide (shown in red), they trigger earthquakes and create nearby volcanoes or build fold mountains. Where plates slide past each other (shown in yellow), earthquakes can also occur.

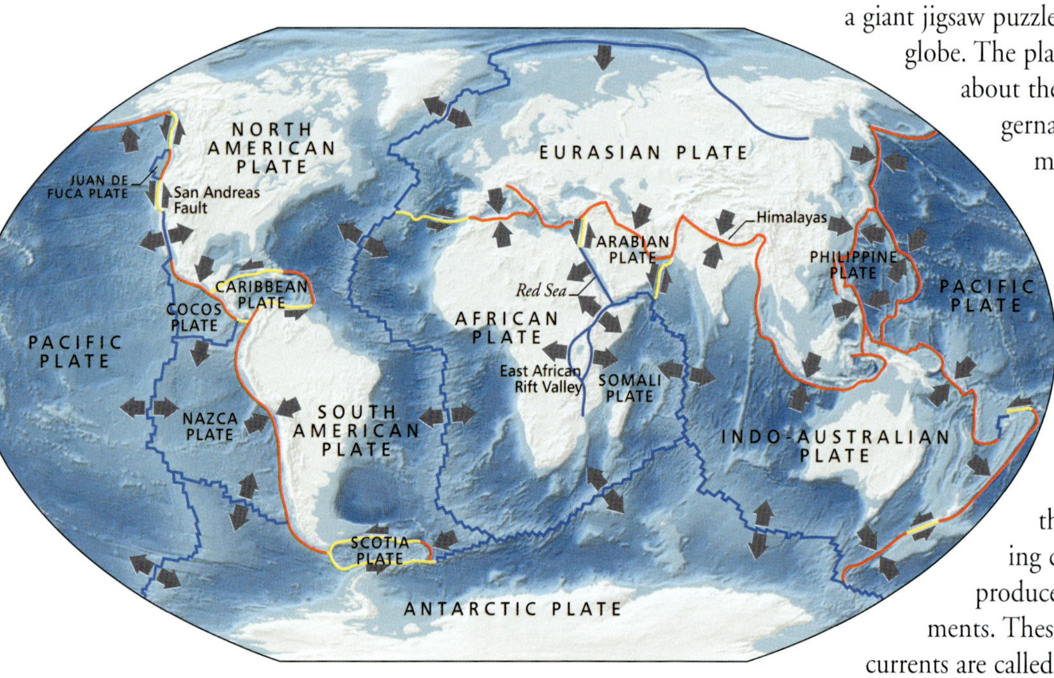

See also **vol 1:** 10–11, 14, 16–17

A satellite view of the Red Sea, seen in the bottom of the picture. Africa is on the left and Arabia is on the right. The Red Sea is widening because of plate movements. Some scientists calculate that, if the Red Sea continues widening at its present rate, in 100 million years time it may reach the size of a great ocean such as the Atlantic Ocean.

They help to circulate the air and sea. Beneath the plates lies a lower part of the mantle called the asthenosphere. "Asthenos" is from the Greek word for "weak", and "sphere" from the Greek for "ball." Scientists believe that the immense heat deep within the Earth causes hot mantle material to rise as it warms, so stirring the mantle and setting up slow-motion convection currents. These currents probably cause the plates to move.

Where plates meet

Where plates meet, zones of earthquake or volcanic activity occur. Where there are cracks, hot molten rock, or magma, can rise up from below to form volcanoes. Where adjacent plates grind or crash together, or even slide past each other, they release shock waves that produce earthquakes.

When two plates collide, one usually slides under the other. Oceanic crust descends and melts to become hot molten rock. This action, known as subduction, creates trenches—deep valleys in the ocean floor. Volcanoes belching lava (magma that erupts onto the Earth's surface) often form above subduction zones. Sometimes, two plates that carry continental landmasses on their leading edges collide. When this happens, one continent cannot sink below the other. Instead, the two plates crash together. As the continents push against each other, the crust buckles. This can create massive mountains. This is how the Himalayas formed.

The San Andreas Fault in California is a site where two plates are sliding past each other. The plates do not slide past each other smoothly. Resistance to slow motion (friction) causes them to lock. When this happens, pressure builds up. Suddenly they unlock, and energy is released as massive vibrations in the rock. This causes earthquakes such as the ones that shake San Francisco and Los Angeles.

When two plates move apart beneath the oceans, magma rises from the crust into the gap that is formed. This feature is called a spreading ridge. It creates new sea floor. Along the length of the Red Sea, the African plate to the west and the Arabian plate to the east are moving apart. As they do so, the Red Sea is gradually becoming wider.

Some 50 million years ago the Indian subcontinent collided with Eurasia to create the Himalayas. The Himalayas are the world's tallest land-based mountains. They are still pushing upward.

NEW SEA FLOOR

The floor of the ocean is constantly in motion. A vast, undersea mountain chain—the mid-ocean ridge—is the birthplace of new sea floor. As new sea floor is forming, old sea floor is destroyed in deep troughs called trenches.

By the 1950s oceanographers were using sonar to map the ocean floor in detail. They began finding ridges running the length of oceans. They named them mid-ocean ridges. A steep ridge runs down the center of the Atlantic Ocean. A gentler ridge lies off the west coast of South America. The ridges and rises are part of an interconnected system—the mid-ocean ridge system—that winds its way through the world's oceans. The ridges and rises show some unusual features. A V-shaped rift often runs along the middle of the ridge. The rock inside the rift, and the water nearby, are warmer than their surroundings. Such findings hinted that fresh volcanic activity was taking place in ridges.

In 1960 American geologist Harry Hess suggested that magma is rising from below at mid-ocean ridges. As it cools and solidifies, it lays down new sea floor. According to Hess, the sea floor is moving slowly, like a creeping conveyor belt. At the same time as new material is being added at ridges, the old sea floor is being removed in trenches. This movement pushes the continents along. They ride on the crust, which is driven by currents in the mantle. Hess called this process sea floor spreading. But where was the evidence?

In the stripes

In the 1960s geologists set out to prove that new sea floor was indeed spreading out from mid-ocean ridges. By 1963 two geologists, Lawrence Morley in Canada and Fred Vine

In the rift running along the middle of a mid-ocean ridge, molten rock rises up from the hot crust and mantle below. It adds to the ocean crust on either side and, as it cools, it shrinks slightly and settles. The main ridge becomes flanked by smaller ridges (called abyssal ridges) on either side. The movement of crust, and the cooling and shrinking of new crust, create tensions that cause the sides of the mid-ocean ridge to crack in places. This produces transform faults that split spreading ridges into shorter sections.

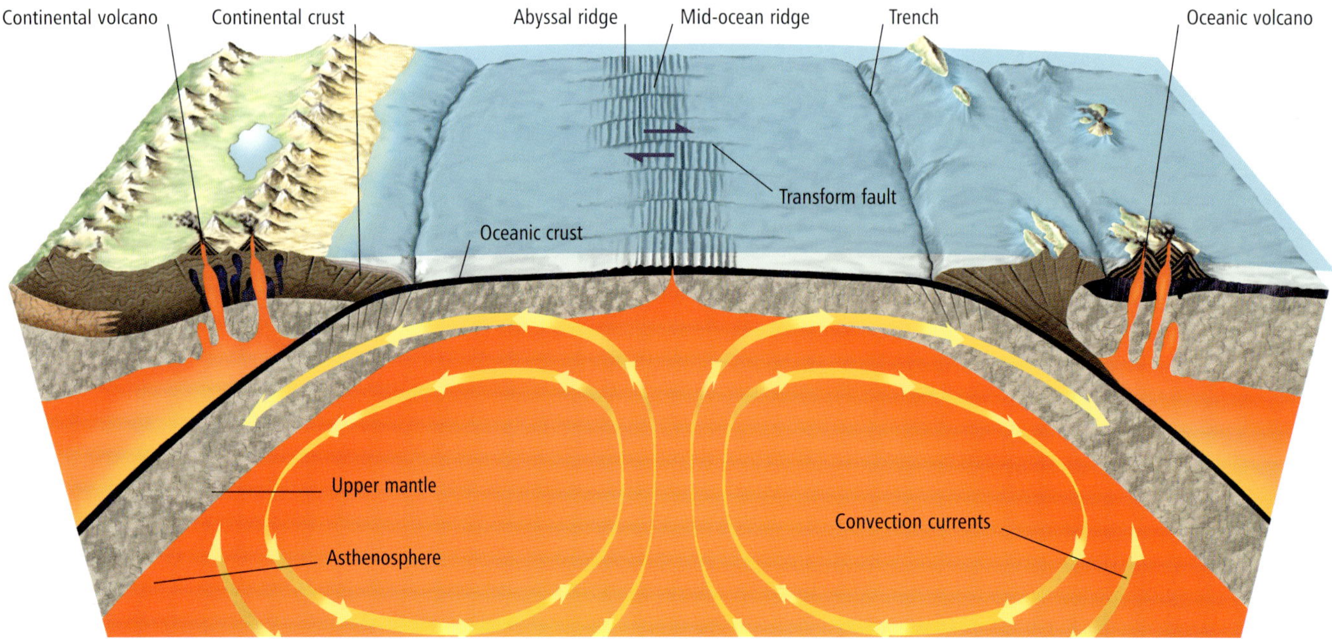

See also **vol 1:** 12–13, 16–17, 37, 38–39, 41

The mid-ocean ridge system winds its way through the five oceans. At 40,000 mi (65,000 km) long, the system is the world's biggest mountain range—more than three times longer than the Andes, Himalayas, and Rockies combined.

in England, came up with an ingenious theory. It involved the Earth's magnetic field.

The Earth is a giant magnet. Every few hundred thousand years its magnetic field (the force that causes magnets to align in a north–south direction) reverses. When new rock is laid down—for instance, at a mid-ocean ridge—the iron inside the rock is trapped in position, and carries the "imprint" of the Earth's magnetic field at the time. When the Earth's magnetic field changes, the imprint in new rock formed after the reversal of the magnetic field matches the new alignment, and scientists can detect this. If new sea floor is laid down at ridges, then the imprint of the Earth's magnetic field should form a series of alternating stripes—known as magnetic reversals—moving away from the ridge. The stripes on either side of the ridge should be a mirror image of each other. And in the mid-1960s this is what scientists found. The theory of plate tectonics (the study of the movement of the Earth's plates) gained a huge boost, and Hess's theory of sea floor spreading was proved correct.

Trenches

Old sea floor is destroyed in ocean trenches when it is pushed down into the mantle and melted. Some of the rock may emerge as magma in nearby volcanoes, or it may become incorporated in the Earth's mantle. If new sea floor was continuously created, but old sea floor was not destroyed, then the Earth's surface would simply grow. However, no oceanic crust is able to survive for more than about 200 million years before being pushed down. Continental crust, on the other hand, can survive for billions of years.

Hydrothermal vents

By 1977 oceanographers in submersibles (research vessels that submerge in water) visited the rifts in mid-ocean ridges to gain a closer view. They discovered that water gushes out of chimneylike vents, or cracks, in the rock. At the high pressures that exist 1.2–1.8 mi (2–3 km) below the sea surface, water from these vents is super-hot (660° F/350° C) but does not boil. Scientists named the chimneylike structures hydrothermal vents. The discovery of hydrothermal vents changed our understanding of the physics, chemistry, and biology of the ocean. For example, all the world's ocean water probably passes through hydrothermal vents every 10 million years or so. The vents "process" the water—adding some chemicals such as sulfur and removing others such as magnesium. The vents play an important role in controlling the chemical makeup of sea water.

Lumps of cooled lava—called pillow lava because they resemble pillows—lie scattered along mid-ocean ridges. They form when fresh lava squeezes out through cracks and then hardens on contact with sea water.

Anatomy of an Ocean

The larger oceans of the world are thousands of miles across and contain many physical features. However, at its simplest, the bottom of the ocean is made up of just two main parts: the continental margin and the ocean floor.

Imagine you could drain the northern Atlantic Ocean and walk across its floor. It would be like a muddy trek across North America, but with very different rocks and "soil" beneath your feet. After leaving the Florida coastline, you cross the coral-dotted continental margin before descending the continental slope off the Bahamas. This brings you to a plain—the Hatteras Plateau—400 mi (650 km) across. This plain is about 2.5–3 mi (4–5 km) lower than the coastline where you began. As you trudge across the plain, you see in the distance an almost unbroken chain of mountains. This is the Mid-Atlantic Ridge, and to cross it you have a 1.2-mi (2-km) uphill climb through a volcanic landscape of ridges. At the crest of the ridge you look down into a hissing, volcanically active rift.

Climbing down the other side of the ridge system—a journey as difficult to undertake as climbing the Rocky Mountains—you find yourself on the welcoming flat of the Porcupine Abyssal Plain. After walking northeastward for hundreds of miles you come to a steep slope—the continental slope. But there are deep valleys—submarine canyons—in the slope, and you must also tackle these as you climb up from a depth of about 2 mi (more than 3 km) to the wide, shallow continental shelf of the British Isles. The sandy trudge across this 500-ft (150-m) deep shelf brings you to Cornwall, the southwesternmost tip of England. Your journey has taken months, and you have traveled 5,000 mi (8,000 km) to achieve it.

Continental margin

The continental margin is the edge of the land that has been covered by sea water. The flattest part, up to the shoreline, is called the continental shelf. Its outer edge—called the shelf break—averages about 500–650 ft (150–200 m) in depth. The overlying water is usually rich in nutrients and well lit by sunlight, and these factors encourage the growth of seaweeds and phytoplankton. Continental shelves make up only 8 percent of the world's total ocean area, nevertheless they are the most food-rich part of the sea.

The seaward edge of the continental shelf gives way to the steep, continental slope—the true edge of the continent. Deep, V-shaped valleys called submarine canyons slice across the continental slope. Underwater avalanches of rocks and sand sometimes sweep through these canyons with enough power to snap thick, submerged telephone cables. Beyond the continental slope is the deep ocean floor.

Ocean floor

Trenches, occurring at depths greater than 3.7 mi (6 km), are the deepest features in the ocean. Large expanses of the ocean floor lie at depths of more than 2.5 to 3 mi (4–5 km). The expanses are almost flat. They are covered with loose

This ocean panorama is many hundreds of miles across. The major features of the ocean floor are shaped by the processes that accompany movements of the Earth's plates. Sediment settling on the sea floor slowly buries or softens some of these features.

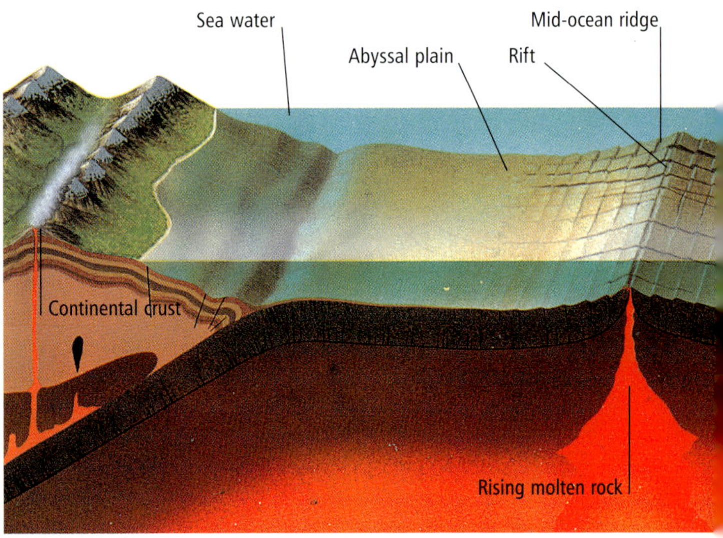

See also **vol 1:** 14–15; **vol 2:** 18–19, 50–51, 56–57, 58–59

Steam rises from boiling sea water as hot volcanic lava meets the ocean at the Hawaiian coast. Lava flows from active volcanoes, such as this one, create new land when they solidify at the surface.

material such as tiny particles of rock and the remains of dead, microscopic plankton that pile up on the ocean floor. This sediment is often about 2,000 ft (609 m) thick but can be more than 1.2 mi (2 km) thick. It covers small geological features so that only large rocky outcrops, such as volcanoes, rise above the level of the plain. Where underwater volcanoes rise to the sea surface, they form islands. Over time, the peaks of these islands may wear away and flatten out.

Abyssal plains, the name of the flat expanses, rise almost imperceptibly toward the mid-ocean ridges. (The word "abyssal" is from the Greek word "abyssos" meaning "without bottom.") Because the sea floor is younger near the ridges, less sediment has settled and more geological features are visible. Abyssal ridges rise stepwise, one after the other, toward the mid-ocean ridge. Running along the center of the ridge is the volcanically active rift—the birthplace of new sea floor.

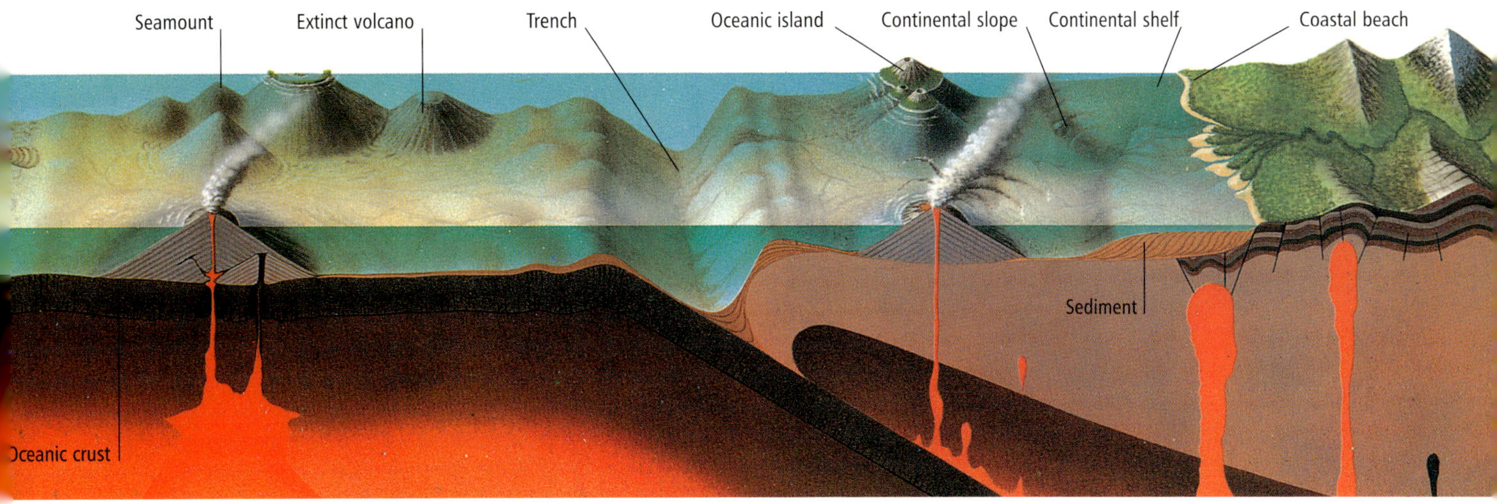

Recycling Rock

The Earth's crust is made of rock. The actions of erosion and deposition, as well as forces deep within the Earth's hot mantle, are continuously forming and reforming rocks. This process is called the rock cycle.

Geologists group or classify rocks into three main types, according to how they are formed. Some rocks, called "igneous," form when magma cools and forms crystals. The word igneous comes from the Latin word meaning "fiery." These rocks can form at the Earth's surface—for example, when volcanic lava cools rapidly. When this happens, it usually forms rock known as basalt. Sometimes magma cools slowly underground—for example, beneath folded mountain ranges. Then it forms granite. Basalt and granite are both types of igneous rock.

Sometimes new rocks are formed from fragments of worn-down rocks, or from the skeletons or shells of dead organisms. All these particles are known as sediment, from which sedimentary rocks get their name. Rivers wash some rock particles into the sea, and winds blow tiny particles of dust from the land onto the sea. The particles may eventually

The rocks of the Devil's Slide, Britain, above, are made of granite. This is an igneous rock, formed when molten rock (magma) cools and forms crystals (crystallizes). The Painted Desert, Utah, on the right, consists of sedimentary rock. This is formed from tiny pieces of rock or minute animal remains that become compacted and cemented together.

See also **vol 1:** 9, 13, 28–29, 40–41

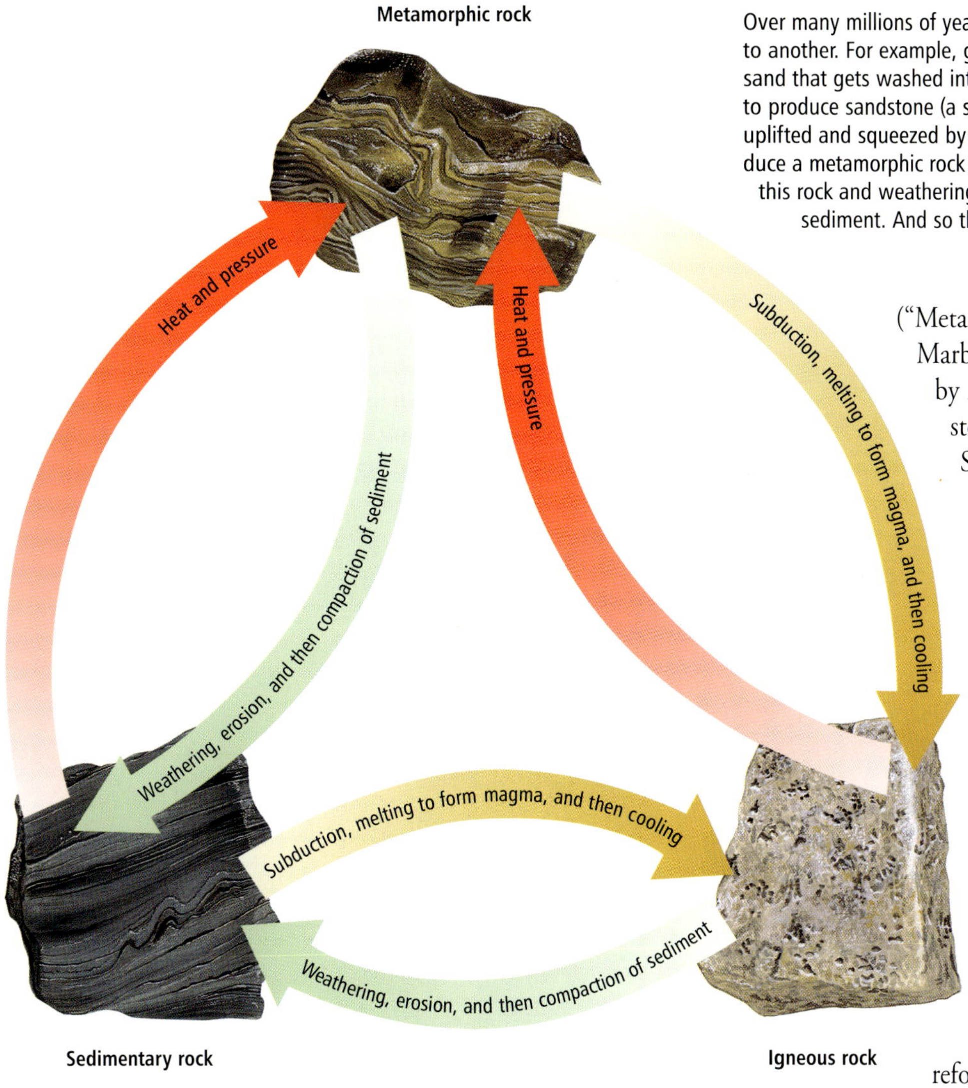

Over many millions of years, rocks are transformed from one form to another. For example, granite on land may be weathered to yield sand that gets washed into the sea. Sand may become compacted to produce sandstone (a sedimentary rock). Sandstone may be uplifted and squeezed by the force of two colliding plates to produce a metamorphic rock such as a schist. Erosion could expose this rock and weathering could fragment it, so producing sediment. And so the rock cycle continues.

("Metamorphic" means "changing shape.") Marble is a metamorphic rock created by igneous rock intruding into limestone, a type of sedimentary rock. Sedimentary rocks may become buried deep underground, and high pressures and temperatures transform them into metamorphic rocks such as slate.

During great movements of the Earth's plates, the force of two colliding plates may fold and squeeze together layers of sedimentary or metamorphic rock and push them upward to form mountain ridges. Old rock returns to the Earth's hot mantle and melts, or squeezes to form a different type of rock deep within the crust. Rocks may also weather and erode to reform as sedimentary rock. All these changes are an ongoing part of the rock cycle.

sink to the bottom of the ocean. As the layers of sediment pile up on top of each other on the seabed, they eventually become squeezed together to form hard sedimentary rocks. The sedimentary rock limestone, for example, can be formed from countless billions of chalky shells of microscopic organisms squeezed together over long periods of time. Sediment from lakes and rivers can produce sedimentary rock on land.

Extreme heat or pressure change existing rock into new types of rock. For example, hot igneous rock, in the form of magma, may force its way up through the Earth's crust. This is known as an intrusion. The intrusion may then melt some of the surrounding rock, creating a new type of rock. Rock formed in this way is called metamorphic rock.

Where sediments settle

In the ocean, most sediment collects on continental margins. Where very large river systems discharge into the sea, river-borne sediments can smother vast areas of the continental shelf and beyond in what is called a "sediment fan." The sediment fan deposited by the Ganges River, India, is the world's largest. It extends about 1,500 mi (2,400 km) beyond the continental shelf.

Some of the strangest sea floor deposits are those that grow mysteriously from metals carried in sea water. In some parts of the ocean, this process produces potato-sized lumps of rock, called manganese nodules, that form gradually over thousands or millions of years.

Islands

There are thousands of islands dotted among the world's oceans. Even small islands, some often only a few miles across or even less, may have great importance both to life on land and in the sea.

Islands are landmasses that are smaller than continents. Even the smallest continents are several times larger than the biggest islands. Islands range in size from uncharted isles a few yards across, to giants such as Greenland, Papua New Guinea, and Borneo—each of these is well over 1,000 mi (1,600 km) long. Most geographers describe three kinds of island in the sea: inshore, offshore, and oceanic.

Inshore islands

Inshore islands lie on continental shelves. Dry land once connected them to the continental mainland, but rising sea levels or sinking land levels have now separated them from

Mykines Island, part of the Faroe Island group, lies in the North Atlantic. The Faroes are offshore islands, formed by volcanoes that emerge close to the Mid-Atlantic Ridge.

Small but important

Oceanic islands may have an important impact on the wildlife of the region. Plants and animals from the mainland that colonize oceanic islands may become isolated and develop in different ways from their mainland relatives, so they become more suited to the new conditions. This process is called evolution. For their size, some oceanic islands contain an unusually large number of unique species. Scientists have so far described about 5,000 species on and around the Galápagos Islands, about 600 mi (960 km) off Ecuador. More than one-third of the species are found nowhere else.

Oceanic islands also have a direct effect on the surrounding ocean. The surface waters of the mid-ocean can be wildlife deserts. Life there can be sparse. The presence of an island deflects deep-sea currents upward. Nutrient-rich water wells up from the deep and encourages marine plants and phytoplankton to grow in the shallows. In warm waters, coral reefs may grow around the island. The waters around oceanic islands often become habitats rich in oceanic life.

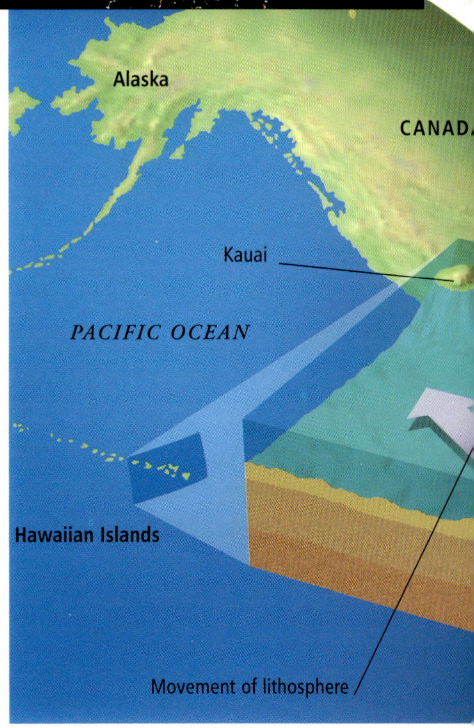

See also **vol 1:** 10, 16–17, 35

The oceanic Galápagos Islands, seen here in a satellite photograph, emerged above a hot spot in the equatorial Pacific Ocean.

their continents. The British Isles are of this type. The British Isles were still connected to continental Europe only 18,000 years ago during the last ice age. As the glaciers that then covered much of northern Europe melted, sea water levels rose and the continental shelf submerged to form what is now part of the North Sea.

Offshore islands

Offshore islands are often found grouped together in clusters called archipelagos. They lie close to a continental landmass but not on the continental shelf itself. Offshore islands have not had a dry-land connection with the continent for many millions of years—if at all. Long ago, some of the plants and animals from the mainland colonized the islands; they may have flown, swam, or drifted there on rafts of vegetation. Ancient offshore islands—such as Madagascar, off the African coast—may harbor animal communities that are different from those on the adjacent mainland. In some cases, animals on the mainland have not survived the competition with other creatures. But the same, or similar species, have survived on the islands.

Oceanic islands

Oceanic islands arise several hundred miles or more from the nearest continent. Most oceanic islands in the Atlantic and Pacific are created by volcanic activity.

In the Atlantic, many islands—including the Faroes and the Azores—have emerged from volcanoes close to the Mid-Atlantic Ridge. Many Pacific islands formed above very hot parts of the mantle, known as "hot spots." A hot spot burns through the overlying crust and creates a volcano that may force lava to the sea surface, creating an island.

In the Indian Ocean, some oceanic islands were created when India broke away from the ancient supercontinent Gondwana. As India moved north, it left giant fragments in its wake. The Seychelles are an example of such fragments.

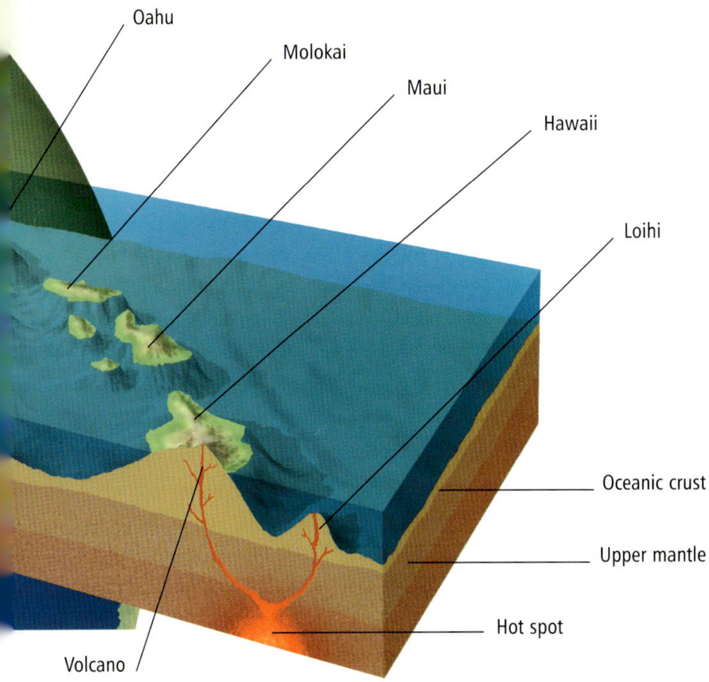

The oceanic Hawaiian Islands and Emperor Seamount chain were formed when the Pacific plate moved slowly over a hot spot. The island of Hawaii formed within the last million years. Southeast of Hawaii, the submarine volcano Loihi is still building and may one day emerge from the sea to create another island in the chain.

Coral Reefs

Day by day, across the world's warmer seas, tiny creatures are building cities of limestone called coral reefs. The world's largest coral reef system, Australia's Great Barrier Reef, is so big that astronauts can see its pale shape from space.

Tiny, anemonelike creatures called coral polyps build coral reefs. These animals are related to jellyfish and sea anemones, but they are smaller. Coral polyps usually grow in colonies of hundreds or thousands to create structures of amazing shape and color—some resembling gray brains and others purple antlers, for example.

There are many kinds of coral polyp. Some, called "soft" corals, are widespread across the world's oceans, but do not build reefs. Reef-building coral polyps are called "hard" corals. Hard corals are unusual because they contain microscopic plants called algae. Reef-building coral polyps and their algae are locked in a partnership for survival. The polyps provide a safe home for the algae. The polyps capture minute sea animals with their tentacles and give nutrients (food) to the algae. The algae provide the polyps with extra nutrients and oxygen when they make their own food. The algae also provide chemicals that enable the polyp to form a hard, cupshaped, limestone skeleton around itself.

Where do reefs grow?

Reef-building coral polyps—and the algae they contain—are fussy about where they live. They grow best in warm tropical and subtropical waters between 68 and 86° F (20–30° C). However, coral polyps die if the water becomes too warm.

The algae that live inside reef-building polyps need plenty of sunlight for photosynthesis. This restricts reef-building polyps to clear, shallow water. Corals will not grow in cloudy waters. If a reef has already formed, and pollution or a change in land drainage causes sea water to become cloudy, the coral may die.

Reef-building corals need to live in sea water that has near-normal levels of saltiness (salinity). Where sea water is

The three main types of coral reef. A fringing reef builds up around an island formed by volcanic activity. If the seabed subsides the island sinks, and a lagoon separates the reef from the island. This is a barrier reef. Eventually, only the coral remains, forming a ring-shaped atoll.

Building a reef

When a hard coral polyp dies, its limestone skeleton is left behind. Reef-building coral polyps, growing in their countless billions, form the "bricks" of a coral reef. New coral polyps grow on top of the skeletons of dead ones, and so the reef extends. Other organisms—such as algae with limestone skeletons and worms that build limestone tubes—live on the coral reef and add more limestone. Together these organisms create the "cement" that helps bind the reef together.

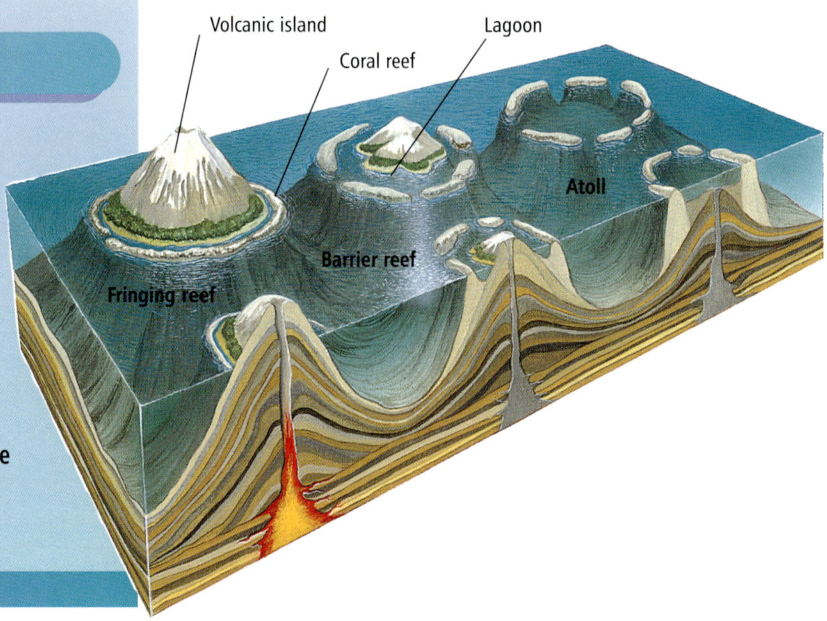

See also **vol 1:** 43; **vol 2:** 52–53

The dark blue water outside a coral atoll contrasts with the lighter, shallower water in the atoll's lagoon.

strongly diluted by fresh water, coral reefs do not form. So coral reefs are not found close to river estuaries, for example.

Types of reef

In 1842 the British naturalist Charles Darwin described three main types of coral reef formation: fringing, barrier, and atoll. Scientists still use this classification today.

A fringing reef grows in shallow water alongside a shore. There may be a small, seawater-filled gap (called a lagoon) between the reef and land. A barrier reef is similar to a fringing reef but grows farther out from the shore. The lagoon is deeper and wider. The Great Barrier Reef, Australia, is probably the world's best-known reef. Its lagoon is at least 20 mi (32 km) wide, and is much wider in places.

Around islands, fringing reefs and barrier reefs sometimes develop to become the third type of reef—a coral atoll. An atoll is a horseshoe- or ring-shaped coral reef that surrounds a lagoon. Charles Darwin suggested that an atoll and its lagoon marked the place where a volcanic island once stood. More than a century later geologists made test drillings on Pacific atolls. They confirmed Darwin's suggestion by showing that some coral atolls are hundreds of feet thick and grow on top of deeply submerged islands.

The red dots on the map below show the location of the world's major coral reefs. Coral reefs grow in warm, shallow waters where the temperature is above 68° F (20° C). Marine biologists calculate that the world's coral reefs harbor between 1 and 3 million different species of organisms.

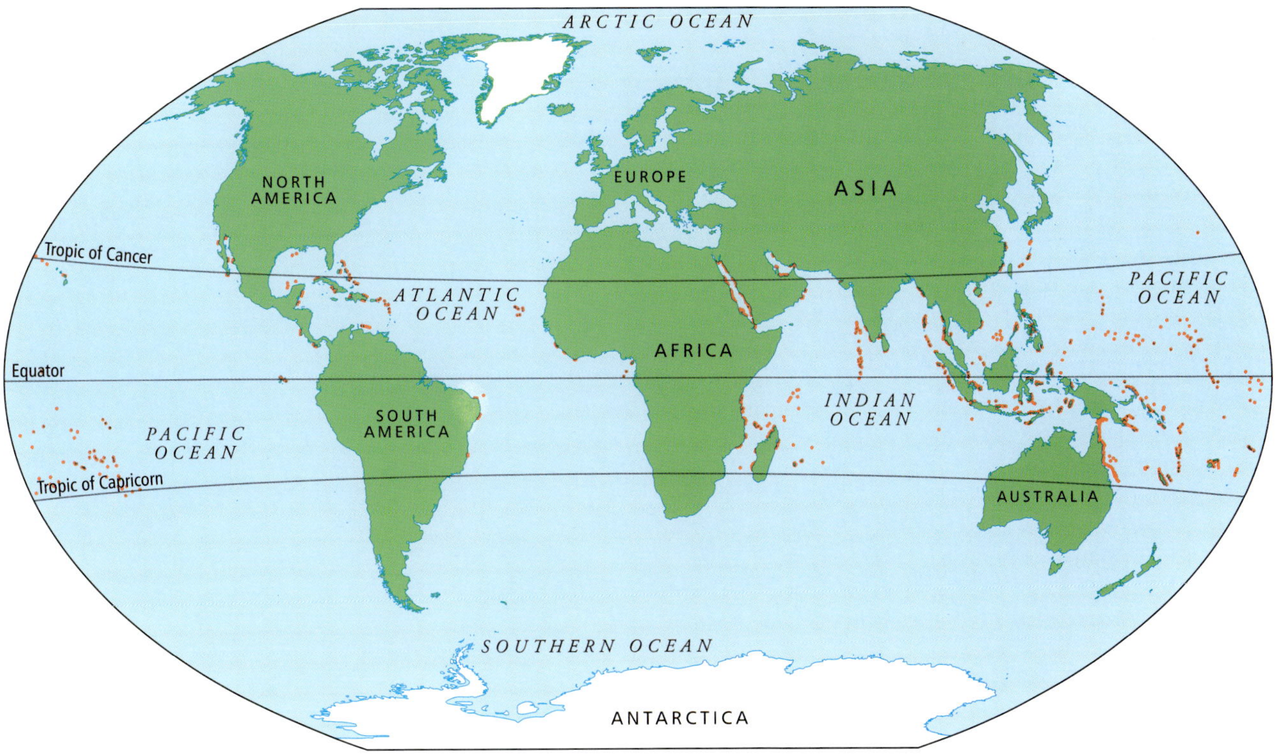

Attack and Defense

The coast is the part of the land that meets the sea. It is a dynamic place that natural forces are constantly reshaping.

The part of the coast that is covered by sea water when the tide comes in, or is splashed by the biggest storm waves, is called the shore. The coastline is the border between shore and coastal land.

The position of a coastline relative to the Earth's plate boundaries greatly affects the coast's features. For example, much of the western Pacific coast of North America lies close to a plate boundary. The coast is being lifted up. In many places, cliffs tower above a rocky shoreline. Below the waterline, the continental shelf is often narrow and the seabed slopes steeply away to depths of thousands of feet.

The eastern North American coastline is quite different. It is a long way from a plate boundary. On the east coast, the land is slowly sinking. The slopes both above and below the waterline are much gentler. In many places, the beaches are wide and sandy, and the continental shelf is broad.

The sea is dramatically eroding the soft rock of the coastline on the left, causing parts of it to collapse onto the beach below. Cliffs liable to break away in this manner can be very hazardous, since the collapse may occur at any time, not just when waves are washing against them.

Altering the coastline

Forces linked to the movement of the Earth's plates (called tectonic forces) establish the overall shape of a coastline. Weathering, erosion, and deposition then alter the coast. Weathering is the gradual breakdown of rock by physical, chemical, and biological action on the Earth's surface. Erosion is the carrying away of rock particles by wind, water, and gravity. Deposition is the laying down of particles in a new place after they have been weathered and eroded.

Waves, currents, and tides are major forces that shape coastlines. On rocky shores, the forces of erosion dominate. On sandy shores, deposition is the major activity.

On a rocky shore, crashing storm waves can slam against cliffs with a pressure similar to the output of a space shuttle's main thrusters. The waves drive compressed air into cracks and then, as the wave retreats, the air expands explosively, enlarging cracks and blasting off small pieces of rock. And storm waves can move boulders weighing hundreds of tons. Meanwhile, sand and gravel suspended in waves scrape against the rock surface and gradually wear it down.

Chemical weathering and erosion occurs, too. When the salt in sea water crystallizes, it expands and forces rocks apart. And sea water weakens and partially dissolves certain types of rock, such as chalk and limestone. A series of 30-ft (9-m) storm waves can erode a soft, chalky cliff by 3 ft (about 1 m) in a single night.

See also **vol 1:** 12–13, 16–17, 26–27, 28–29, 37

The sea unloads sediment (mainly sand and mud) in sheltered parts of coastlines where the action of waves, currents, and tides is less strong. Sandy beaches are common along the Atlantic coast of the United States, for example.

Waves rarely strike a sandy beach "head on." Instead, they hit the beach at an angle. Prevailing winds blow from a particular direction at certain times of the year, so that the sand on a beach tends to be swept along the coast in the direction of the prevailing winds. This feature is known as longshore drift.

Fighting the sea

Coastal engineers design defenses to withstand the immense power of the sea. For sea defenses to be successful, they need to be designed with great care. The people living in most seaside towns view sandy beaches as a valuable place for recreation, and so they will take steps to reduce the impact of longshore drift. Building groins—barriers at right angles to the beach—is one way of doing this. Next time you visit a beach, notice that sand gathers on the upcurrent side of a groin and is scoured away from the downcurrent side.

A sloping beach is a natural wave break. Creating artificial beaches is a costly process and can go disastrously wrong if the effects of waves, winds, and currents are not taken into account. In 1982 at Ocean City, New Jersey, local residents spent more than $5 million replacing the sand on their local beach. However, the cost and effort were wasted because the sand was swept away within three months.

Building concrete seawalls is one way to protect coastal towns from storm damage. However, a seawall does not gradually disperse the sea's energy as a natural beach would. Instead, the seawall may receive the full force of the sea and can erode severely. One way to reduce this effect is to place concrete boulders at the base of the seawall.

Artificial barriers like these groins can reduce erosion locally, but because sand is no longer swept along at the same rate, downcurrent of the beach erosion may be greater.

Rocky Shores

Rocky shores occur where coastal land is being lifted up by forces linked with the movement of the Earth's plates. Such coastlines often feature dramatic cliffs, and rocky features such as caves and pinnacles.

The size and shape of a sea cliff depend on the nature of the underlying rock together with the force of the waves that attack it and wear it away. For example, soft, sedimentary rock such as limestone erodes much faster than hard, volcanic rock such as granite.

A rocky coastline usually has headlands that jut out into sea, with sheltered bays between them. The headlands bear the main force of the waves, so that they are eventually cut back. As rock is worn from the headlands, it creates suspended sediment that is carried away and settles in the bays, forming a beach at the base of the cliffs. In time, as the headlands are cut back, the shape of the coastline becomes smoother. Eventually, the gently sloping beach may become wide enough to absorb the impact of most waves, so that few, if any, reach the cliffs. When this happens, cliff erosion slows down dramatically.

When a rocky coastline is cut back by weathering and erosion, distinctive features form. Wave action at the cliff base undercuts the face. The upper part may then collapse, spreading rubble across the base of the cliff. The softer parts of headlands are eroded more quickly, creating sea caves. If the cave breaks right through the headland, it creates a sea arch. Further erosion may cause the arch to collapse, leaving pinnacles of rock called stacks.

Lost down the slope

In southern California, the damming of rivers is changing the shape of the nearby coastline. Along the coast between Santa Barbara and La Jolla lie four "beach compartments." Each compartment has three parts: one or more rivers, a sandy beach, and a submarine canyon offshore. The rivers add sand to the coastal

The rising Pacific coast

The North American Pacific coast is slowly rising as the Pacific plate slides under or alongside the North American plate. In some places, marine terraces—wavecut platforms at the base of old sea cliffs that are now raised high and dry—show evidence of this happening. In the Palos Verdes hills of southern California, marine terraces rise in a series of tiers from the present coastline to 1,300 ft (400 m).

Most of the Pacific coast is exposed to large storm waves—up to 20 ft (6 m) high in winter. All year round, exposed areas of coastline are washed by smaller but high-energy waves that have traveled far across the ocean. Exposed beaches, such as Boomer Beach in La Jolla, California, have their sand swept away by storms in winter, exposing pebbles and boulders. In summer, smaller waves refill the beaches with sand. Some parts of the Alaskan coast are eroding rapidly—at the rate of nearly 8 ft (about 2.4 m) a year—where soft rocks are exposed to strong wave attack.

See also **vol 1:** 18–19, 28–29; **vol 2:** 44–45

On some rocky beaches, a rugged platform of exposed rock at the foot of the cliffs largely protects the land behind from constant wave attack.

water. Longshore currents spread the sand along the nearby beach. Eventually, the sand from the beach finds its way to the head of a submarine canyon downstream. The sand becomes lost from the coastal system when it flows down the canyon and spreads onto the ocean floor below. As fast as sand flows down the submarine canyon, rivers upstream replace the sand that is lost from the beach compartment.

In recent years, however, local authorities have begun damming rivers and lining the riverbanks with concrete. This has massively reduced the amount of sediment entering the sea. The sand in beach compartments is no longer being replaced at the same rate, and the beaches are becoming starved of sand.

The dams provide communities with hydroelectric power and offer some protection against flooding, so it is very unlikely that the authorities will remove the dams. Over time, if the sand is not replaced, the shoreline will become more and more rocky. And replacing the sand artificially, in the huge volumes required, would cost several million dollars a year for each affected beach compartment.

Wave action has eroded these limestone cliffs, creating dramatic scenery. On the left, the cliff has been almost completely washed away leaving an isolated pinnacle called a stack. On the right, the erosion has created an arch.

SANDY SHORES

The combined forces of the wind and the sea continually shape and reshape all types of sandy beaches—from those on tropical coral islands to the ones surrounding many continental coastlines.

Sandy beaches form when waves, ocean currents, and tides carry materials such as clay particles and tiny rock fragments to parts of the coastline quicker than they carry them away. Sandy beaches are common on the Atlantic and Gulf coasts of the United States, as well as along parts of the Californian coast of the Pacific Ocean.

The sediment that settles on beaches comes from several sources. Sandy and gravelly beaches contain coarse mineral fragments that erode from cliffs and coastal mountains. Mud comes from clay and other tiny particles emptied into the sea by rivers. On tropical and subtropical shores, where there are no nearby mountains or rivers, the shells and skeletons of marine organisms may be the main source of beach material. Beaches also gather sand blown from inland.

The stunning, tropical white beaches of the Seychelle Islands in the Indian Ocean are composed of coral sand and shell fragments.

Beach formation

Particles of varying sizes settle in different places on a coast according to the way ocean currents and waves carry them. The size of the particles deposited determines the nature of the beach.

The receding Atlantic coast

The Atlantic coast of the United States is currently sinking, relative to sea level, at the rate of about 1 ft (0.3 m) every 100 years. At the same time, the coast is being eroded, on average, at the rate of about 2–3 ft (0.6–0.9 m) every year. Each year, the Atlantic coastline is moving westward by a distance roughly equal to the length of your legs. Atlantic coast real estate is slowly disappearing.

See also **vol 1:** 35, 44–45; **vol 2:** 46–47

Muddy beaches, which have the smallest particles, form on the most sheltered shores, such as those in and around river estuaries. Sandy beaches have medium-sized particles. Shingle (water-worn pebbles) and gravel—which have the largest particle sizes—collect on shores exposed to strong currents and persistent battering by large waves. Whatever the composition of the beach particles, however, the action of waves, sea currents, and tides can shift them almost constantly, even when they settle.

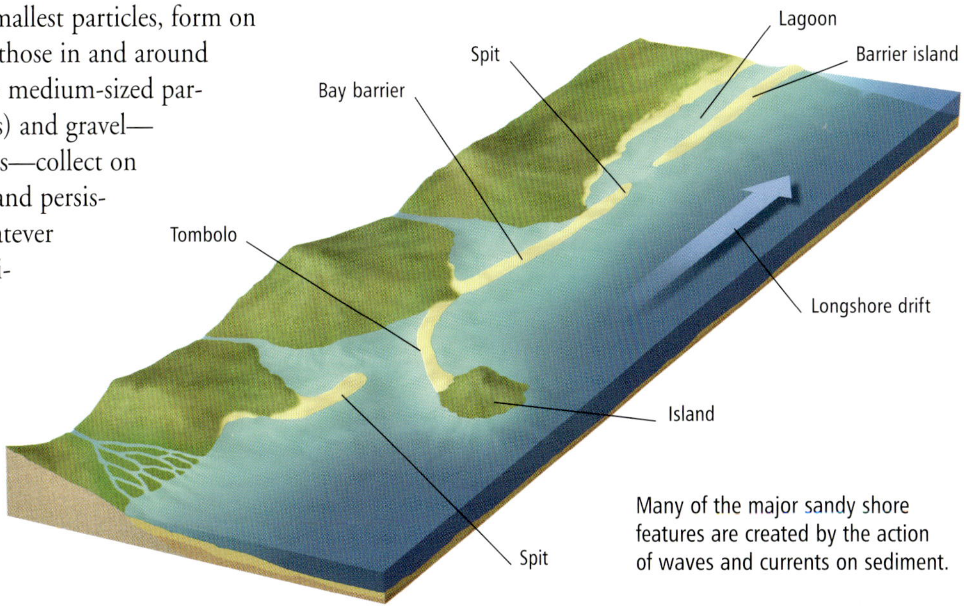

Many of the major sandy shore features are created by the action of waves and currents on sediment.

Coastal features

On both the Atlantic and the Pacific coasts of North America, waves arriving at an angle to the shore tend to move coastal sediments in a southerly direction. The movement creates features typical of shores where deposition (the laying down of particles) dominates. These features are particularly obvious where there are bays sheltered from strong waves and currents by headlands, and the water is fairly shallow.

On the downstream side of a headland—in other words, in the direction of the current—a stretch of sand known as a spit sometimes forms roughly parallel to the shore. The spit may grow large enough to almost or totally close off a bay or harbor. Then it becomes known as a bay barrier or bar.

Barrier islands

Along the shoreline of the United States, between Long Island, New York, and southern Texas, there are dozens of low-lying, sandy islands that lie roughly parallel to the coast. These are barrier islands. Scientists still argue about how these barrier islands, so-called because they are a barrier to storm waves, were formed. However, most scientists agree that the islands formed after the last ice age (a time when ice covered much of the Northern Hemisphere).

Since then, many glaciers and ice sheets have melted, raising sea levels by some 300 ft (90 m). The barrier islands may have been created when sea water flooded low-lying coastal land, and glacier meltwater washed large amounts of sediment into coastal waters. Longshore currents would have distributed river sediment along the coast to form spits and bars. As sea levels continued to rise, sea water would have flooded the ends of spits and bars, making them into islands.

Regardless of how they were formed, Atlantic or Gulf barrier islands are prime pieces of real estate. Miami Beach, Florida, and Corpus Christi, Texas, are typical examples. However, barrier islands have certain drawbacks. They are at risk of flooding and, unless they are properly protected by sea defenses, they are unstable.

Sometimes sand settles between a near-shore island and the mainland and creates a connection, or causeway, called a tombolo. This one is in Taormina, Sicily.

Estuaries and Deltas

Estuaries are places where rivers flow into the sea, and where fresh water mixes with sea water. Estuaries are an important part of the connection between the ocean and the land.

Some estuaries are small and shallow, but others are huge and deep. A few estuaries form giant natural harbors that are the site of the some of the world's greatest cities—Bombay, Buenos Aires, New York, and Tokyo, for example. Larger estuaries are usually classified according to their shape and how they were formed. Chesapeake Bay, between Maryland and Virginia, is the drowned mouth of a river valley. Rising sea levels have engulfed the valley since the peak of the last ice age some 18,000 years ago. Chesapeake Bay is known as a coastal estuary.

Some estuaries are created when the land sinks. San Francisco Bay is an example of such an estuary. Its irregular shape was created by movement and sinking along the San Andreas Fault. Pamlico Sound, North Carolina, on the other hand, is a shallow estuary separated from the open ocean by sand bars or barrier islands. A fjord, by contrast, is a steep-sided, U-shaped coastal valley gouged out by huge, moving sheets of ice called glaciers. Fjords are common in mountainous, coastal regions in higher latitudes, such as on the west coasts of Canada, Alaska, and Norway.

Returning nutrients to the land

Estuaries play a largely unseen role in recycling nutrients (substances that provide nourishment) between sea and land. Nutrients washed into the estuary by rivers find their way into mud, where they are consumed by microbes and burrowing creatures. Other animals then eat these creatures, or their waste, and so absorb the nutrients. Finally, the nutrients end up in the stomachs of seabirds that visit the rich feeding grounds of the estuary. When birds shed their droppings on land, they fertilize it and return nutrients to the soil.

Deltas

Deltas are vast, flat plains of sediment that accumulate at the mouths of some rivers. They form where rivers dump mud and sand at their mouth, but currents flowing parallel to the shore (longshore currents) fail to spread the sediment immediately. The delta often forms a triangular or fan shape as the sediment spreads over low-lying land or fills an indented coastline. As the river flows over the delta, it usually breaks up into several distributaries or branched channels.

Salt and nutrients

The range of creatures that live in an estuary is not large, because environmental conditions are harsh. The saltiness of water in an estuary changes with the state of the tide and

In some places, estuaries are dredged to provide deep-water channels for vessels, and in others they are filled to "reclaim" land for building. All kinds of development—industrial complexes, oil refineries, docks, and marinas—are sited in and around estuaries. The features that make estuaries so productive and attractive for development, also make them very liable to pollution.

See also **vol 1:** 12–13, 29, 44–45; **vol 2:** 48–49

An infrared satellite photograph of the Mississippi delta. Offshore, the seabed is subsiding, preventing the buildup of a typical fan-shaped delta. Instead, low mounds of sediments (called levees) accumulate around river channels to create a "bird's foot" shape that extends into the Gulf of Mexico.

with the amount of fresh water flowing into the sea. Many aquatic creatures cannot cope with such drastic changes in salt levels. Nor can they cope with the vast amounts of sediment that could smother them, or the lack of oxygen in the water or mud. However, the rivers that wash into estuaries bring plenty of nutrients, so finding food is not generally a problem.

Although estuaries do not have a great variety of animal life, the species that do live there may be found in vast numbers. Several hundred burrowing clams may be found in just 1 sq ft (0.1 sq m) of mud on some mudflats. Birds such as ducks, geese, and waders—as well as flamingos—gather in their hundreds or even thousands to harvest food in, and on, the mudflats.

Living on estuaries and deltas

Deltas are highly fertile places for farming. One of the earliest civilizations in history, ancient Egypt, depended on the rich soils of the Nile delta to grow food for its population. But because the land is low-lying, flooding can be a serious hazard. Most of the population of Bangladesh (more than 120 million people) lives in the delta of the Ganges and Brahmaputra Rivers in the Bay of Bengal. Swollen rivers and storm surges frequently cause floods in Bangladesh, killing tens of thousands of people.

Many estuaries formed important routes for people traveling between the sea and inland destinations. The control of an estuary was therefore a source of economic and political power. This is one of the reasons why so many major towns and cities have grown up beside them.

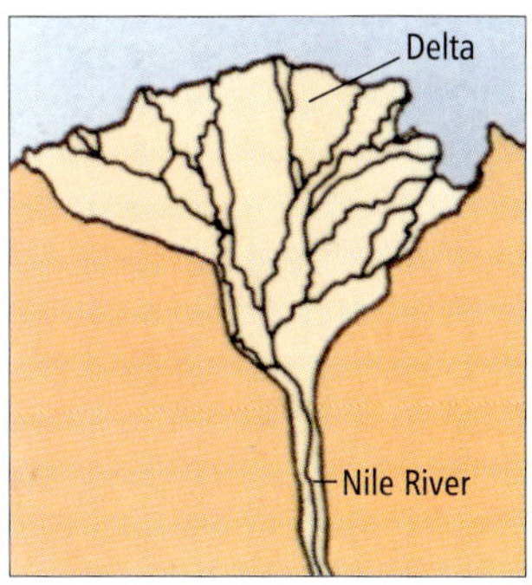

The Nile delta is a classic fan-shaped delta. In 1964, the Aswan Dam was built on the Nile River about 700 mi (1,120 km) from the sea to provide Egypt with hydroelectricity. As a result, less sediment is being carried downstream to the river estuary, and the delta is now being eroded.

Coastal Wetlands

Coastal wetlands are waterlogged strips of land. Certain salt-tolerant terrestrial plants grow on them. In cool parts of the world, these wetlands are usually called saltmarshes. In warmer places, they are called mangrove swamps.

Along some gently sloping, sheltered shores, where the water is usually calm, tiny particles settle as mud on the sea bottom. Gradually the deposits build up into mudflats, which grow high enough to be exposed to the air for long periods at low tide. Then they become wetlands.

Under threat

Wetlands, like the estuaries with which they are commonly linked, are often threatened by human activities. Many wetlands are vital and even unique havens for wildlife. But being so close to the sea, wetlands are also prime targets for land filling or draining to create coastal resorts. When this happens, vital nursery beds for commercially important fish and shellfish are lost. Ducks, geese, and other migratory water birds must seek new feeding or nesting grounds.

The particles in the mud of wetlands are small and closely packed together. Water is trapped in the gaps between them, and oxygen has difficulty getting through. The lack of oxygen, and regular covering by sea water, limit the types of animals and plants that live in the mud. The main wetland plants in cool parts of the world are salt-tolerant marsh grasses and rushes, but in warmer climates plants called mangrove shrubs and trees grow instead.

Safe havens

Like the mudflats of estuaries, wetlands accumulate mud that is rich in organic waste (mostly the decaying parts of plants and animals). This provides rich sources of food for animals and plants. The wetlands are safe nurseries for the young of many species of fish and other marine animals that breed there. Large predators are likely to flounder in the shallow water and mud, and marsh grasses, rushes, or mangroves offer numerous hiding places for small creatures such as snails and crabs.

Mangrove shrubs and trees often grow profusely in tropical and subtropical wetlands. Mangroves have branching roots to hold them firmly in the mud and special roots that grow upward out of the mud to take in air.

See also **vol 1:** 29, 30–31; **vol 2:** 54–55

This lagoon at Cancun, Mexico, forms part of a beach resort. Coastal lagoons are unusual because they often have connections with both the sea and with rivers.

Scrubbing clean

Beneath the surface of wetland mud, remarkable processes are at work, "cleaning" the water of pollutants, natural and manmade, that drain off the land. For example, mud particles trap heavy metals such as lead. Marsh grasses absorb nitrate fertilizers that wash off agricultural land. The effect is to reduce the amount of polluting chemicals that find their way into the sea. When wetlands are destroyed, this natural "water scrubbing" facility is also lost.

In Egypt, the United Nations Development Program has established a scheme to do artificially what wetlands do naturally. Engineers have created beds of marsh plants to clean the Bahr El Baqar drain, a waterway that carries human and other waste from Cairo to the coastal lagoon of Lake Manzala. "Cleaning" the water reduces the health hazard to people living and working near the lake. Once the water has been cleaned, it is used to rear high-value fish on fish farms.

In the southeastern United States, saltmarshes and their creeks are nurseries for the young of more than half the locally caught fish species. Fishers also harvest clams, oysters, and scallops from the marshes, along with crab, eel, and smelt.

Coastal lagoons

Shallow lakes, or lagoons, sometimes form behind the wetland shoreline. They contain a mixture of fresh water and sea water. The saltiness of the water may vary widely across the lagoon according to the tides and seasons of the year. Rain water and fresh water draining into the lagoon from the land dilute the sea water on the landward side of the lagoon. But on the seaward side, the lagoon may get fresh supplies of sea water when the tide comes in. If the heat of the Sun evaporates water from the surface more quickly than it is replaced by fresh water, concentrations of salt build up. The water on the seaward side may become more salty than normal sea water.

CHANGING SEA LEVEL

The level of the sea is never constant. It rises and falls as the tide goes in and out. The global sea level also changes over time.

Tides are not the only physical factor that influence sea levels. Weather conditions also affect local sea levels. Onshore winds push the sea upward slightly toward the coast and raise the local sea level. Offshore winds have the reverse effect: they locally lower the sea level. A combination of high tides and coastal storms can massively raise coastal sea levels, causing the sea to surge inland.

Measuring sea level

Sea level is the average level between high and low tide. The global sea level is the average sea level throughout the world. To measure sea level, scientists use a vertical post with a measurement scale (known as a tide gauge) and then make adjustments for local rises and falls in the level of the land. Scientists have developed new technologies to make sea level measurements more accurate. For example, they bounce radar waves onto the sea from a satellite. The time it takes for the reflection to return tells them the sea level.

Longterm change

Global sea level changes often over long periods of time. When the Earth's climate cools—for example, during an ice age—rain tends to freeze and stay on land rather than run off into the sea. So the sea level drops. During the last ice age, the sea level probably fell by 560 ft (170 m). During warmer periods of the Earth's history, ice melts off the land and enters the sea. Also, sea water expands as it warms, contributing to a rise in sea levels. Since the thaw of the last ice age, beginning 18,000 years ago, global sea level has risen by 300–400 ft (about 90–120 m).

Ups and downs

Upheavals in the Earth's crust due to forces created when the continental plates move can cause large portions of a continental landmass to rise or fall. Sometimes these changes can be very rapid. An earthquake can fracture the landscape, triggering a drop in one place and a rise in another, over the

A raised beach is a shelf of coastal land above sea level. It is formed when the sea level drops or the land rises. Four different levels are visible here, each indicating an ancient beach formed by a previous sea level.

See also **vol 1:** 12–13, 29, 54–55; **vol 3:** 29

The North Sea lies between Britain and the European mainland. About 18,000 years ago, near the peak of the last ice age, the North Sea was not a sea at all. About two-thirds of it was land, over which mammoths and other cold-climate creatures roamed. Since then, the north European ice sheets and glaciers have retreated, and sea levels have risen. The land became flooded to a depth of about 165–330 ft (50–100 m) to create today's North Sea. Where mammoths once roamed, fish now swim.

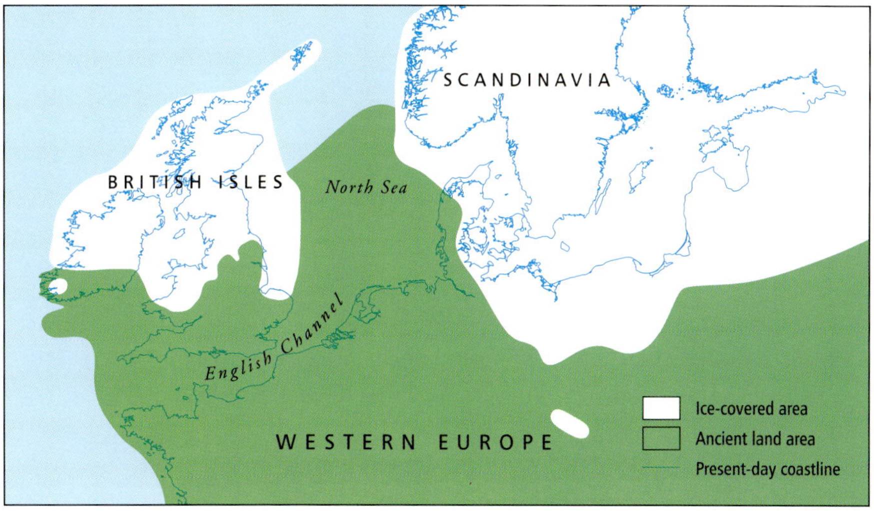

space of a few minutes. However, most changes occur almost imperceptibly over thousands or millions of years.

The landmasses also move because they are floating on the Earth's crust and upper mantle. For example, when a big covering of material such as ice presses down on the landscape, the landmass sinks slightly in that location and tilts and rises in another. When the ice sheet melts, or the rock layer weathers and erodes, the weight is slowly lifted and the landmass in the area rises slightly. These rises and falls usually occur over many thousands of years. Tilted landmasses are still rising in some places, and falling in others, because of the melting of ice from the last ice age.

Accelerating change

Human impact on the environment may be causing sea level to rise. The city of Venice in Italy frequently floods, but in the mid-1960s it began to experience more floods than ever. Geologists suspected that water taken from natural underground reservoirs for people to drink was causing the land to sink. By 1970 some locations had subsided by 6 in (15 cm).

Over the last century the average sea level throughout the world has probably risen by about 6 in (15 cm). This could be the result of an average rise in temperatures across the world (called global warming), which is causing the sea to expand slightly as it warms up.

People use boardwalks to cross this square in Venice in winter floods when the sea covers parts of the city. Venice is slowly submerging, partly because it is built on medieval wooden piles that are slowly sinking into the ground. At the same time, the Adriatic Sea—which flows around Venice—is rising.

The Ocean Waters

The planet's water is constantly on the move. It is shunted between land, sea, and air as part of the water cycle. Scientists also call this the hydrological cycle.

In the Sun's heat, water evaporates from the ocean. The rising water vapor forms clouds, which the wind blows across the planet. The clouds unload water in the form of rain, hail, or snow. The precipitation falls back into the sea or onto land. If it falls on land, the water refills lakes and rivers or soaks into the ground (when it is known as groundwater). Eventually, most groundwater finds its way into the sea, and so the cycle continues. The distribution of water on land governs where land animals and plants live. Most plants and animals are at least 65 percent water.

A unique liquid

Water (H_2O—a form of chemically bonded hydrogen and oxygen) exists naturally on Earth in all three physical states—solid, liquid, and gas, but especially liquid. Water is unusual because, unlike most other liquids, it becomes less dense when it freezes. Ice floats and helps separate the water below from the cool air above, allowing the water to remain liquid. When the water surface freezes, heat can still circulate below. If this did not happen, the water at the poles would freeze solid, and only the warmer regions of the world would have liquid sea water. Ocean circulation carries heat from the equator to the poles. This causes the tropics to be less hot, and the poles to be less cold, than they would be if the ocean did not circulate heat around the globe. Without the ocean to regulate the Earth's temperature, our planet would be difficult to live on.

Water also stores heat very well. It takes a large amount of heat energy to warm sea water. But once sea water is warm, it stores a lot of energy. Temperatures on land rise and fall much more than they do in the sea. When land warms up in the heat of the day, and then cools in the chill of the night, the temperature rise and fall in the nearby sea is much

About 97.5 percent of the water on the Earth's surface lies in the oceans. Of the remaining 2.5 percent, about three-quarters is ice and a quarter is groundwater. Lakes, rivers, and living organisms together contain only a tiny fraction of the world's water—much less than 1 percent.

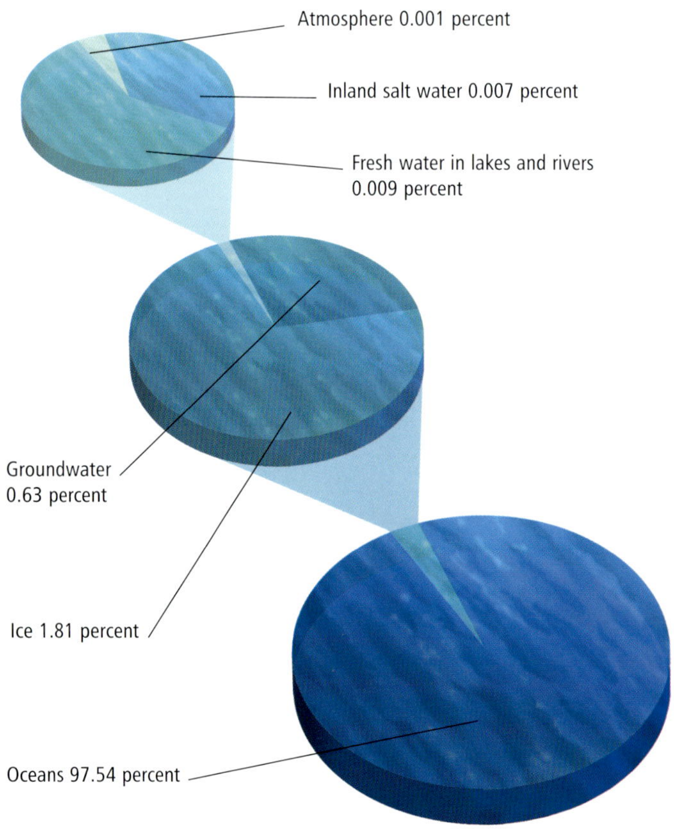

What is salt?

Sea water is 96.5 percent pure water. The other 3.5 percent is made up of dissolved solid materials that chemists often lump together and call "salts." Common salt, or sodium chloride, is by far the most frequently found salt. It makes up more than 60 percent of the salts in sea water.

See also **vol 1:** 38–39, 42, 48–49, 50–51

Water evaporates from the oceans and land and forms clouds that unload their water as precipitation. Water from the land finds its way back into the sea through rivers and groundwater flow.

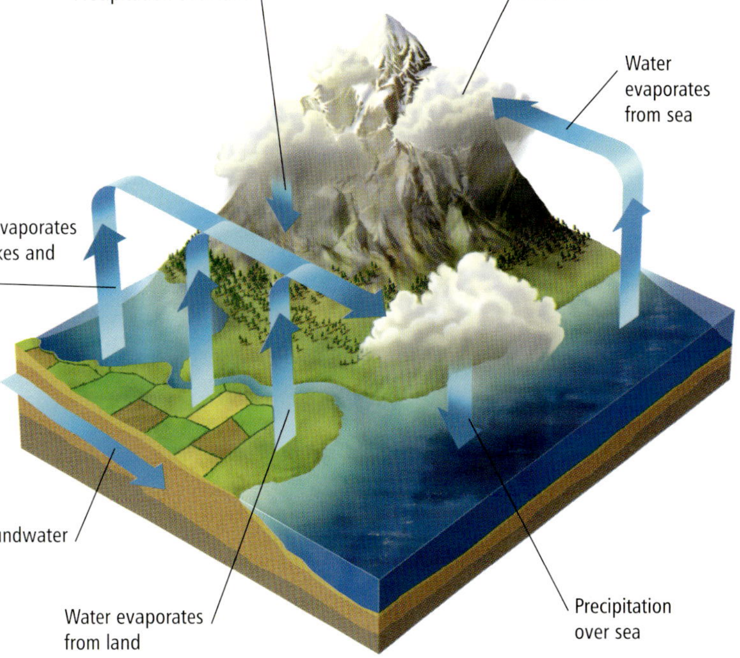

less. Summer and winter temperature differences are also smaller in the sea than they are on land.

Why is the sea salty?

According to an old Norwegian legend, the sea is salty because in the ocean depths lies a salt mill that grinds away endlessly. Not true, of course! The real "salt mills" are the processes of weathering and erosion. During weathering and erosion, chemical substances in coastal and inland rocks and soil become dissolved in rain water or get washed into rivers. Eventually, many of these substances end up in the sea.

Chemicals are also added to sea water in other ways. Volcanoes on land, for example, produce gases such as sulfur dioxide that dissolve in rain water. Volcanic vents under the sea add chemicals, too. Pollutants on land (such as fertilizers) and in the air (such as chimney smoke), also dissolve in water and find their way into the sea.

Saltpans at Lanzarote, Canary Islands, give an idea of how much salt the sea contains. Shallow ponds are dug to hold sea water. As the water evaporates in the heat of the Sun, the salt is left behind. Workers scrape the salt into piles ready for collection.

RECYCLING CHEMICALS

The chemicals contained in surface sea water are the same the world over. They are constantly recycled between land, sea, and air.

Scientists believe that the saltiness of the world's oceans has remained about the same for millions of years. Chemical substances are forever being washed into the sea from the land, but the sea does not seem to get saltier. This is because chemicals do not accumulate in one place forever. Chemicals from the sea eventually find their way back onto land again.

Sometimes the process can be extremely slow. Chemicals washed from the land attach themselves to particles of sediment on the sea floor. The particles become buried as more particles pile on top, trapping chemicals and removing them from the sea. Eventually, the compressed particles form sedimentary rock. Millions of years later, the sea floor may be lifted onto the land when two of the Earth's plates collide. The rock is then exposed to the air. Weathering by wind and rain eventually releases the trapped chemicals. They enter rivers, lakes, and groundwater, and wash back into the sea.

Nature recycles some elements, such as sulfur and carbon, quite quickly. One of the ways they find their way from the ocean to the land is through the biological activities of living organisms, such as feeding.

The sulfur cycle

The element sulfur, present in the form of sulfates (a kind of salt), is an important plant nutrient. Sulfates are weathered from rocks and washed out from soils. Eventually, they enter the ocean.

Volcanoes also contribute to the sulfur cycle by releasing large amounts of the gas sulfur dioxide into the air when they erupt. Sulfur dioxide also enters the air when people burn fossil fuels (gasoline, coal, and natural gas). The sulfur dioxide dissolves in water in the atmosphere to produce acid rain (rain that contains sulfur dioxide and other chemicals). Acid rain falls onto the ocean, and onto the land, where it replaces some of the sulfur lost as sulfates, but not all. Somehow, sulfur must be finding its way back onto land from the sea. But how?

In the early 1970s English scientist James Lovelock suggested that the substance dimethyl sulfide (DMS) might be the link. Some of the microscopic plants called phytoplankton produce DMS. Research in the 1980s showed that DMS given off by phytoplankton enters the atmosphere, where it reacts with oxygen to produce droplets of sulfuric acid. Winds carry the droplets over land. The droplets then fall to the ground in rain, so renewing the land's sulfur supplies. Remarkably, these acid droplets also help form rain clouds.

The carbon cycle

All plants and animals are carbon-based. This means they are partly made up of proteins, carbohydrates, and fats, which all contain the element carbon. Plants absorb the gas carbon dioxide from the air when they photosynthesize. During photosynthesis, plants trap the energy from sunlight to help

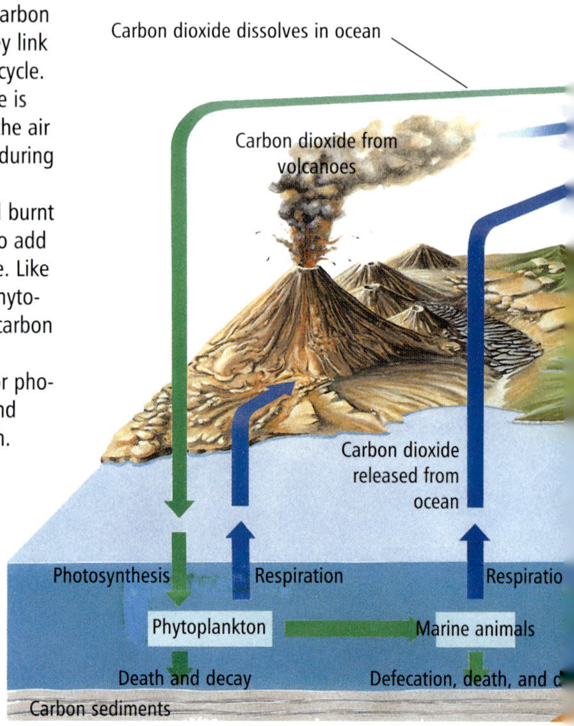

Atmospheric carbon dioxide is a key link in the carbon cycle. Carbon dioxide is released into the air by organisms during respiration. Volcanoes and burnt fossil fuels also add carbon dioxide. Like land plants, phytoplankton use carbon dioxide in the atmosphere for photosynthesis, and release oxygen.

See also **vol 1:** 18–19; **vol 2:** 12–13, 16–17

Giant blooms of phytoplankton, like this one off the coast of Namibia, can reach hundreds of miles in length. These microscopic plants alter the chemical composition of sea water by taking up nutrients such as nitrates and iron. The activities of phytoplankton also help recycle chemicals between sea, air, and land.

them make the foods they need in order to live, such as the carbohydrate known as starch.

On land, trees and other plants take in carbon dioxide and give out oxygen, which animals breathe in order to live. In the sea, phytoplankton do this. Phytoplankton also pass carbon between land, sea, and air. Carbon dioxide from the atmosphere dissolves in sea water. Marine phytoplankton take in carbon dioxide when they photosynthesize. They release carbon dioxide when they break down foods to obtain energy. This process is called respiration.

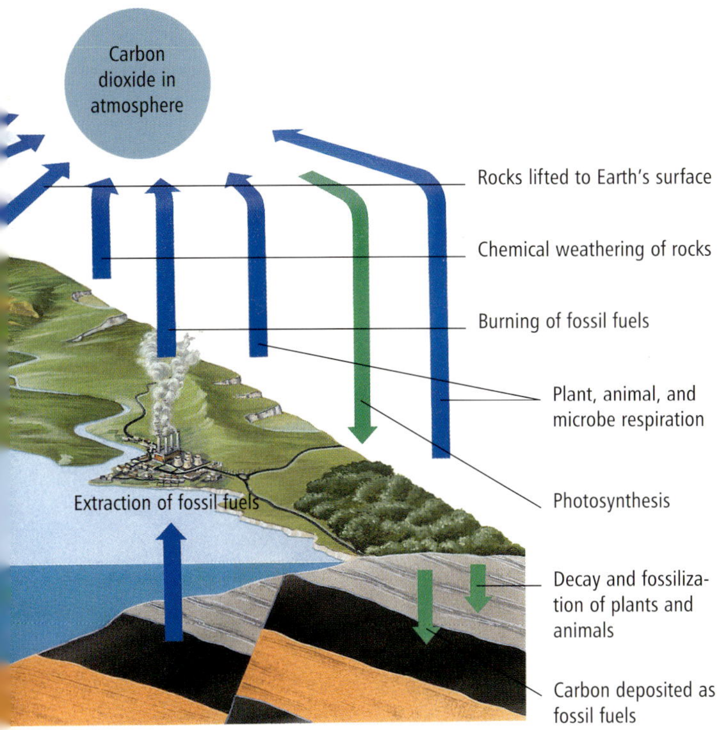

The carbon sink

Some phytoplankton use carbon to build their skeletons. When phytoplankton die, their skeletons often settle on the seabed, where they become buried and compacted to form limestone. In this way, carbon from carbon dioxide becomes "locked up" in rock. This carbon may take millions of years to be recycled and enter the atmosphere again as carbon dioxide. The carbon from dead organisms trapped in limestone and other carbon-rich rocks is part of the carbon sink—carbon removed from circulation for long periods. Some scientists believe that if they can find ways to increase phytoplankton photosynthesis, the process may take more carbon dioxide out of the air. This may lower atmospheric carbon dioxide levels and help counter global warming.

BURIED TREASURE

The oceans contain buried treasure in the form of precious minerals. Geological processes have created these minerals naturally over millions of years.

There is enough gold in the sea for each person on Earth to have a piece about the size of an orange. But the gold is usually spread thinly throughout the oceans, and it is only found in large quantities in a few places. The same is true for many of the ocean's other mineral resources.

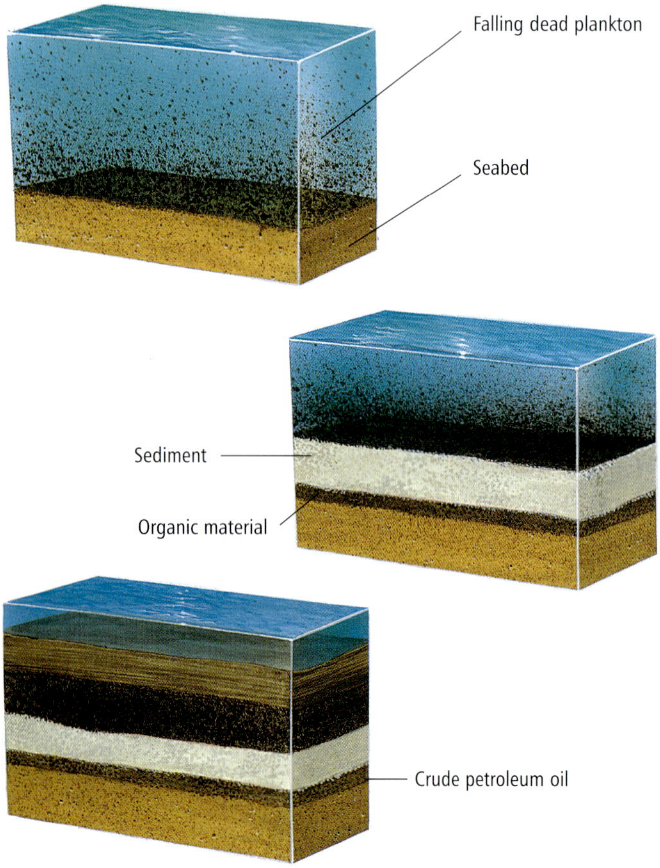

Fossil fuels can form when tiny marine plants and animals (marine plankton) are buried rapidly on the seabed by sediment, and the organic (carbon-rich) parts are preserved before they decay. Eventually, as more sediment piles on top, the remains become squeezed and heated. The carbon-based matter turns to crude petroleum oil. As the process of breakdown continues, petroleum oil turns to natural gas.

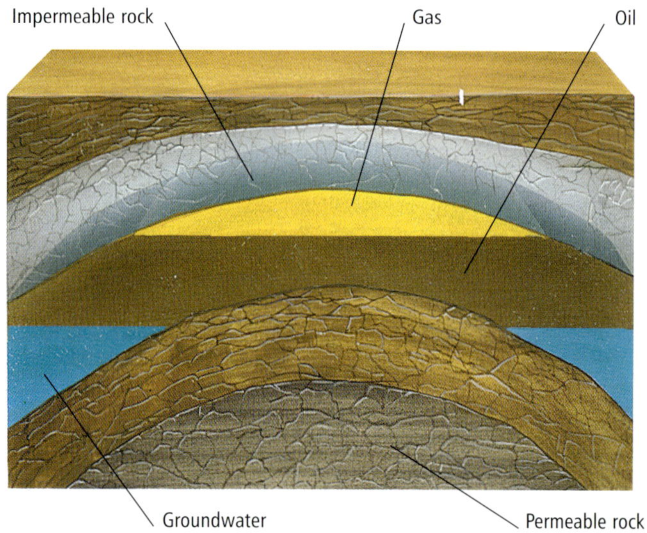

Most petroleum oil and natural gas deposits are found trapped in rock at depths of between 500 and 25,000 ft (150–7,600 m) below the ground. Many of these deposits are in rocks under the sea. Gas collects at the top of the trap and petroleum oil at the bottom.

Fossil fuels

Much of the world's wealth comes from dead marine plankton that sank to the bottom of ancient seas millions of years ago. The carbon in their bodies has become converted to petroleum oil and natural gas, used to fuel our industries, heat our homes, and power our cars.

One way these fossil fuels form is when marine plankton is buried rapidly at the bottom of the sea—often in shallow water. Over millions of years the organic (carbon-rich) matter becomes squeezed and heated several thousand feet below the sea floor. It breaks down into substances called hydrocarbons. Hydrocarbons are made up of carbon and hydrogen. They are the raw ingredients of crude petroleum oil. Eventually, under continuing pressure, petroleum oil breaks down still further to natural gas.

See also **vol 1:** 15; **vol 2:** 12–13, 20–21; **vol 3:** 46–47

The oil and gas travel through holes and cracks in rock, usually seeping upward with the groundwater. Sometimes, a layer of rock through which nothing can seep (known as impermeable rock) blocks their path to the surface. When this happens, the oil and gas may collect beneath the impermeable rock in what is known as a trap. Oil prospecting companies use sound waves to tell them which rocks contain deposits, before drilling to reach them.

Metallic nodules, also called manganese nodules, form in parts of the deep ocean. The lumps of metal are usually about 4 in (10 cm) across, and they grow very slowly. As well as manganese, the nodules contain metals like nickel, copper, and platinum. This gives them high value, but bringing nodules up from depths of 2.5 mi (4 km) and more is currently too expensive to make it worthwhile.

Gas hydrates

On the continental slopes and rises at the edge of continental shelves, cool temperatures and high water pressures prevent the conditions that produce petroleum oil. However, decomposing remains on the sea floor do produce natural gas. Under these cold, high-pressure conditions, the gas combines with water to form unusual crystals called gas hydrates. The crystals are fragile, and if they were raised to warmer, shallower water, they would quickly disintegrate and release their gas. Gas hydrates contain methane, which is a greenhouse gas (one that traps infrared radiation and contributes to atmospheric warming). If the gas hydrates broke down—perhaps because of ocean warming—this could release vast quantities of methane gas into the atmosphere, which might speed up global warming. Before we try to use gas hydrates as fuel, we need to learn how to do so without harming the environment.

Sand and gravel

Until the mid-20th century American and European prospectors dug up seashore sand and gravel to use for building materials. Then people realized that keeping sand and gravel on a beach would not only encourage leisure visitors, but the deposits would also act as natural sea defenses.

Today, sand and gravel (called "aggregates") are mined from beneath the sea. Many shallow-water, marine aggregates are sediments from ancient shores. They have become covered by sea water since the last ice age. Construction companies use marine aggregates in vast quantities for buildings and roads. In bulk, aggregates are surprisingly valuable. In fact, in yearly value, they are second only to oil and gas as offshore mineral resources.

The chimneys that form at some deep-sea hydrothermal vents are a potential source of rare metals. In 1998 an Australian mining company bought the rights to explore fields of hydrothermal vents off Papua New Guinea. The company plans to develop technology to mine precious metals such as gold and mercury in water 1 mi (1.6 km) deep.

Heat, Light, and Sound

Heat causes sea water to circulate. The temperature, light levels, and even the sound-transmitting qualities of water also have a major effect on marine life.

When warm water meets cool air, as here in this bay, sea mist forms. Usually, the mist burns off in the heat of the day.

The oceans absorb and store a great deal of heat energy from the Sun, while only rising slightly in temperature. When the oceans cool down slightly, they give out large amounts of heat. The heat energy released from the oceans plays a vital part in controlling land and air temperatures around the globe.

When sea water warms up, it becomes lighter and rises. When sea water cools, it becomes slightly heavier and sinks—unless it becomes ice, when it becomes lighter again. The rising warm water and the sinking cool water power ocean currents below the surface and create a pattern of water circulation around the globe. Meanwhile, currents at the sea surface warm the air and feed clouds with moisture, thus driving the world's weather systems.

Distribution of marine life

The temperature of the sea is one of the main factors that determines where marine creatures live. Creatures found in warm surface waters may be quite different from those

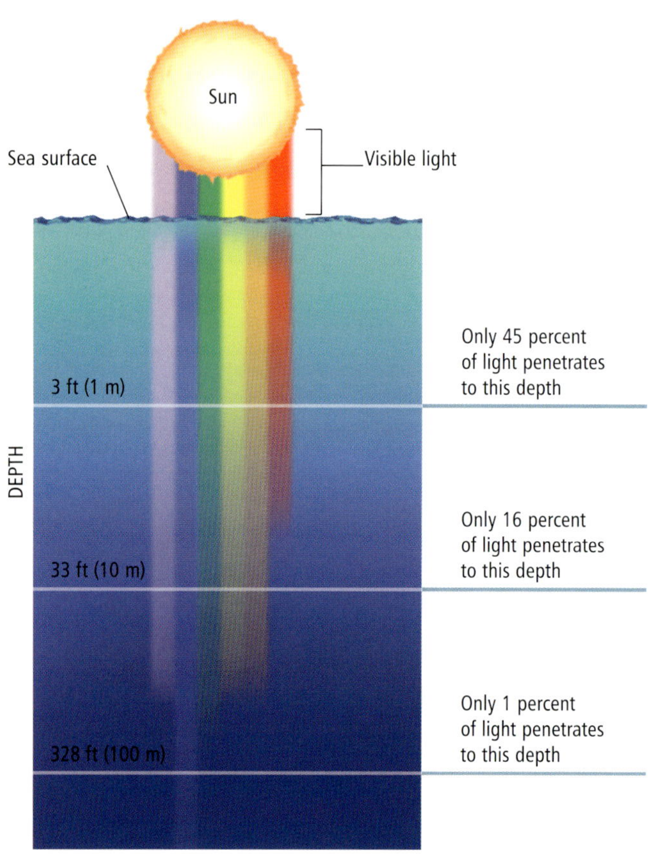

Water absorbs light. It filters out some wavelengths (colors) of light better than others. Reds, oranges, and yellows are filtered out first. Even in clear, tropical sea water, about 99 percent of all light is absorbed within about 330 ft (100 m).

See also **vol 1:** 48–49, 50–51; **vol 2:** 16–17, 18

A good sound transmitter

The undersea world is anything but silent. In shallow waters, among plentiful marine life, the sea is alive with the sounds of creatures. Dolphins and whales produce clicks and whistles, while populations of clams or shrimp make a sound like sizzling bacon. Some fish grunt or click by using their gas-bladders.

Sound travels farther and faster in water than in air. Very loud, low-pitched sounds can travel thousands of miles around the world through the oceans. So good is water at transmitting sounds, that scientists often use sound, rather than light, to explore the ocean depths. One of the most common methods is to bounce sound waves off the sea floor or objects in mid water using sonar.

Whales use sound waves for different purposes, according to the species. Baleen whales, like the humpback whale, use sound to communicate. Toothed whales, such as the sperm whale, use sound to navigate as well as to communicate.

found in cooler waters. Coral reefs occur in some warm, shallow waters where temperatures are 68° F (20° C) or higher. Many nutrients are recycled as a result of the close association between coral polyps and their algae, and reefs support a huge variety of animals. In the cold surface waters of polar seas, where temperatures are in the range 32–40° F (0–5° C), fewer types of organisms are found.

In temperate regions (the parts of the Earth between the tropics and the polar regions), the near-surface waters become three-layered in summer. The upper layer is warm and the deep layer is cool. Between the two there is a layer where the temperature changes by as much as 18° F (10° C) from top to bottom. This layer is the thermocline, and it acts like a physical barrier. Many small marine creatures have difficulty crossing the thermocline, and it keeps many deep-sea creatures separate from those living in the surface waters.

Life-giving light

Light is essential for life in the oceans. Without light, seaweeds and other marine plants could not make their food by photosynthesis. Because marine plants need light, they cannot grow at depths where light does not penetrate. Throughout the world's oceans, phytoplankton bloom as they absorb sunlight and use it to photosynthesize. Millions of tons of shrimplike creatures called krill then feed on the plankton. Other marine animals then eat the krill. Therefore, most marine organisms ultimately depend on the light, because it encourages the growth of phytoplankton—the first link in the ocean food chains.

And many marine creatures, of course, need light to find their way about, to attract others for mating, and to find their food.

Frozen Seas

Sea ice is frozen sea water. In the northern winter, the area of polar sea ice doubles and almost covers the Arctic Ocean. In the southern winter, the area of floating antarctic ice increases fivefold to cover about half the Southern Ocean.

When sea water freezes, a thick, continuous surface layer of ice forms. This is pack ice. Each year, as the polar summer ends, sea water freezes afresh over parts of the Arctic and Antarctic Oceans. First, ice crystals form near the sea surface and, as they gather together, the water takes on a greasy appearance. This grease ice thickens and, in calm conditions, forms smooth, clear ice (called nilas). In choppy waters, the sea surface becomes scattered with round plates of ice called pancake ice. Usually, the soupy sea water between the plates freezes as well to create a complete covering of pack ice.

In the Arctic Ocean, fresh pack ice reaches about 6 ft (2 m) thick in the first winter. In the Antarctic Ocean, first-year pack ice is closer to 10 ft (about 3 m) thick. If the pack ice does not melt away in the following summer, it may become thicker in subsequent years. This multi-year ice can reach more than 18 ft (5 m) thick in the middle of the Arctic Ocean and is impassable to all but the most powerful ice-breaking ships.

More or less salty

When sea water freezes, the salt dissolved in it separates out from the water. The ice crystals that form grease ice contain almost pure water. The salt they leave left behind becomes trapped in tiny pockets of salty liquid as the ice thickens. The tiny pockets are known as brine cells. Microscopic channels in the ice allow the salty liquid to flow out into the sea water below. As a result, pack ice contains very little salt, and where sea water freezes, the sea water below the ice is even saltier than usual.

See also **vol 1:** 36

Icebergs

Icebergs are floating chunks of ice that have slipped into the sea from the land. Some icebergs break off the edges of ice sheets. The ice sheet covering most of Antarctica grows outward each winter. As it does so, ledges of ice called ice shelves extend across the ocean surface from the continent. At their outer edges, the ice shelves shed icebergs. Glaciers also shed icebergs. Glaciers are rivers of ice fed by snowfall on high ground. They flow slowly down steep-sided valleys, and icebergs break off from their ends. Glaciers in

Pack ice may be buckled by sea swells and broken into large chunks called floes. In some places, the pack ice freezes solid again; in others, currents keep the floes separate.

Several thousand icebergs break off Antarctica's ice shelves each year—particularly from the Ross Ice Shelf and in the Weddell Sea. The icebergs often drift for months on end and over thousands of miles before melting away in warmer waters.

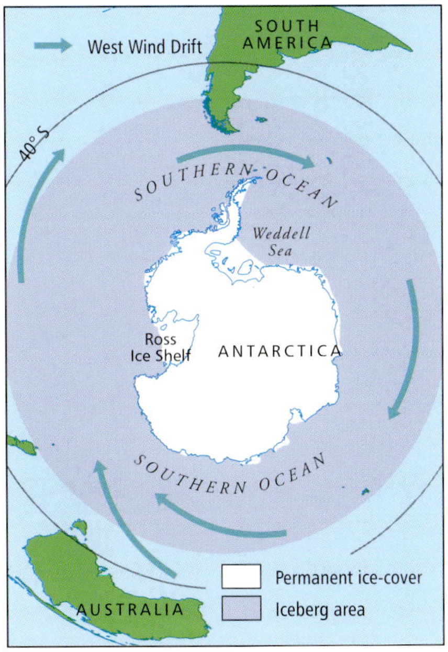

Most arctic icebergs break off the Greenland icecap and drift south and west to Canadian and even U.S. waters. They gradually melt as they travel south.

Greenland, Alaska, and Antarctica shed thousands of icebergs every year. Glaciers in landmasses around the Arctic Ocean alone shed as many as 10,000 icebergs a year.

Arctic icebergs are usually smaller, deeper, and narrower than antarctic ones. Even so, arctic icebergs can be 1,000 ft (300 m) or more thick and can weigh over 1,000,000 tons (900,000 tonnes). Typically, only about 15 percent of an arctic iceberg reveals itself above the sea surface. Each year, a few hundred arctic icebergs reach the busy shipping lanes that cross the North Atlantic. It was one such iceberg that struck the *Titanic* on its maiden voyage in 1912 and caused it to sink with the loss of more than 1,500 lives. Today, the International Ice Patrol (IIP) uses radar and aerial reconnaissance to monitor northwest Atlantic icebergs. Since the IIP was set up in 1914, no more lives have been lost due to iceberg collisions in this part of the Atlantic Ocean.

In 1957 the crew of the icebreaker USS *Glacier* spotted one of the largest antarctic icebergs on record. They estimated it to be about 60 mi (97 km) wide and 208 mi (335 km) long—more than twice the area of Connecticut.

Antarctic icebergs tend to be wide but shallow. Many are as much as 5 mi (8 km) across and about 500 ft (150 m) thick. Less than 20 percent of an antarctic iceberg is visible above the waterline.

Global Winds

Hot air rising and cool air sinking generate wind systems over the oceans. Throughout history, seafarers have depended upon these winds to fill their sails.

When the Sun's rays strike the Earth, it warms the atmosphere—the layer of air that envelops the planet. Warm air rises, and cold air sinks down to replace it. The air movements create wind systems that flow around the Earth. Air at the equator and tropics is warmer than air at the poles. There are several reasons for this.

At the equator and tropics the Sun is directly overhead, or nearly so, and the Sun's rays strike the Earth at a steep angle. Because of this, much of the Sun's heat is absorbed. At the poles, however, the rays hit the Earth on the slant, and more of the Sun's heat is reflected into space. Any area at the poles receives a lower intensity of light (and heat) than the equivalent area at the equator. Furthermore, because the ice-covered poles are pale, they reflect more light than the blue seas of the tropics or the brown, green, or yellow hues of tropical lands. Darker colors absorb more light.

If the tropics absorb more of the Sun's heat than the poles, why don't the tropics simply get warmer and warmer? The reason is that both the atmosphere and the oceans carry heat away from the tropics to the poles. As warm air at the tropics rises, cooler air farther away from the equator moves in to replace the air that has risen. This sets up a movement of warm air toward the poles at high altitude, and a movement of cool air toward the equator at low altitude.

The horse latitudes

Geographers divide the Earth horizontally into imaginary lines called lines of latitude. These are drawn parallel to the equator, which is at 0° latitude.

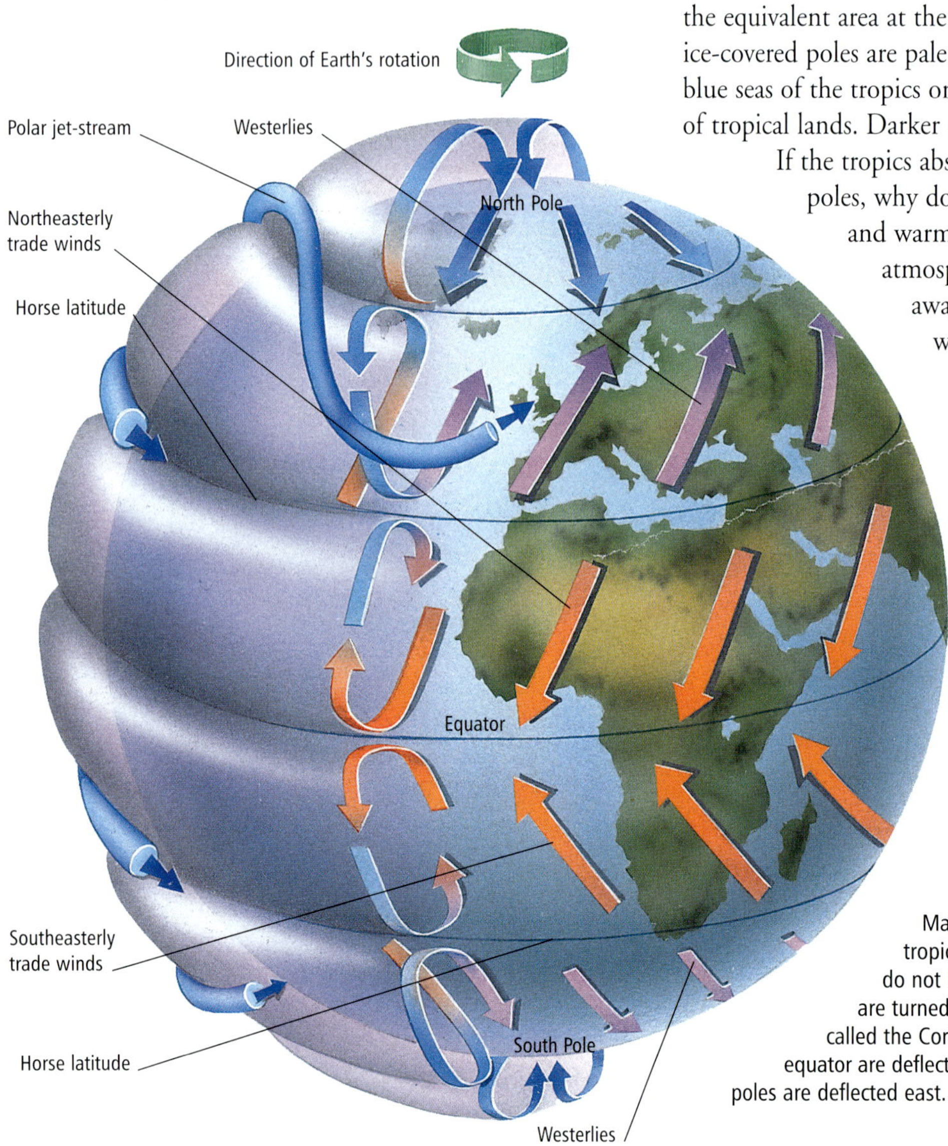

Major winds carrying warm air from the tropics to the poles, and cool air back again, do not blow directly north–south. Instead, they are turned by the Earth spinning on its axis; this is called the Coriolis effect. Winds blowing toward the equator are deflected west; winds blowing toward the poles are deflected east.

See also **vol 1:** 49; **vol 3:** 11

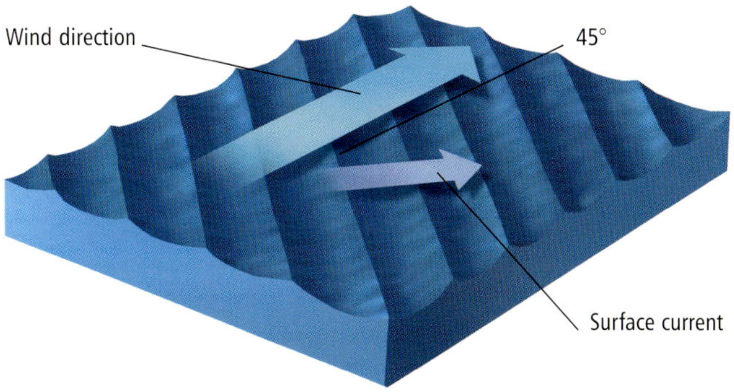

The Coriolis effect can also turn currents. Many surface ocean currents do not move parallel to the wind. Instead, the currents move at a 45° angle to the wind direction.

The North and South Poles are at 90° N and 90° S respectively. The air that rises near the equator, and travels toward the poles, gradually cools and descends to the Earth's surface at about latitudes 30° N and 30° S. These are the horse latitudes. The descending air produces a high-pressure zone that has characteristically warm, dry conditions. On land, most of the world's hot deserts lie at these latitudes. In the days of sail, ships crossing these latitudes often encountered calm conditions that prevented them from moving. When the ship's water supplies ran low, any horses on board were the first to be sacrificed—hence the name sometimes used today to describe these latitudes.

Trade winds

Air currents moving from the high-pressure zones at 30° N and 30° S, and traveling toward the equator, produce the steadiest and most reliable winds on Earth. They are the "trade winds," so called because for many hundreds of years the world's seafaring traders depended upon them to sail across oceans. In the Northern Hemisphere, the trade winds are northeasterlies (they blow from the northeast). In the Southern Hemisphere, they are southeasterlies (they blow from the southeast).

The trade winds die out near the equator, in the region called the doldrums (from an old English word for "dull"). Mariners could become stuck there and die of thirst, as this old engraving shows.

Why Christopher Columbus missed North America

At the time when Columbus set out on his voyages of discovery, European navigators often sailed "along the parallel." This means they sailed along a line of latitude when crossing open ocean. Had Columbus sailed due west of Spain—his starting point—he would have landed somewhere in, or near, present-day North Carolina. However, to do so, he would have had to battle against unfavorable winds and currents. Instead, Columbus sailed with the northeast trade winds to the Canary Islands before sailing west across the Atlantic on a more southerly course. This way, he missed North America, and actually landed in the Bahamas.

Ocean Currents

A current is a flow of water, usually in a fixed pattern. Currents are constantly moving around the oceans. Currents help bring warm weather to cold regions and help lower the temperature of hot areas.

The Earth's major currents are like giant rivers carrying vast volumes of water across the globe. Some ocean currents carry warm water, and others carry cold water. There are about 40 major ocean currents. They shift much more water than rivers on land. The Gulf Stream, a warm-water current that comes from the Gulf of Mexico and flows up across the North Atlantic toward northwest Europe, alone carries several times more water than all the world's freshwater rivers put together.

Surface currents

Winds drive the major surface currents of the world. Underwater obstructions (such as submarine mountains) and landmasses, as well as the rotation of the Earth, may deflect ocean currents. These combined effects deflect major surface currents right (or clockwise) in the Northern Hemisphere. The same effects deflect surface currents left (or counterclockwise) in the Southern Hemisphere. This creates massive spirals of circulating water called gyres. Gyres flow clockwise in the Northern Hemisphere and flow counterclockwise in the Southern Hemisphere.

Near the equator, the trade winds slacken, and the northern and southern gyres create westward-flowing equatorial currents. Between them, weak equatorial countercurrents flow east. The only surface current that flows right around

The main ocean currents and circulation patterns. Warm-water currents are shown in red, and cold-water currents are shown in blue.

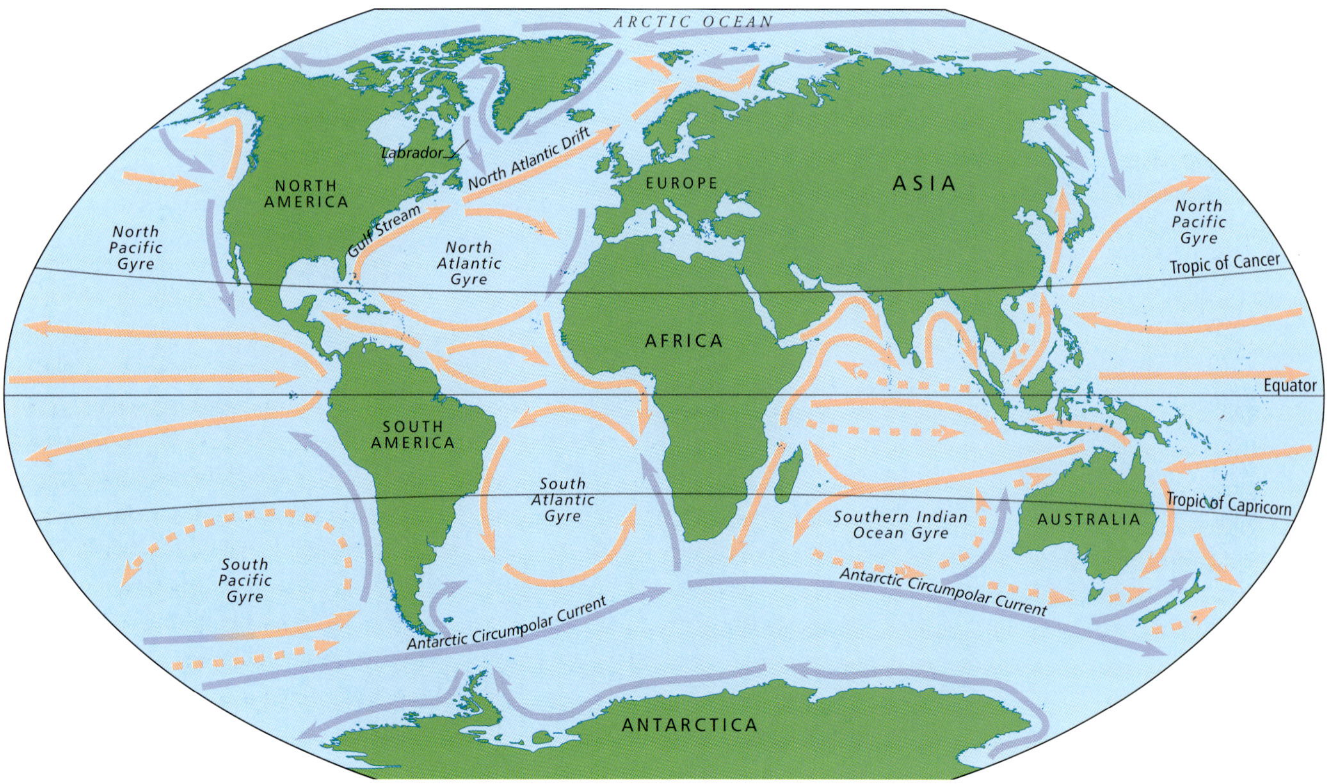

See also **vol 1:** 46, 50–51

Shoes ahoy!

For anyone shipwrecked on a small tropical island, floating a message in a bottle is one way to summon help. Even if the castaway does not know exactly where he or she has been swept ashore, a mariner with a good knowledge of ocean currents might be able to trace the bottle back to its source.

In May 1990 a storm swept 21 cargo containers off the ship *Hansa Carrier* as it sailed across the North Pacific. Among the lost cargo were about 39,000 pairs of a well-known brand of shoes. Within six months of the spill thousands of shoes washed ashore between Oregon and British Columbia, Canada. The shoes had floated about 1,500 mi (2,400 km) east from their start point on the North Pacific Current. Most of the shoes were in good condition, but they were not in pairs. Advertisements began appearing in newspapers, as new shoe owners tried to match single shoes with their partners.

the world, unblocked by landmasses, is the Antarctic Circumpolar Current. Westerly winds drive this current eastward and keep warm surface currents away from Antarctica, which is one reason why the continent is so cold.

In the days of sail, successful seafarers needed a sound knowledge of ocean currents, as well as winds. Portuguese traders sailed a figure-eight route to reach South Africa and then return home. They discovered that they got to the southern tip of Africa faster if they sailed across the Atlantic and then back again, rather than sailing directly along the African coast. On the outward voyage, ships followed the trade winds and "hitched a ride" on the Canary Current—the eastern part of the North Atlantic gyre. They then sailed westward along the Equatorial Current, crossing the Atlantic, before picking up the South Atlantic Gyre, which brought them back across the Atlantic to Africa. On the return journey, they sailed northwest with the South Atlantic gyre, and then west of the Canary Current, before turning for home.

Shifting the heat

Warm surface currents carry the Sun's heat far beyond the tropics, while cool surface currents carry the chill far beyond the poles. The warm Gulf Stream, for example, merges with the North Atlantic Drift that flows toward the Arctic. The North Atlantic Drift keeps the climate of western Europe warmer than would otherwise be expected. Winter temperatures in London, England, are about 18° F (10° C) warmer than in Labrador—lying at a similar latitude but cooled by the Labrador Current flowing south from subpolar waters.

A helpful chart

The Gulf Stream fascinated the 18th-century statesman Benjamin Franklin. When he was postmaster general of the American colonies, he noticed that mail ships took about two weeks less time to sail to England than to return. By questioning sea captains, he discovered that the eastward flow of the Gulf Stream slowed ships returning from England. Franklin published the first detailed chart of the Gulf Stream, together with instructions on how to avoid it on the return voyage. Franklin's chart was soon to prove invaluable to hundreds of transatlantic traders.

Subsurface Currents

Surface currents form only part of the flow of sea water around the world. More than three-quarters of the ocean's water moves in currents flowing at greater depths. These are the subsurface currents.

Currents flow at all levels, from the surface to the deepest depths. The Gulf Stream is a rare example of an ocean current that extends from the surface almost to the bottom of the ocean in places. Most currents do not. It is quite common for a current at one depth to flow in the opposite direction to a current at another depth.

Subsurface currents

Differences in water density produce the major subsurface currents. For example, when sea ice forms, most of the salt trickles out of the ice into the sea water beneath. Sea water becomes denser, or heavier, as its saltiness increases. So, the water beneath new sea ice sinks to the sea bottom. In the ocean depths, this cold, salty water moves toward the equator. The water that sinks at the poles is replaced by water arriving at higher (but still deep) levels from warmer parts of the ocean. Thus there is a flow toward the equator of cold, salty bottom water, and a return flow—toward the poles—of deep, less salty, warm water. This flow is called the thermohaline circulation, from the Greek words "therm" meaning "heat," and "hals" meaning "salt."

The bottom and deep-water currents flow more slowly than surface currents. For example, bottom water traveling from the far North Atlantic to antarctic waters takes several hundred years to make the journey. When it rises to the surface in the Southern Ocean, it is rich in nutrients. These nutrients are the result of the steady, downward cascade of dead organisms and solid waste.

Upwelling

The rise of bottom water to the surface is called upwelling. It happens where surface currents move apart, or where winds and surface currents move water away from coastlines. The sideways movement of surface water creates room for

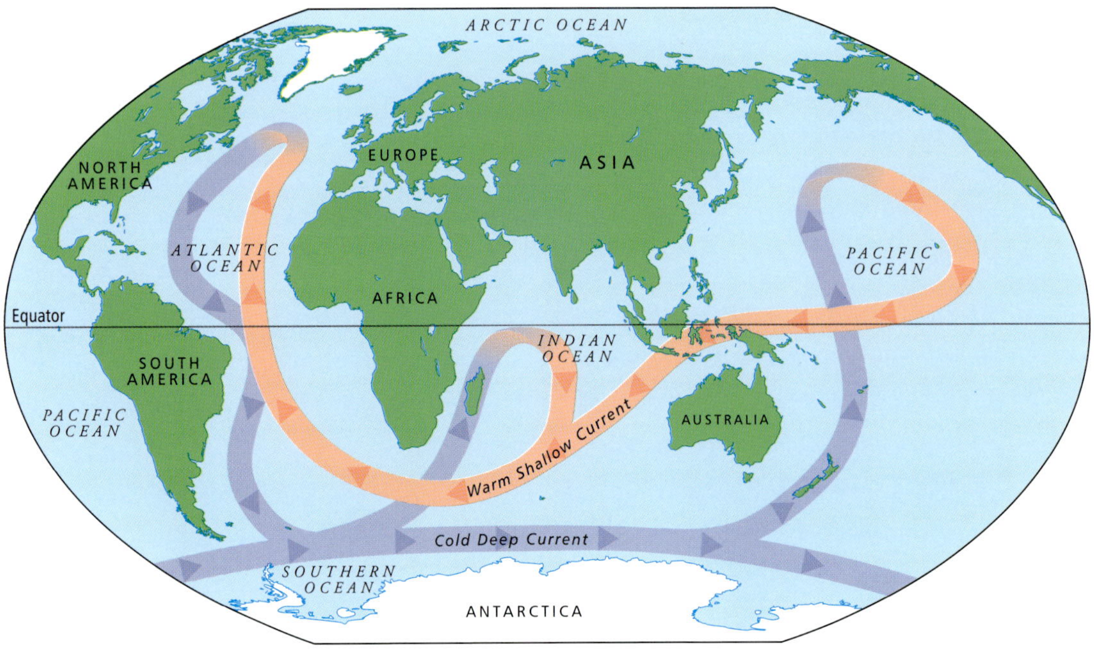

Deep ocean currents carry cold water to the tropics and subtropics. Water sinks to the depths in the North Atlantic and moves southward to join antarctic water. The water flows into the Indian and Pacific Oceans, and then rises closer to the surface as it travels back to the North Atlantic in a giant conveyor belt pattern.

See also **vol 1:** 44–45, 48–49, 52–53

Regulating the world's climate

More than three-quarters of the sea water in the open ocean lies below the thermocline (the thick layer of subsurface water with a temperature difference of several degrees between its top and bottom). The currents in the deeper layers are isolated from surface waters except where the surface waters descend (called downwelling) or where the deep layers rise to the surface (called upwelling). Because the currents carry so much water, they have a major influence on world climate.

If global warming melted most of the Arctic Ocean's sea ice in summer, it would probably have a massive effect on thermohaline circulation. The ice-making "machine" would be temporarily halted, interrupting the creation of Atlantic bottom water. This might effect the circulation of sea water in the North Atlantic. For example, the North Atlantic gyre might shrink southward, causing the Gulf Stream and North Atlantic Drift to move southward. If this happened, northwest Europe might lose its warm current, and parts of the continent could have temperatures 9–18° F (5–10° C) cooler than now.

deep water to rise up and take its place. Upwelling is important because it brings nutrients to the sunlit surface where phytoplankton thrive. The nutrients and sunlight fuel the rapid growth of phytoplankton, on which many other marine creatures depend.

Important areas of upwelling lie on the eastern sides of the Atlantic and Pacific Oceans. There, the combination of winds and surface currents pulls surface water away from the coast. Water then rises up from below to fill the space. Along most of the Pacific coast of the Americas, from Canada to Argentina, the waters are full of plankton-feeding fish such as sardines, anchovies, and herrings. In the Atlantic, similar rich fishing grounds lie off Portugal and off the northwest and southwest coasts of Africa. The strength of coastal upwelling varies with the seasons and from year to year.

Galápagos penguins feeding on fish called anchovies. Upwelling in the eastern equatorial Pacific fuels an explosion of plankton. Anchovies arrive to eat the plankton, and penguins then feast on the anchovies.

El Niño

Winds and ocean currents can combine to produce rapid and far-reaching changes in climate. The El Niño climatic event takes place every few years. When it occurs, its impact is felt across the world.

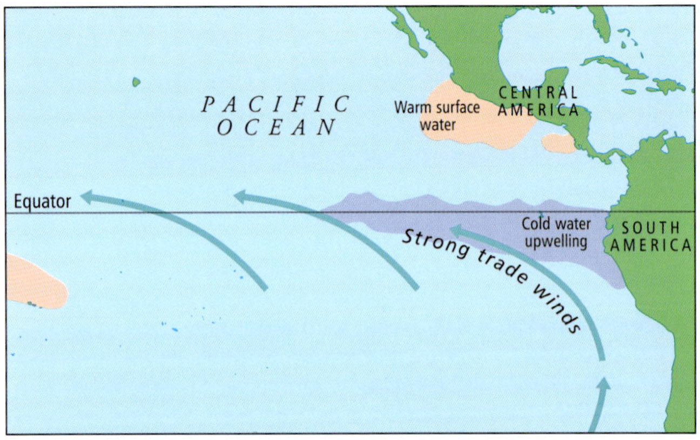

Normally, in the eastern equatorial Pacific, trade winds drive warm surface water to the west, allowing cold water in the Peru Current to well to the surface, forming a cool tongue that extends from the South American coast (top). In an El Niño year, trade winds slacken, or even reverse (bottom), and the upwelling is suppressed for months.

El Niño is the term traditionally used by fishermen from Peru and Ecuador to describe a slackening of the trade winds that usually occurs near Christmas. The words "El Niño" are Spanish for "the male child," and are a reference to the Christ child. The Peruvian upwelling supports a very large anchovy fishery. However, when the trade winds slacken in December or January, the upwelling declines. Nutrient-laden cool water no longer reaches the surface, and the growth of phytoplankton slows. The cool waters off the Peruvian coast become warm. The anchovies no longer have a concentrated food supply, and they scatter. This change marks the end of the main fishing season.

El Niño also refers to a much larger climatic event. In some years, the decline in southeast Pacific trade winds happens much earlier than normal. The upwelling is lost for months, and the explosion in phytoplankton growth does not take place at all. Such years, when the slackening of southeast Pacific trade winds comes early, is called an El Niño year. Many people simply call it an El Niño.

The effect of an El Niño year on fishermen and wildlife along the Peru–Chile coast can be dramatic. Following the El Niño of 1957–58, the numbers of local seabirds fell from about 28 million to only 6 million. The El Niño of 1982–83

See also **vol 1:** 50–51

was the biggest for a hundred years, and the anchovy catch dropped to well below 1 percent of normal.

Wider impacts

An El Niño event is part of a wider weather pattern called the El Niño-Southern Oscillation (ENSO) that links the southeast Pacific with the eastern Indian Ocean. As part of the ENSO, an El Niño heralds many changes in weather. The buildup of warm water in the southeast Pacific affects wind patterns at many levels of the atmosphere, including narrow belts of high-altitude winds called jet-streams. Jet-streams help set the weather patterns over thousands of miles. An El Niño causes shifts in weather patterns across North and South America, Africa, Southeast Asia, and Australia.

The 1997–98 El Niño

From their observations of changing patterns, oceanographers are able to spot the buildup to an El Niño event six months before it happens. Scientists studied in detail the 1997–98 El Niño, the largest one since 1982–83. The U.S. National Oceanic and Atmospheric Administration (NOAA) used satellites in space and buoys in the equatorial Pacific Ocean to measure temperature, currents, and winds, and forecast the arrival of El Niño. Its arrival caused droughts in Australia, India, and Southeast Asia.

Torrential rain triggered by the 1997–98 El Niño caused extensive flooding in parts of the United States. Here, at Grand Forks, North Dakota, the Red River burst its banks.

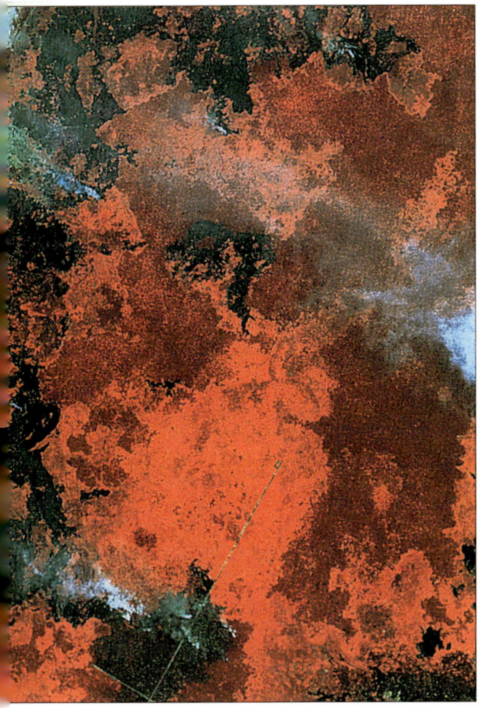

In 1997 Indonesian farmers lit fires to clear forest areas for plantations, but El Niño caused them to burn out of control. Smoke traveled hundreds of miles over Southeast Asia. Thick smog caused a passenger jet to crash on the island of Sumatra. In this infrared satellite view, the green areas show land that has been burnt by the fires. The red color is forest and other vegetation.

While the 1997–98 El Niño was causing drought in some areas, its effects were battering other places with storms and torrential rain. Floods ravaged places as far apart as California, Florida, Brazil, and Kenya. In 1997, between October and December, parts of Kenya's dry grassland, or savanna, received five times more rain than usual. The floods took months to subside, and many local people died of the contagious diseases dysentery and cholera.

El Niños are not all bad news, however. During the 1982–83 El Niño, for example, there were almost no anchovies available to support the Peruvian anchovy fishery. But warm-water fish such as skipjack tuna and yellowfin tuna appeared elsewhere off the Pacific coast of South America, providing a new source of income for fishermen. Scientists can provide a service by warning people worldwide of the onset of an El Niño. People can then prepare for its worst effects. Farmers can alter their crop-planting, and fishermen can change their fishing techniques in order to catch other species.

Rhythm of the Tides

The gravitational pull of the Moon and the Sun causes coastal waters on Earth to rise and fall once or twice each day.

Sea levels rise and fall in a rhythmic pattern over about a 24-hour period. These rises and falls are called tides. Tides effect the lives of marine organisms in several ways—particularly those that live on the shore or in the shallow water over continental shelves. A falling (ebb) tide can leave marine organisms high and dry on a shore, while a rising (flood) tide submerges them again a few hours later. Many marine creatures time their feeding or breeding to coincide with particular states of the tide.

The cause of tides

Tides occur when pulses, or bulges, of water travel across the Earth's surface. Tidal surges build up along coastlines and funnel through bays and estuaries. The gravitational forces that the Moon and Sun exert on the oceans cause the tides. Although the Moon is much smaller than the Sun, it is four hundred times closer to the Earth and therefore exerts a more powerful gravitational pull on the Earth's surface. This force pulls the Earth's water slightly toward the Moon and produces a bulge of water on the side closest to the Moon. At the same time, a counterbalancing bulge forms on the opposite side of the Earth. As the Earth turns on its axis, the two tidal bulges move around the Earth. In between the tidal bulges are slight depressions where the water has been drawn away. The bulges are the places where high tides form, and the depressions are the positions of low tides.

The Earth turns once on its axis each 24-hour day, so we might expect any place on Earth to experience two high tides and two low tides in that period. However, because the Moon orbits the Earth slightly faster than the Earth rotates, there are two high tides and two low tides every lunar day—about every 24 hours and 50 minutes.

Furthermore, the Earth is not covered in sea water to a uniform depth. The ocean basins vary in size and depth, and they are interspersed with landmasses. Tidal bulges cannot move freely over the Earth's surface. They are deflected and channeled by obstructions such as submarine mountains and by peaks and troughs in the ocean floor. Tides are also deflected by the Earth's rotation, and they bounce off landmasses in their path.

Tidal bores

When a tidal surge enters an estuary or bay, the water can sometimes form a moving "wall," or tidal bore that advances upstream. This is most likely to happen around the time of the highest spring tides of the year. The estuary of the Chientang River in northern China probably experiences the world's largest bore. It reaches nearly 25 ft (7.5 m) high. The Brazilian Amazon's tidal bore, the pororoca, reaches 16.5 ft (5 m) with speeds of 13.5 mi per hour (22 km/h).

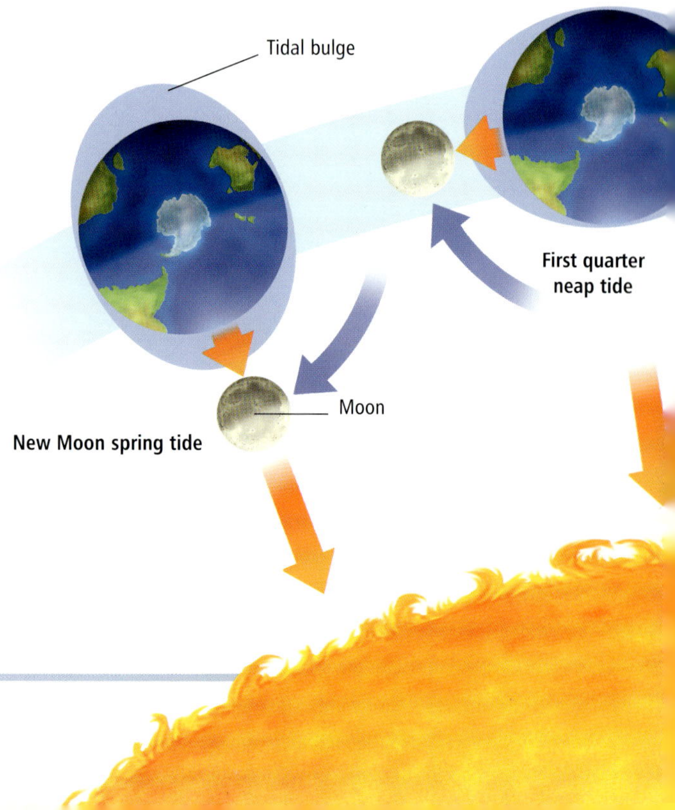

See also **vol 1:** 16–17, 32, 47; **vol 2:** 44–45, 46–47, 48–49

Low and high water (left and right) in Canada's Bay of Fundy. The world's largest tidal ranges (the difference between low and high tides) are found there. The range can be as great as 60 ft (18 m). In most parts of the Mediterranean Sea, on the other hand, the difference is less than 3 ft (0.9 m).

As a result, tidal patterns are quite complicated, although they are predictable. Many places—including most Atlantic coasts—have two tides of similar range each lunar day. The high tides are about 12 hours and 25 minutes apart. Other places—including parts of the Gulf of Mexico—have only one tide each lunar day. Some places have two tides a day, but the tides differ in height.

Spring and neap tides

What role does the Sun play in tides? The Moon orbits the Earth once every 29.5 days—a lunar month. Therefore, twice each lunar month the Sun and Moon are in alignment. These are the periods of Full Moon and New Moon. Then, the Sun's gravity adds to that of the Moon's, and the tides are larger than normal. These are called spring tides because they "well up," or "spring." Halfway between the spring tides are the neap tides. These are the smallest tides of the lunar month. At these times, the Sun and Moon are at right angles to the Earth. Their gravitational attractions have opposing effects, thus lowering the level of the high tide and raising the level of the low tide.

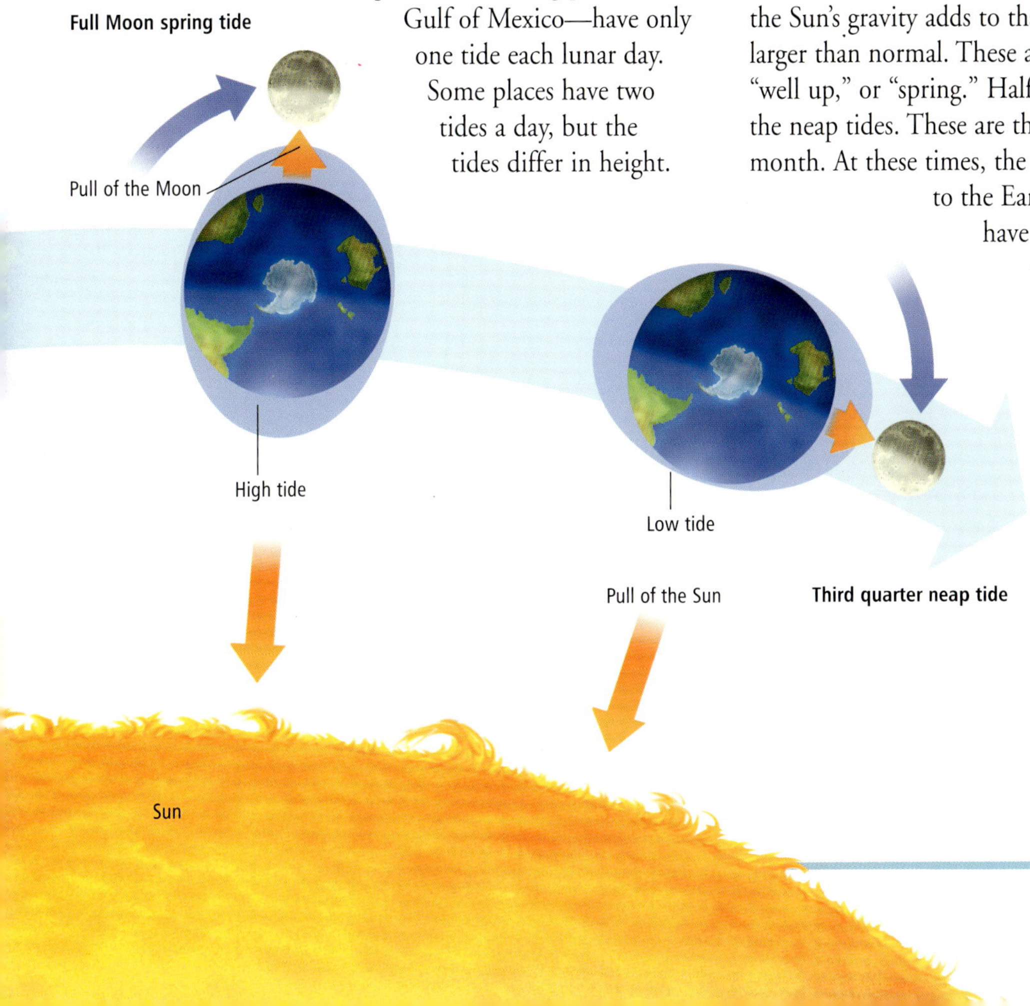

The height of the Earth's tides varies according to the alignment of the Sun and the Moon.

Making Waves

Waves are among the most familiar features of oceans. There are several kinds, and their movement is more complex than appears at first sight.

Winds cause most ocean waves. The stronger the wind, and the longer it blows, the larger the waves it normally creates. The tallest storm wave on record was observed in 1933 in the Pacific Ocean. A ship's officer measured it at about 112 ft (34 m). Even in hurricane winds, with speeds up to 106 mi per hour (170 km/h), it is rare for storm waves to reach more than 43 ft (13 m) high.

Breaking ocean waves like this one are sought out by surfers. Such waves form on exposed, gently sloping beaches.

Sea waves and swells

At sea, where the wind whips up the water, waves begin life with sharp peaks. As the waves move away from their place of origin, their shapes smooth to that of a swell. The pattern of waves at sea is rarely straightforward. Often, waves of different sizes arrive from different directions and interfere with each other to produce a choppy sea. Where two wave crests merge, the wave grows, and where a crest and a trough cross, they temporarily help cancel each other out.

Breaking waves

Winds and currents push objects along at sea, but sea waves do not. Instead, waves cause floating objects to bob up and down. It is only when waves reach shallow water and "break" that they drive objects forward, as any surfer can confirm.

When a wave reaches shallow water, it "touches bottom" and its circular motion is disrupted. The bottom part of the wave slows, but the top part keeps traveling at the same speed. As a result, the crest topples forward and the wave breaks, crashing to form surf. Energy is released onto the shore, and the wash carries loose objects up the beach.

An ocean wave is an up-and-down movement of water that travels along the sea surface. Anything floating on the surface is raised up by the wave crest and moved forward, but then lowered by the wave trough and shifted back again in a circular motion. The up-and-down wave movement mixes water near the surface and circulates oxygen and nutrients.

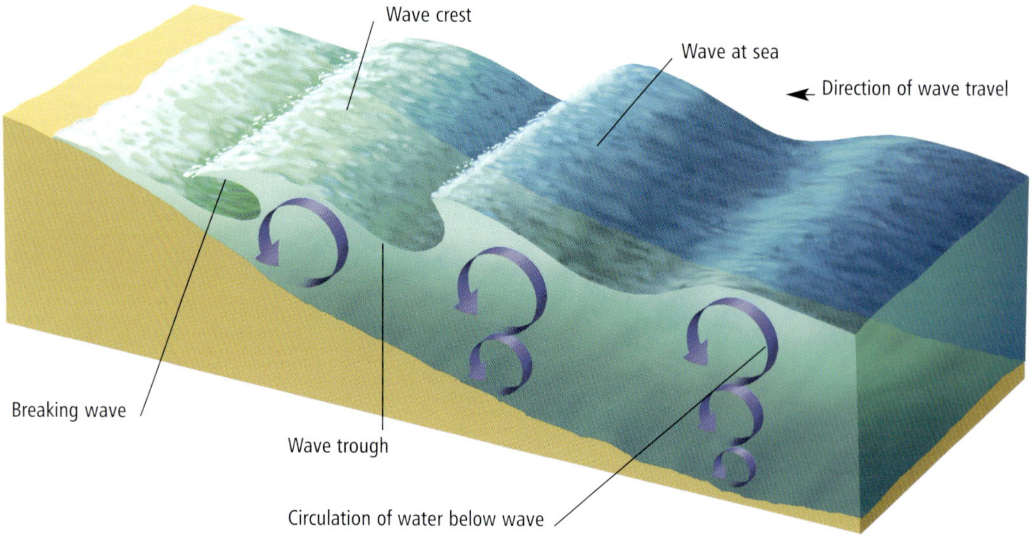

See also **vol 1:** 46–47, 58–59

Wave power

The shape of the shoreline determines the way in which the wave unleashes its power. On shores with a gentle gradient, or slope, the waves break gradually over a long distance, sometimes creating good conditions for surfers. On steeply shelving shores, the waves break suddenly and violently.

On sandy shores, waves striking the shoreline at an angle shift sand and pebbles along the shore by the process known as longshore drift. On rocky shores, storm waves crash into the base of cliffs and erode them.

The biggest waves

The world's biggest waves build in shallow water and are not created by winds at all. Called seismic sea waves or tsunamis (from the Japanese word meaning "harbor waves"), they are sometimes mistakenly called tidal waves—because they look like a tidal surge.

Tsunamis are triggered by the sudden displacement of water by earthquakes, landslides, or volcanoes. At sea, tsunamis are only a few feet high and perhaps 40 mi across, but they move very rapidly at up to 435 mi per hour (700 km/h)—the speed of an airplane. As they approach land, tsunamis slow down and build massively. They then sweep ashore, carrying buildings, vehicles, and people before them. In July 1998 an earthquake off Papua New Guinea produced tsunamis up to 49 ft (15 m) tall. The waves swept away whole villages and claimed more than 2,000 lives.

Tsunamis are most common in the Pacific Ocean. The wide circle of volcanic and earthquake activity around the Pacific rim in the area known as the "Ring of Fire" produces, on average, two life-threatening tsunamis each year. The Pacific Tsunami Warning Center (PTWC), with headquarters in Hawaii, issues warnings of impending tsunamis. People have limited time—a few hours or less—to move away from a threatened shore if a tsunami is due.

Rogue waves

In some places, bigger-than-expected storm waves build up. This happens when two wave crests merge, or when a series of storm waves move against an oncoming current. Off the southeast coast of Africa, the Agulhas Current flows south and meets storm waves moving north from the Southern Ocean. Steep "rogue" waves build to 65 ft (20 m) and swamp boats. Ocean-going tankers are regularly lost in these waters.

An artist's impression of a tsunami destroying a small harbor. In 1883 the volcanic cone of Krakatoa, an island off the coast of Java in Indonesia, erupted. It produced 115-ft (35-m) high tsunamis that claimed the lives of 36,000 people around the Indian Ocean.

Ocean Weather

Some of the most dangerous weather systems build up over the oceans.
Moist, rising air from tropical seas fuels these storms.

The huge amounts of heat energy and moisture that the oceans store can fuel violent weather. When moisture evaporates rapidly from the sea surface and then condenses to form clouds, it transfers heat energy from the sea to the clouds. This can create strong updrafts and the beginnings of a storm system. A storm is an atmospheric disturbance accompanied by wind, thunder, and precipitation.

Storms at sea play a major role in shaping coastlines. In winter and fall, storms sweep away sandy beaches and erode soft cliffs, reshaping the coastline. Winter storms over temperate seas produce storm waves that endanger lives at sea.

Tropical storms

A tropical storm at sea begins to spin under the influence of the Earth's rotation. It may then grow and become a tropical cyclone, or hurricane. The winds of a hurricane spin at over 74 mi per hour (119 km/h) around a calm, low-pressure center. Hurricanes wreak havoc with wind and water. Tropical cyclones regularly ravage Asian countries bordering the Bay of Bengal, killing many thousands of people.

The North Atlantic hurricane season occurs in late summer and early fall. Violent tropical storms sometimes strike the Caribbean and the United States. This map shows the path of Hurricane Georges in 1998.

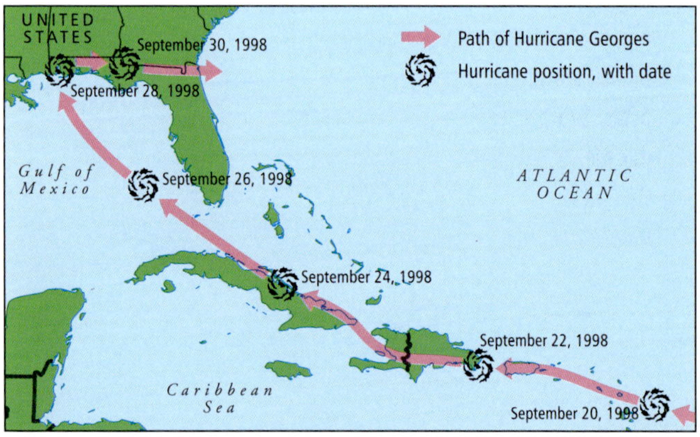

A cross-section through the spiraling storm clouds of a hurricane. In the center of the hurricane is a region of calm weather called the eye. Once a hurricane moves over cooler water, or onto land, it loses its source of heat and moisture, and subsides. Even so, it can still unload torrential rain and give birth to tornadoes (destructive, narrow, spiraling updrafts of air).

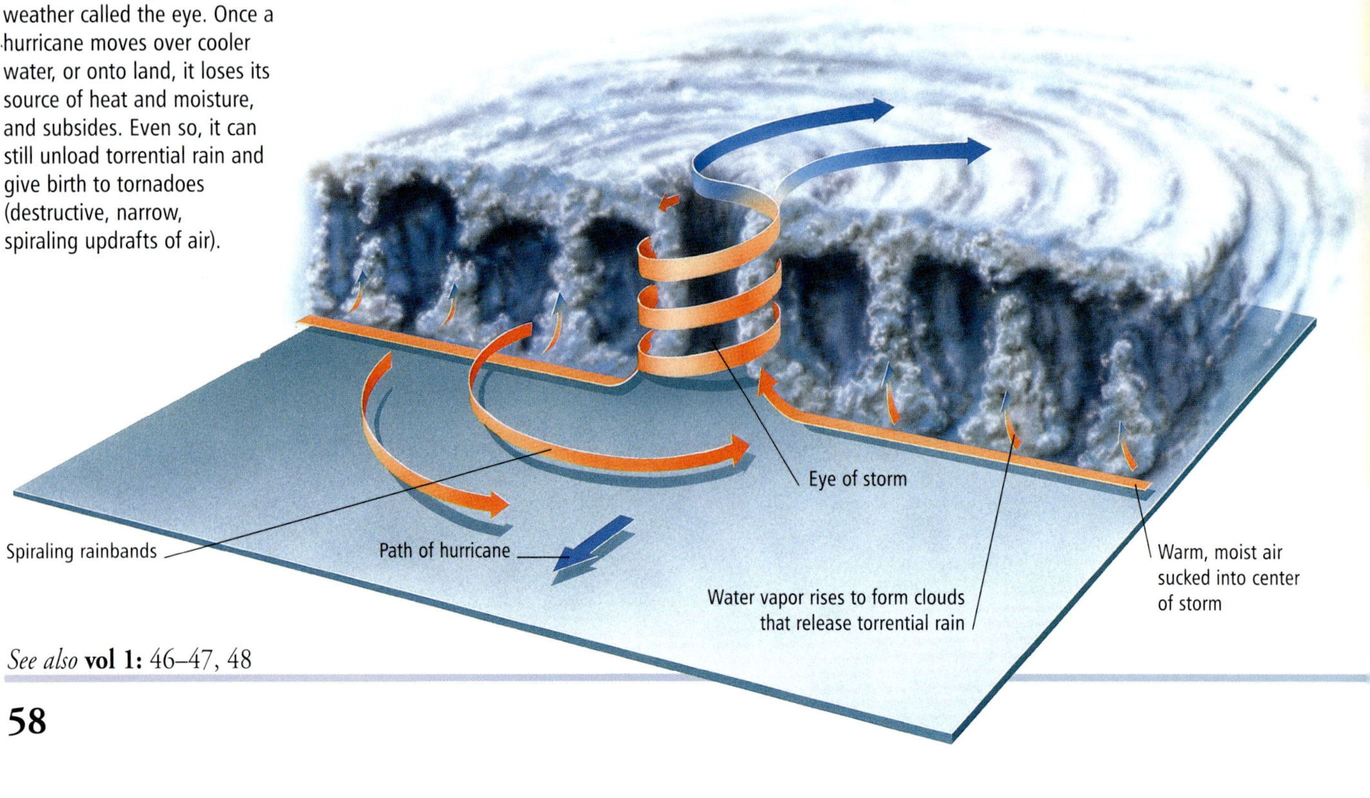

See also **vol 1:** 46–47, 48

Sea fog

Where warm and cold ocean currents mix, sea fogs form. This happens on the Grand Banks off Newfoundland, Canada, for example, where the warm waters of the Gulf Stream meet the cold Labrador Current. Warm, moist air suddenly cools, and the moisture condenses into tiny droplets to create fog. At certain seasons, the Grand Banks has notoriously poor visibility. Over the centuries, hundreds of vessels have collided there in the fog.

The worst storms

- **In the Great Storm of 1703, winds and storm waves sank more than 100 vessels sheltering off the coast of southeast England. Nearly 10,000 naval men were drowned—about one-third of the British Navy.**
- **In November 1970 a tropical cyclone in the Bay of Bengal produced a storm surge that killed more than 300,000 people in Bangladesh.**
- **On October 12, 1979 the northwest Pacific's Typhoon Tip maintained wind speeds of 190 mi per hour (305 km/h).**

In September 1998 Hurricane Georges lashed Florida and the Caribbean. It uprooted trees, toppled buildings, and hurled boats ashore. In total, this hurricane killed about 600 people.

A waterspout is a tornado over water. It looks like a spinning column of water between sea and storm cloud. In fact, the spout is mostly water vapor. But it is still a hazard when it collapses. The spout can unload several tons of water—enough to swamp a boat.

Climate Change

The 1990s were the Northern Hemisphere's warmest decade since records began. Many of the scientists who study the Earth's climate believe this may be a sign of increasing global warming.

In 1990, 1995, and 2000 teams of scientists belonging to the Intergovernmental Panel on Climate Change (IPCC) met to report on global warming (a rise in average temperatures across the world). They considered whether or not global warming was happening. And if it was, what were the causes and likely effects, and what could be done about it.

In their 2001 report, the IPCC made a "best" estimate that the Earth's surface would warm by nearly 5.4° F (3° C) in the next hundred years. If so, this could cause global sea levels to rise by nearly 20 in (about 50 cm). Greater melting of ice in polar waters would cause part of the rise, but most of the change would result from ocean water expanding slightly as it warmed.

The enhanced greenhouse effect

If the Earth's surface is gradually warming, what is the cause? When sunlight strikes the Earth's surface and atmosphere, much of it is absorbed and warms up the planet. However, some of the absorbed heat is emitted back into space as infrared radiation (lightwaves not visible to the human eye).

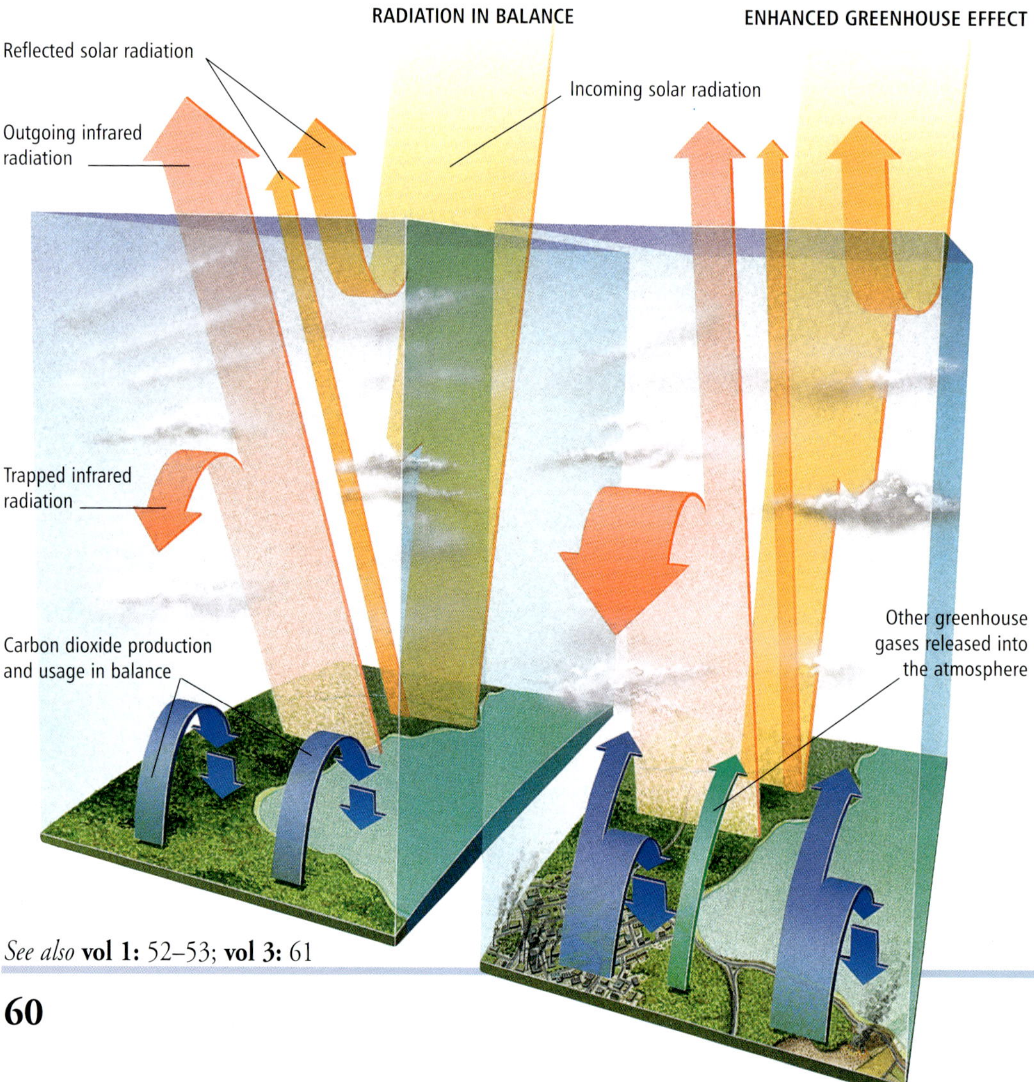

The global climate has been fairly stable for the last few thousand years. Solar radiation absorbed by the planet is in balance with heat energy radiated back into space, as shown far left. But it is possible that "greenhouse gases" released by the burning of fossil fuels are creating an enhanced greenhouse effect, shown left, with more carbon dioxide being produced and more heat energy being trapped in the atmosphere. The heat energy radiated back into space may be less than the amount absorbed, so the planet is warming.

See also **vol 1:** 52–53; **vol 3:** 61

Global warming and marine life

In 1998 many coral reefs in the Indian Ocean ejected their algae and turned white. This is called bleaching, and it leads to the death of the coral polyps. The cause appears to have been a rise in water temperatures of only 1.8–3.6° F (1–2° C) above normal, triggered by a strong El Niño—perhaps enhanced by global warming. Some scientists suspect that El Niños may be becoming more frequent because of global warming.

In 1999 a report by three Canadian scientists described how polar bears in Hudson Bay were losing weight and having fewer cubs. They claimed that global warming was causing pack ice to break up earlier in the season than usual. Without solid sea ice as a platform, the polar bears could not hunt for seals and were coming ashore hungry.

Chemical analysis (examination) of ice drilled from the antarctic ice cap tells scientists how the Earth's climate has changed since the time the ice was formed thousands of years ago.

Certain gases in the atmosphere—mainly carbon dioxide, water, and methane—absorb some of the infrared, trapping heat energy in the Earth's atmosphere. These gases are called "greenhouse gases" because, like the glass in a greenhouse, they allow most wavelengths of sunlight to pass through them, but they trap outgoing infrared radiation. This has a warming effect.

The greenhouse effect is entirely natural and has been operating for billions of years. In fact, without the greenhouse effect, the Earth would be much cooler. However, human actions seem to be enhancing or heightening the greenhouse effect, and many scientists believe this is a major cause of global warming. By burning fossil fuels (oil products, natural gas, and coal) in vehicles, factories, and homes, people are adding carbon dioxide to the atmosphere. This is trapping more infrared radiation and causing the planet to get hotter. In the last 150 years alone, atmospheric carbon dioxide levels have risen by more than a quarter.

Future prospects

If humans continue to pour carbon dioxide and other greenhouse gases into the atmosphere unchecked, weather will probably become more extreme and less predictable. Some regions will experience heavier rainfall and others worse droughts. Mighty ocean currents may shift direction slightly, bringing unexpected heat and moisture to some landmasses and withdrawing it from others. A sea-level rise of 20 in (50 cm) would threaten many low-lying coastal areas.

Fighting global warming with iron

The best way to counter the threat of global warming is to reduce greenhouse gases. Some scientists have been investigating a new way of trying to do this. Near the equatorial Pacific, and around Antarctica, the absence of dissolved iron seems to prevent phytoplankton growing fast. Scientists have therefore been adding dissolved iron to surface water in the hope of boosting phytoplankton growth. This would encourage greater photosynthesis, which would cause the phytoplankton to extract more carbon dioxide from the air and might help counter global warming. Experimental results have been promising so far, but scientists still need to do more research on the project.

Page numbers in *italics* refer to illustrations

abyssal plains *16*, 17
abyssal ridges *14*, 17
acid rain 38
aerial reconnaissance 45
Africa 10
African Plate *12*
aggregrates 41
Agulhas Current 57
air, temperature 4, 46
Alaska 45
  coast 26
algae 22, 43
algal bloom *39*
Amazon, tidal bore 54
amphibians 4
anchovies 51, *51*, 52, 53
animals, water content 36
Antarctic Circumpolar Current 49
Antarctic Ocean 44
  icebergs 45, *45*
Antarctic Plate *12*
Antarctica 10, *10*, 45, 49
Arabian Plate *12*
Arctic Ocean 5, 6, 7, 44
  depth 6
  icebergs 45, *45*
  size 6
asthenosphere 13
Atlantic Ocean 5, 6
  barrier islands 29
  birth 10
  changes 11
  coastline 24, 28
  floor 16
  size 7
  tides 55
  upwelling 51
atmosphere 8, 46
atolls, coral *22*, 23, *23*
Australia 10, *10*
avalanches, underwater 16

bacteria 9
Bahamas 47
Bahr el Baqar drain 33
Bangladesh 31, 59
bar 29
barrier islands 29
barrier reefs 22, *22*, 23
basalt 18
bay barrier 29
bays 5, 26
beach compartments 26–7
beaches
  artificial 25
  below cliffs 26
  coastal *17*, 24
  formation 28–9
  muddy 29
  raised *34*
  sandy 25, 29
  starved of sand 27
Bengal, Bay of 5, 31, 58, 59
Bengulea Current *49*
birds
  droppings 30
  effect of El Niño on 52
  of mudflats 31
  of wetlands 32
Boomer Beach, La Jolla, California 26
Borneo 20
Brahmaputra River, delta 31
brine cells 44
British Navy 59

Canary Current *49*
Cancun, Mexico *33*
canyons, submarine 16, 26–7
carbohydrates 38, 39
carbon 38, 40
carbon cycle 38–9, *38–9*
carbon dioxide 4, 38–9, *38–9*, 61
carbon sink 39
Caribbean, Hurricane Georges *58*, *59*
Caribbean Plate *12*
Caribbean Sea 5
causeways *29*
caves 26
chalk 24
chemical weathering 24
chemicals 37
  recycling 38–9, *38–9*
Chesapeake Bay 30
Chiengtang River, tidal bore 54
chimneys *41*
cholera 53
clams 31, 43
cliffs, coastal 26, *27*
climate 4
  change 60–1
  effects of currents on 51
  El Niño 52–3, *52*
clouds 4, 8, 36, 38
coal 61
coastal estuaries 30
coastal lagoons 33, *33*
coastal wetlands 32–3
coastline, changes to 24–5, *24*
Columbus, Christopher 47
comets 8, *8*
continental drift 12
continental margin 16
continental shelf 16, *17*, 19, 24
  islands on 20
continental slope 16, *17*
convection currents 12–13
copper *41*
coral *28*
coral atolls *22*, 23, *23*
coral polyps 22
coral reefs 20, 22–3, *22*, 43
  bleaching 61
  classification *22*, 23
  locations *23*
Coriolis effect *46*, *47*
crabs 32
crust
  continental 15
  oceanic 15
currents
  air 47
  convection 12–13
  effect on coastline 24
  equatorial 48
  longshore 27, 29, 30
  ocean 48–51, *48*, *49*, *50*, 61
  subsurface 50–1, *50*
  surface 48–9
  temperature and 42
  winds and *47*
cyclones 58, 59

dams 27
Darwin, Charles 23
days, lunar 54, 55
deltas 30–1, *31*
deposition 24, 29
deserts 47
Devil's Slide, Britain *18*
dimethyl sulfide (DMS) 38
Dinosaurs, Age of 10
doldrums *47*
dolphins, communication 43
downwelling 51
ducks 31, 32
dysentery 53

Earth
  area of water on 4
  asthenosphere 13
  core 5
  crust 12
  formation 8, *9*
  lithosphere 12
  magnetic field 15
  plates 12–13, *12*
  upper mantle 12
earthquakes 13, 34–5
  causing tsunamis 57
ebb tide 54
Egypt
  farming 31
  water cleaning 33
El Niño 52–3, *52–3*, 61
El Niño Southern Oscillation (ENSO) 53
Emperor Seamount chain *21*
equatorial currents 48, *49*
erosion 24, *24*, *26–7*, 28, 37
estuaries 6, 30–1, *30*
Eurasia 10, *10*, 11
Eurasian Plate *12*
Europa 9
Europe 11
Everest, Mt 7, *7*
evolution 20
exploration
  beneath ocean,
    early methods 4–5
    modern methods 5

farming 31
Faroe Islands *20*
fathoms 7
fertilizers 33, 37
fish
  communication 43
  plankton-feeding 51, *51*
  of wetlands 32–3
fjords 30
flamingos 31
floes *44*
flood tide 54
flooding
  caused by El Niño 53, *53*
  protection from 27
fogs 59
fossil fuels 40–1, *40*, 61
  formation *40*
Franklin, Benjamin 49
friction 13
fringing reef *22*, 23
Fundy, Bay of *55*

Galápagos Islands 20, *21*
Ganges River 19
  delta 31
gas, natural 40–1, *40*, 61
  location *40*
gas hydrates 41
geese 31, 32
geologists 12
Gibraltar, Straits of 11
Glacier, USS 45
glaciers 30, 44–5
global warming 39, 51, 60–1, *60*
gold 40
Gondwana 10
Grand Banks 59
Grand Forks, North Dakota *53*
granite 18, *18*, 26
grasses, marsh 32, 33
gravel
  on beaches 28, 41
  erosion by 24
  extracting 41
gravity 54, 55
grease ice 44
Great Barrier Reef 22, 23
Great Rift Valley *11*
greenhouse gases 41, *60*, 61
Greenland 20, 45, *45*
groins *24–5*, 25
groundwater 36, *37*, 41
Gulf (U.S.),
  barrier islands 29
  coast 28
Gulf Stream 48, 49, 50, 51, 59
gulfs 5
gyres 48

Hatteras Plateau 16
Hawaiian Islands *17*, *20–1*

headlands 26, 29
heat 42–3
Hell's Gate, Kenya *11*
herrings 51
Hess, Harry 14, 15
high-pressure zones 47
Himalayas 13, *13*
horse latitudes *46*, 47
Hudson Bay 61
Hurricane Georges *58*, *59*
hurricanes *58*
hydrocarbons 40
hydroelectricity 27
hydrogen 40
hydrological cycle 36
hydrothermal vents 5, 15, *41*

ice 9, 36, 44–5, *44*
ice ages 29, 34, 35
ice shelves 44
icebergs 44–5, *45*
igneous rock 18, 19, *19*
impermeable rock 41
India 10, 11
Indian Ocean 5, 6
  changes 11
  size 7
Indo-Australian Plate *12*
Indonesia, forest fires *52–3*
industry, on estuaries *30*
infrared radiation 60–1, *60*
Intergovernmental Panel on Climate Change (IPCC) 60
International Ice Patrol (IIP) 45
intruson 19
iron 61
islands 17, *17*, 20–1
  barrier 29
  inshore 20–1
  connection to mainland *29*
  oceanic 20, 21, *21*
  offshore 20, 21
  volcanic *22*

jet-streams 53
Juan de Fuca Plate *12*
Jupiter (planet) 9

krill 43

La Jolla, California 26
Labrador 49
Labrador Current 49, 59
lagoons *22*, 23, *23*
  coastal 33, *33*
lakes 36
land
  reclamation *30*
  temperature 4
landmass, rise and fall 34–5
landslides, causing tsunamis 57
Lanzarote, Canary Islands *37*
latitudes 46–7
Laurasia 10
lava *9*, 13, *15*
  cooling 18
lead 33
levees *31*
light *42*, 43
  on water *42*
limestone 19, 22, 24, 26
lithosphere 12
Loiho (volcano) *21*
London, England 49
longshore currents 27, 29, 30
longshore drift 25, 57
Los Angeles, earthquakes 13
Lovelock, James 38
lunar day 54, 55
lunar month 55

magma 13, 14, 15
  cooling 18
magnesium 15
magnetic field 15
manganese nodules 19, *41*
mangrove swamps 32, *32*
Manzala, Lake 33
marble 19
Mariana Trench 7
marine life, effects of tides on 54
marine organisms 4
marsh grasses 32, 33
Mauna Kea *7*
Mediterranean Sea 5, 11
  tidal range *55*
message in a bottle 49
metallic nodules *41*
metals, deposits 19
metamorphic rock 19, *19*
meteorites 8
methane 41, 61
Mexico, Gulf of 5, 48, 55
Miami Beach, Florida 29
microbes 8, 30
Mid-Atlantic Ridge 16, 21
mid-ocean ridges 14, *14*, *15*, 16, *16*
Milky Way 9
minerals, in sea 40–1
Mississippi delta *31*
mist *42*
Moon
  orbiting Earth 54, 55
  tides and 54, *54–5*, 55
Morley, Lawrence 14–15
mountains
  growth 13
  submarine 5, 7, *7*, *15*, 16
mud, on beaches 28
mudflats 31
Mykines Island *20*

NASA 8, 9
National Oceanic and Atmospheric Administration (NOAA) 53
Nazca Plate *12*
neap tide 55
Newfoundland 59
nickel *41*
nilas 44
Nile, delta *31*
nitrates 33
North America 10, 11
North American Plate *12*
North Atlantic Drift 49, 51
North Atlantic gyre *49*, 51
North Pole 47
North Sea *35*
nutrients 43, 50, 51
  recycling 30

Ocean City, New Jersey 25
oceanographers 5
oceans
  anatomy 16–17, *16–17*
  changes in 10–11, *11*
  circulation 36
  color 6
  depth 6–7, 16, 17
  effects 4
  floor 5, 16
    changes 12, 13
    formation 13
    mapping 5, 7
  formation 8–9, *11*
  layers 43
  names 5
  physical structure 4
  seas in 5
  size 6, 7
oil 40–1, *40*, 61
  location *40*, 61

Pacific Ocean 5, 6, *6*, 7
  birth 10–11
  breaking waves 56
  changes 11
  coastline 24
  depth 6, 7
  rising coast 26
  size 6, 7
  trough in 7
  upwelling 51
  waves 56–7, *56*
Pacific Plate *12*
Pacific Tsunami Warning Center 57
pack ice 44, *44*
Painted Desert, Utah *18*
Palos Verdes hills, California 26
Panama, Isthmus of 11
pancake ice 44
Pangaea 10
Panthalassa 10
Papua New Guinea 20, *41*
penguins, Galápagos *51*
petroleum oil 40–1, *40*
  location *40*
Philippine Plate *12*
photography
  satellite 5, *5*
  underwater 5
photosynthesis 4, 22, 38–9, *38*, 43, 61
phytoplankton 4, 16, 20, 38, 39, *39*, 51, 52, 61
Piccard, Jacques 7
pillow lava *15*
Pimlico Sound, North Carolina 30
pinnacles 26
planets
  formation 8
  life on 8–9
  water on 8–9
plankton *39*, 40
  see also phytoplankton
plants
  marine 20
  respiration *38*, 39
  water content 36
plate tectonics 15
plates 12–13, *12*, 19
  effects on coast 24
  sea levels and 34–5
platinum *41*
polar bears 61
polar seas 43
pollution 37
  of estuaries *30*
  of wetlands 33
Porcupine Abyssal Plain 16
pororoca 54
precipitation 36, *37*
proteins 38

radar altimetry 7, 34
radar reconnaissance 45
radiation 4
  infrared 60–1, *60*
  solar *60*
rain 8
  acid 38
Red River, North Dakota *53*
Red Sea 11, *11*, 13, *13*
respiration *38*, 39
ridges 14, *14*, *15*, 16, 17
Rift Valley *11*
rifts 14, 15, 16, *16*
Ring of Fire 57
rivers 6, 19, 36, *37*
  damming 27
  sediment from 27, 28
rock
  arch of *27*
  on coastline 26–7, *27*
  cycle 18–19, *19*
  igneous 18, 19, *19*
  impermeable 41
  metamorphic 19, *19*
  molten (liquid) 8, *9*, 13
  sedimentary 18–19, *18*, *19*, 26, 38
  types 18
  weathering 24

salt 36, 44
  concentrations of 33
  effect on currents 50
  in estuary 30–1
  in pack ice 44
  plants tolerant of 32
  sources 37
saltmarshes 32, 33
saltpans *37*
San Andreas Fault *12*, 13, 30
San Francisco, earthquakes 13
San Francisco Bay 30
sand
  on beaches 25, 26–7, 28–9, *28*, *29*, 41
  erosion by 24
  mining 41
sandstone *19*
sardines 51
Sargasso Sea 5
satellite altimetry 7, 34
satellite photography 5, *5*
sea defenses 25, *25*
sea floor spreading 14, 15
sea fogs 59
sea levels
  changes 34–5
  longterm changes 34
  measuring 34
  rises 61
seabed
  changes 12, 13, 14–15, *14*

formation 13
mapping 5, 7
seamount *17*
seas
color 6
definition 5
frozen 44–5
seawalls *25*
seaweeds 16
sediment 17, 18–19, 25, 26
on beaches 28
levees *31*
river-borne 19, 27
sediment fan 19
sedimentary rock 18–19, *18*, *19*, 26, 38
seismic sea waves 57
Seychelles *28*
shelf break 16
shingle 29
shoes, floating on ocean 49
shores 24
rocky 26–7
sandy 28–9, *28*, *29*
waves breaking on 57
shrimp 43
single-celled organisms 4
Sinus Borealis 10
sky, color 6
slate 19
snails 32
snowballs 8
solar radiation *60*
Somali Plate *12*
sonar surveys 5, 7, 43
sound 42, 43
made by whales 43
soundings 7
South America 10, 11
South American Plate *12*
South Atlantic gyre *49*
South Pole 47
Southern Ocean 5, 6, 44
size 7
spit 29
spreading ridge 13
spring tide 55
stacks 26, *27*
starches 39
steam 8
storm waves 56, 57
storms 58
effects on sea levels 34
worst 59
subduction 13
submersibles 5
sulfates 38
sulfur 15, 38
sulfur cycle 38
sulfur dioxide 37, 38
sulfuric acid 38
Sun
heat from 46
tides and 54, *54–5*, 55
sunlight 4, 6
superocean 6
surfing *56*
swells 56

Taormina, Sicily *28–9*
tectonic forces 24
temperate regions 43
temperature
changes 36–7
and marine life 43
ocean's effects on 4
Tethys Sea 10, 11
thermocline 43, 51
thermohaline circulation 50, 51
tidal bores 54
tidal bulges 54
tidal surges 54
tide gauge 34
tides 34
in Bay of Fundy *55*
cause of 54–5
effect on coastline 24
effect on marine organisms 54
frequency 54–5
neap 55
ranges *55*
rhythm of 54–5
spring 55
Titanic 45
tombolo *29*
trade winds 47, *47*, 48, *52*
transform faults *14*
trenches 7, 13, 14, *14*, 15, 16–17, *17*
Trieste (deep-diving vessel) 7
tsunamis 57, *57*
tuna 53

upwelling 50–1
valleys 16
Venice 35, *35*
vents, hydrothermal 5, 15, *41*
Vine, Fred 14–15
volcanoes 8, *9*, 13, 15, 37
activity 14, *17*
causing tsunamis 57
extinct *17*
island 17
oceanic *14*

waders 31
Walsh, Don 7
water
area covered 4
in atmosphere 61
chemical makeup 15
chemical recycling 38–9
circulation 36, 42
cleaning 33
evaporation 4, 36, *37*
heat storage 36–7
light on *42*
on planets 8–9
properties 36
source of 8
water vapor 8
waterspout *59*
waves 56–7
absorbing impact 25, *25*, 26
biggest 57
conflicting 56
effect on coastline 24, *27*
effects of wind on 56
height 56
hitting beach 25
movement of water *56*
power of 57
rogue 57
sea waves 56, *56*
storm waves 56, 57
tsunamis 57, *57*
winds and 56
weather 4
ocean 58–9
weathering 24, 37
Wegener, Alfred 12
wetlands, coastal 32–3
whales, communication 43, *43*
wildlife
of estuary 30–1
of oceanic islands 20
sea temperature and 42–3
use of light 43
of wetlands 32
winds
Coriolis effect *46*, *47*
currents and *47*
deflection *46*
effects on sea levels 34
global 46–7, *46*, *47*
offshore 34
onshore 34
waves and 56
worms 5, 22

Picture Credits

Photographs

Abbreviations

A Ardea; C Corbis; FLPA Frank Lane Picture Agency; NHPA Natural History Photographic Agency; SPL Science Photo Library

1 Douglas Peebles/C; 3 Michael S. Yamashita/C; 5 Thomas Ligon/SPL; 6–7 Tom Van Sant, Geosphere Project/Planetary Visions/SPL; 8 Jack Finch/SPL; 9 Peter Menzell/SPL; 11 Piers Cavendish/A; 13t C; 13b Simon Fraser/SPL; 15 B. Murton/Southampton Oceanography Centre/SPL; 17 G. Brad Lewis/SPL; 18l Bob Gibbons/A; 18r Bernhard Edmaier/SPL; 20 Silvestris Fotoservice/FLPA; 21 C; 23 Kurt Amsler/A; 24 Su Gooders/A; 24–25 Trevor McDonald/NHPA; 25 Richard Bickel/C; 26–27 Elio Ciol/C; 27 N.A. Callow/NHPA; 28 Norbert Wu/NHPA; 28–29 Tony Waltham/Geophotos; 30–31 David Woodfall/NHPA; 31 C; 33 Danny Lehman/C; 34 Angela Murphy/SPL; 35 Tony Waltham/Geophotos; 37 Don Smith/FLPA; 39 Digital Image © 1996 Corbis; Original image courtesy of NASA/C; 41t P. Morris/A; 41b Ralph White/C; 42 R. Sorensen & J. Olsen/NHPA; 44 Stephen Krasemann/NHPA; 45 Kevin Schafer/NHPA; 47 Hulton Archive; 51 D. Parer & E. Parer-Cook/A; 52–53 CNES, 1997 Distribution Spot Image/SPL; 53 Adrian Warren/A; 55l, 55r François Gohier/A; 56 Douglas Peebles/C; 57 Chris Butler/SPL; 59l Warren Faidley/Oxford Scientific Films; 59r H. Hoflinger/FLPA; 61 CSIRO/Simon Fraser/SPL; cover photograph: Powerstock.

Illustrations

Advocate Ltd/Alex Pang 19, 22, 40; Andromeda Oxford Ltd 4–5, 10, 11, 12, 14, 15, 16–17, 23, 31, 32, 35, 38–39, 43, 45, 46, 48, 49, 50, 52, 58, 60, 61; Maltings Partnership 20–21, 29, 36, 37, 42, 47, 54–55, 56; Tony Truscott Designs 7.